SPECTROSCOPY

New Uses and Implications

SPECTROSCOPY

New Uses and Implications

Roy H. Williams

Astronomer and educator, Kopernik Observatory and Science Center, Vestal, New York, U.S.A.

Apple Academic Press

Spectroscopy: New Uses and Implications

First Published in the Canada, 2011
Apple Academic Press Inc.
3333 Mistwell Crescent
Oakville, ON L6L 0A2
Tel. : (888) 241-2035
Fax: (866) 222-9549
E-mail: info@appleacademicpress.com
www.appleacademicpress.com

The full-color tables, figures, diagrams, and images in this book may be viewed at www.appleacademicpress.com

ISBN 978-1-926692-82-1

Roy H. Williams

Printed at Thomson Press (I) Ltd.
9 8 7 6 5 4 3 2 1

Cover Design: Psqua

Library and Archives Canada Cataloguing in Publication Data
CIP Data on file with the Library and Archives Canada

CONTENTS

INTRODUCTION

Spectroscopy is a technique utilizing the interaction of light with matter. In many cases, this interaction causes the substance to become ionized, creating an emission or absorption of light at various wavelengths of electromagnetic radiation, called a spectrum. The instrument that splits light into its component wavelengths is called a spectrograph or spectrometer. The assortment of different wavelengths that is recorded by a spectrograph is called spectra. The discovery of unique spectra of different elements was a starting point in the development of quantum mechanics. It was by discovering the discrete energy levels of electrons and their subsequent release or absorption of photons that correlate to energy levels that allowed the formulation of modern physics. These energy levels correspond to a "fingerprint" that is unique to that particular element. It is the comparison of known spectra or fingerprints that allows scientists to identify the chemical composition of a substance.

Spectroscopy utilizes the absorption, emission, and scattering of light as it interacts with matter creating a large variety of spectra. Analysis of various spectra can yield important physical characteristics, including, but not limited to, chemical composition, temperature, luminosity, and mass. Fields of study include astronomy, medicine, analytic chemistry, and material science.

Spectroscopy as a research tool reveals the chemical composition and temperature of elements, molecules, and compounds. Traditionally, spectroscopy developed out of the need of scientific fields such as astronomy, physics, and chemistry.

However, nearly all scientific fields are now using this powerful technique. For example, geologists use mass spectrometers for mineralogical analysis of rock composition. These mass spectrometers are even used on robotic space missions such as the Mars Exploration Rovers (MER) and the Mars Global Surveyor (MSG). Both of these spacecraft used a Thermal Emission Spectrometer (TES) and discovered an abundance of the mineral hematite on Mars. This finding supports the hypothesis that Mars at one time had an abundance of standing water.

In astronomy, spectrographs are widely used to determine the temperatures and compositions of stars. In fact, helium was first discovered by French astronomy Pierre Janssen in 1868. Janssen used a primitive spectroscope during a solar eclipse, which allowed him to examine the spectrum of the sun's corona. Spectral analysis of stars led to their classification. This scheme is displayed on the famous Hertzsprung-Russell diagram that uses the relationship of luminosity versus temperature. These discoveries led astronomers to develop a clear understanding of the lifecycle of stars.

In addition, Earth orbiting satellites that utilize remote sensing can detect mineral deposits, types of vegetation, pollution, and temperature of various surfaces. The most famous of these Earth orbiting satellites is Landsat. Since 1972, there have been six of these satellites orbiting the Earth in polar orbits, using a spectrum imager called a Thematic Mapper. This instrument is able to map coastal waters, differentiate between soil and vegetation, and discern biomass content, soil moisture, vegetation heat stress, and the weathering of different clays. It is the clay content and its relative ratio with aluminum silicates that helps geologists identify different rock types.

In the medical sciences, spectroscopy is applicable in a large variety of ways. Magnetic Resonance Imaging (MRI) uses a strong magnetic field to find subtle differences in magnetic properties within various tissues. This allows for detailed imaging of the tissues, organs, and joints without the need for surgery, CT scans, or X-rays. Mid-infrared and Raman spectroscopy can identify tissue abnormalities, diseased tissue, and cells. This makes it useful for detecting cancer.

Nuclear Magnetic Resonance (NMR) resolves the location and variety of various structures and dynamics of molecules at the atomic level. This high resolution makes it very useful for the diagnosis of brain lesions, tumors, and metabolic liver disease. The main advantage of NMR is the ability to diagnose diseases of the heart, liver, and kidneys without using the potentially harmful ionizing radiation found in CT scans. In the future, NMR may be used to study metabolic processes and blood vessel flow.

Careers in spectroscopy are extremely wide open due to the fact that all scientific fields have embraced this method. Spectroscopists work in the fields of archaeology, art history, geology, astronomy, medicine, analytical chemistry, and physics. In addition, private industry is always seeking technicians, researchers, and physicians that have experience in spectroscopy. According to a 2005 survey by *Spectroscopy Magazine*, the average industrial spectroscopist earned a salary of $78,170, academic spectroscopists averaged $54,640, and government spectroscopists averaged around $66,500.

Although spectroscopy is over one hundred years old, researchers are constantly discovering new applications. Huge technical leaps might be achieved as spectroscopy merges with material sciences, namely superconductivity and nanotechnology. In addition, materials will be tested under actual operating conditions, such as stress, temperature, magnetic, and electric fields. This may allow for materials to be greatly improved at the atomic level. In addition, advances in superconductivity will not only bring about better spectroscopy hardware but could encourage research on superconductivity and the ability to fabricate nano-scale electronic circuits using organic molecules. This could lead to huge scientific breakthroughs, such as superconducting power lines and true quantum computers.

— Professor Roy H. Williams

Collision-Induced Infrared Absorption by Molecular Hydrogen Pairs at Thousands of Kelvin

Xiaoping Li, Katharine L. C. Hunt, Fei Wang, Martin Abel and Lothar Frommhold

ABSTRACT

Collision-induced absorption by hydrogen and helium in the stellar atmospheres of cool white dwarfs causes the emission spectra to differ significantly from the expected blackbody spectra of the cores. For detailed modeling of radiative processes at temperatures up to 7000 K, the existing H_2–H_2 *induced dipole and potential energy surfaces of high quality must be supplemented by calculations with the* H_2 *bonds stretched or compressed far from the equilibrium length. In this work, we describe new dipole and energy surfaces, based on more than 20 000 ab initio calculations for* H_2–H_2*. Our results agree* *well with previous ab initio work (where those data exist); the calculated rototranslational absorption spectrum at 297.5 K matches experiment similarly well. We further report the calculated absorption spectra of H_2–H_2 for frequencies from the far infrared to 20,000 cm^{-1}, at temperatures of 600 K, 1000 K, and 2000 K, for which there are no experimental data.*

Introduction

It is well known that dense gases of infrared inactive molecules such as H_2 absorb infrared radiation. Absorption continua range from the microwave and far infrared regions of the spectrum to the near infrared and possibly into the visible. Collisionally interacting pairs of hydrogen molecules possess transient electric dipole moments, which are responsible for the observed absorption continua [1, 2]. Planetary scientists understood early on the significance of collision-induced absorption (CIA) for the modeling of the atmospheres of the outer planets [3, 4]. More recently, it was shown that the emission spectrum of cool white dwarf stars differs significantly from the expected blackbody spectrum of their cores: CIA in the dense helium and hydrogen atmospheres suppresses (filters) the infrared emissions strongly [5–10]. Detailed modelling of the atmospheres of cool stars with proper accounting for the collision-induced opacities is desirable, but it has been hampered heretofore by the highly incomplete or nonexisting theoretical and experimental data on such opacities at temperatures of many thousands of kelvin.

Quantum chemical calculations of the induced dipole surfaces of H_2–H_2, H_2–He and other complexes have been very successful [11–14]. Based on such data, molecular scattering calculations accounting for the interactions of the molecular complexes with photons have been under-taken which accurately reproduced the existing laboratory measurements at low temperatures (T ≤ 300 K or so) [2]. At higher temperatures, virtually no suitable laboratory measurements of such opacities exist, but reliable data are needed. We therefore decided to extend such quantum chemical calculations of the induced dipole (ID) and potential energy surfaces (PES) of H_2–H_2 complexes to highly rotovibrationally excited molecules, as encountered at high temperatures (up to 7,000 K) and photon energies up to ~2.5 eV.

Ab Initio Calculations of the Induced Dipole and Potential Energy Surfaces

At the temperatures characteristic of cool white-dwarf atmospheres, the CIA spectra depend on transition dipole matrix elements with vibrational quantum numbers up to v≈7. To evaluate these matrix elements, we have determined the induced dipoles and interaction energies of pairs of hydrogen molecules with bond lengths ranging from 0.942 a.u. to 2.801 a.u. (1 a.u. = a0 = 5.29177249·10-11 m). For comparison, the vibrationally averaged internuclear separation in H_2 is 1.449 a.u., in the ground vibrational state. We have used MOLPRO 2000 [15] to calculate the PES for H_2–H_2 and to calculate the pair ID by finite-field methods, at coupled-cluster single and double excitation level, with triple excitations treated perturbatively [CCSD(T)]. In this work, we have employed MOLPRO's aug-cc-pV5Z(spdf) basis, consisting of (9s 5p 4d 3f) primitive Gaussians contracted to [6s 5p 4d 3f]; this gives 124 contracted basis functions for each of the H_2 molecules. The basis gives accurate energies and properties [16]; yet it is sufficiently compact to permit calculations on H_2 pairs with 28 different combinations of H_2 bond lengths, at 7 different intermolecular separations, in 17 different relative orientations (the orientations listed in Table 1), and at a minimum of 6 different applied field strengths for each geometrical configuration.

In the calculations, the centers of mass of the two H_2 molecules are separated along the Z axis by distances R ranging from 4.0 to 10.0 a.u. The vector R joins molecule 2 to molecule 1. The molecular orientations are characterized by the angles (θ_1, θ_2, $\phi 12$), where θ_1 is the angle between the Z axis and the symmetry axis of molecule 1, θ_2 is the angle between the Z axis and the symmetry axis of molecule 2, and $\phi 12$ is the dihedral angle between two planes, one defined by the Z axis and the symmetry axis of molecule 1 and the other defined by the Z axis and the symmetry axis of molecule 2.

Calculations were performed first for two molecules with bond lengths of r_1 = r_2 = 1.449 a.u., the ground-state, vibrationally averaged internuclear separation. The interaction energies were evaluated in the absence of an applied field; then the pair dipoles were obtained from finite-field calculations, grouped into three sets of 40. Within each of the sets, the fields were confined to the XY, XZ, or YZ planes, and the two components of the applied field were selected randomly, in the range from 0.001 a.u. to 0.01 a.u., for a total of 120 calculations. For each fixed set of the bond lengths, orientation angles, and intermolecular separation, the total energies were fit (by least squares) to a quartic polynomial in the applied field F:

$$E = E_0 - \mu_\alpha F_\alpha - \left(\frac{1}{2}\right)\alpha_{\alpha\beta} F_\alpha F_\beta$$

$$-\frac{1}{6}\beta_{\alpha\beta\gamma} F_\alpha F_\beta F_\gamma \tag{1}$$

$$-\frac{1}{24}\gamma_{\alpha\beta\gamma\delta} F_\alpha F_\beta F_\gamma F_\delta - \ldots,$$

where the Einstein convention of summation over repeated Greek subscripts is followed. The coefficients of the linear terms were selected from each fit, to obtain the Cartesian components of the induced dipole moments μ_X, μ_Y, and μ_Z. In Table 1, our results for the components of the pair dipole are given for pairs with $r_1=r_2=1.449$ a.u.

Table 1. Cartesian components μX, μY, and μZ of the H2–H2 dipole in a.u. (multiplied by 106) with bond lengths $r_1=r_2=1.449$ a.u.

R (a.u.)	4.0	5.0	6.0	7.0	8.0	9.0	10.0
$(\theta_1, \theta_2, \varphi_{12})$				μ_X			
(π/12, π/6, π/3)	48	150	103	63	39	25	16
(π/12, π/4, π/6)	-3675	-2393	-1393	-804	-480	-299	-195
(π/12, π/3, π/6)	-2790	-1791	-1044	-607	-366	-230	-150
(π/12, 5π/12, π/6)	-144	-28	-17	-16	-15	-11	-8
(π/6, π/4, π/3)	1417	1294	806	471	280	175	115
(π/6, π/3, π/4)	399	562	365	210	121	74	49
(π/6, 5π/12, π/3)	2065	1922	1196	695	411	255	167
(π/4, π/3, π/6)	1109	879	528	302	177	109	72
(π/4, 5π/12, π/6)	3481	2432	1424	815	481	299	195
(π/3, 5π/12, π/6)	2555	1804	1062	611	363	226	148
(7π/12, π/12, π/6)	-7979	-5226	-3027	-1740	-1037	-648	-424
(7π/12, π/6, π/4)	-9089	-5973	-3462	-1988	-1184	-740	-484
(7π/12, π/4, π/6)	-11040	-7181	-4151	-2381	-1417	-885	-580
(7π/12, π/3, π/6)	-9759	-6345	-3669	-2107	-1255	-785	-515
(π/2, π/12, π/6)	-3628	-2337	-1341	-765	-452	-282	-184
(π/2, π/6, π/3)	-3575	-2303	-1322	-754	-447	-278	-182
(π/2, π/4, π/6)	-7071	-4535	-2606	-1489	-884	-551	-361
				μ_Y			
(π/12, π/6, π/3)	-6236	-4288	-2519	-1453	-864	-539	-352
(π/12, π/4, π/6)	-3691	-2635	-1566	-908	-542	-338	-221
(π/12, π/3, π/6)	-2801	-2088	-1257	-733	-440	-275	-180
(π/12, 5π/12, π/6)	-1060	-952	-599	-355	-214	-135	-88
(π/6, π/4, π/3)	-5443	-4082	-2455	-1429	-854	-533	-349
(π/6, π/3, π/4)	-2847	-2414	-1496	-880	-529	-332	-217
(π/6, 5π/12, π/3)	-1427	-1430	-916	-545	-330	-207	-136
(π/4, π/3, π/6)	-1008	-1209	-793	-473	-286	-179	-117
(π/4, 5π/12, π/6)	419	-239	-224	-145	-90	-57	-37
(π/3, 5π/12, π/6)	404	-211	-199	-127	-78	-49	-32
(7π/12, π/12, π/6)	-997	-877	-543	-317	-189	-118	-77
(7π/12, π/6, π/4)	-2588	-2187	-1342	-780	-464	-290	-189
(7π/12, π/4, π/6)	-2416	-1890	-1142	-662	-394	-246	-161
(7π/12, π/3, π/6)	-2415	-1757	-1045	-604	-360	-225	-147
(π/2, π/12, π/6)	-835	-798	-501	-294	-175	-109	-71
(π/2, π/6, π/3)	-2521	-2367	-1481	-865	-515	-321	-210
(π/2, π/4, π/6)	-1734	-1570	-976	-569	-339	-211	-138
				μ_Z			
(π/12, π/6, π/3)	-15702	-5371	-2141	-1026	-568	-345	-223
(π/12, π/4, π/6)	-35330	-12145	-4900	-2374	-1322	-808	-525
(π/12, π/3, π/6)	-53105	-18342	-7486	-3664	-2053	-1258	-820
(π/12, 5π/12, π/6)	-65061	-22550	-9278	-4573	-2574	-1580	-1032
(π/6, π/4, π/3)	-19683	-6793	-2764	-1349	-755	-463	-301
(π/6, π/3, π/4)	-37478	-13007	-5355	-2641	-1486	-914	-597
(π/6, 5π/12, π/3)	-49514	-17248	-7156	-3553	-2008	-1237	-810
(π/4, π/3, π/6)	-17837	-6231	-2596	-1293	-731	-451	-296
(π/4, 5π/12, π/6)	-29903	-10485	-4400	-2205	-1253	-774	-509
(π/3, 5π/12, π/6)	-12057	-4257	-1805	-913	-522	-323	-213
(7π/12, π/12, π/6)	65301	22600	9286	4573	2573	1580	1032
(7π/12, π/6, π/4)	49757	17294	7161	3553	2008	1237	810
(7π/12, π/4, π/6)	30125	10528	4404	2206	1253	775	510

Table 1. *(Continued)*

R (a.u.)	4.0	5.0	6.0	7.0	8.0	9.0	10.0
$(7\pi/12, \pi/3, \pi/6)$	12133	4272	1807	914	522	323	214
$(\pi/2, \pi/12, \pi/6)$	69310	24042	9916	4898	2761	1697	1109
$(\pi/2, \pi/6, \pi/3)$	53764	18746	7796	3879	2196	1354	887
$(\pi/2, \pi/4, \pi/6)$	34212	11998	5042	2532	1441	891	58

In earlier work on the polarizabilities α for H_2–H_2 [16], we conducted several tests of this fitting procedure: we compared results from quartic fits with 120 different field strengths, quartic fits with 200 different field strengths, and quintic and sixth-order fits with 200 field strengths (at one set of orientation angles and an intermolecular distance of 2.5 a.u., where the differences between the calculations were expected to be magnified); we found excellent agreement among the results from all of the fits. We also compared the results from the random-field calculations with the values obtained analytically, based on calculations with 6 or 8 selected values of the field strengths, for fixed orientation angles and the full range of intermolecular separations. The field values were grouped into the sets $\{f, 2^{1/2}f, 3^{1/2}f, -f, -2^{1/2}f, -3^{1/2}f\}$, $\{f, 2^{1/2}f, 5^{1/2}f, -f, -2^{1/2}f, -5^{1/2}f\}$, and $\{f, 2^{1/2}f, 3^{1/2}f, 5^{1/2}f, -f, -2^{1/2}f, -3^{1/2}f, -5^{1/2}f\}$, with f = 0.001, 0.002, 0.003, and 0.004 a.u. At the shortest intermolecular distance (R=2.5 a.u.), the results for f = 0.001 a.u.—0.003 a.u. were affected by numerical imprecision in the hyperpolarization contributions; at larger R, they agreed well with the random-field results. Agreement between the random-field results and the results obtained with f = 0.004 a.u. was excellent for all R values. On this basis, we have used random-field fits in the work with r1 = r2 = 1.449 a.u., but we have used analytic fits with 6 different field values for the computations with r_1 or $r_2 \neq 1.449$ a.u. In [16], we also compared the results obtained via analytic differentiation at the self-consistent field (SCF) level using Gaussian 98 versus the results from our SCF calculations, for the full range of intermolecular separations and three different relative orientations, again with excellent agreement. Basis set superposition error (BSSE) has been shown to be negligible [16], as tested by function counterpoise ("ghost-orbital") methods. BSSE occurs when the pair basis provides a better representation of H_2–H_2 than the single-molecule basis provides for an isolated H_2 molecule. In these calculations, BSSE has been suppressed by the large size of the single-molecule basis.

The interaction mechanisms that determine the induced dipole include classical multipole polarization, van der Waals dispersion, and short-range exchange, overlap, and orbital distortion. At long range, the leading term in the collision-induced dipole comes from quadrupolar induction, which varies as R-4 in the separation R between the molecular centers [2]. The next long-range polarization term is of order R-6; it results both from hexadecapolar induction and from the effects of the nonuniformity of the local field gradient (due to the quadrupole moment of the collision partner). The magnitude of the latter term depends on the dipole-octopole polarizability tensor E. At order R-7, back-induction [17, 18] and dispersion [17–21] affect the pair dipole. Back-induction is a static reaction field effect: the field from the permanent quadrupole of molecule 1 polarizes molecule 2, which sets up a reaction field that polarizes molecule 1 (and similarly, with molecules 1 and 2 interchanged). The van der Waals dispersion dipole results from dynamic reaction-field effects, combined with the effects of an applied, static field [21], via two physical mechanisms. (1) Spontaneous, quantum mechanical fluctuations in the charge density of molecule 1 produce a fluctuating field that acts on molecule 2; then molecule 2 is hyperpolarized by the concerted action of the field from 1 and the applied field F. This sets up a field-dependent dynamic reaction field at molecule 1, giving a term in the van der Waals energy that is linear in the applied field F. (2) The correlations of the fluctuations in the charge density of molecule 1 are altered by the static field F acting on 1; molecule 2 responds linearly to field-induced changes in the fluctuations of the charge density of 1, again giving a term in the van der Waals energy that is linear in the applied field F. The precise functional forms of the short-range exchange, overlap, and orbital-distortion effects on the dipole are not known; however, these contributions are expected to drop off (roughly) exponentially with increasing R [2].

The dipole moment of the pair can be cast into a symmetry-adapted form, as a series in the spherical harmonics of the orientation angles of molecules 1 and 2 and the orientation angles of the intermolecular vector:

$$\begin{aligned}\mu_1{}^M(R,r_1,r_2) &= (4\pi)^{3/2}3^{-1/2}\sum A_{\lambda_1\lambda_2\Lambda L}(R,r_1,r_2)\\ &\times Y_{\lambda_1}{}^{m_1}(\Omega_1)Y_{\lambda_2}{}^{m_2}(\Omega_2)Y_{\lambda_L}{}^{M-m}(\Omega_R)\\ &\times\langle\lambda_1\lambda_2 m_1 m_2\,|\,\Lambda m\rangle\langle\Lambda L m(M-m)\,|\,1M\rangle,\end{aligned} \tag{2}$$

where the sum runs over all values of λ_1, λ_2, m_1, m_2, Λ and m; M=1, 0, or -1, corresponding to the dipole components,

$$\begin{aligned}\mu_1{}^1 &= -\left(\frac{1}{2}\right)^{1/2}(\mu_X + i\mu_Y),\\ \mu_1{}^0 &= \mu_Z,\\ \mu_1{}^{-1} &= \left(\frac{1}{2}\right)^{1/2}(\mu_X - i\mu_Y).\end{aligned} \tag{3}$$

In (2), Ω_1 and Ω_2 denote the orientation angles of molecules 1 and 2, that is, the orientation angles of the z axes of the molecule-fixed frames, ΩR is the orientation angle of the vector R (note that R runs from molecule 2 to molecule 1, in this work), and the quantities $\langle \lambda_1\lambda_2 m_1 m_2|\Lambda m\rangle$ and $\langle \Lambda LM(M-m)|1M\rangle$ are Clebsch-Gordan coefficients. Equation (2) follows immediately from the fact that the collision-induced dipole of H_2–H_2 is a first-rank spherical tensor, which is obtained by coupling functions of r_1, r_2, and R. Therefore λ_1, λ_2, Λ, L, and the magnitudes of r1, r2, and R completely determine the dipole expansion coefficients $A_{\lambda1\lambda2\Lambda L}(R,r_1,r_2)$.

The dipole coefficients arising from various long-range polarization mechanisms are categorized in Table 2, through order R^{-7}. In this table, Θ denotes the molecular quadrupole moment; $\bar{\alpha}$ is the trace of the single-molecule po-

Table 2. Long-range dipole induction mechanisms that contribute to the coefficients $A_{\lambda\lambda'\Lambda L}$ of (2) for a pair of molecules A and B [17, 18].

Induction mechanism	Power law	Properties	Coefficients
Quadrupolar field	R^{-4}	$\Theta, \bar{\alpha}$	A_{2023}, A_{0223}
		$\Theta, \alpha_\parallel - \alpha_\perp$	$A_{22\Lambda3}, \Lambda = 2,3,4$
Hexadecapolar field	R^{-6}	$\Phi, \bar{\alpha}$	A_{4045}, A_{0445}
		$\Phi, \alpha_\parallel - \alpha_\perp$	$A_{42\Lambda5}, \Lambda = 4,5,6$
			$A_{24\Lambda5}, \Lambda = 4,5,6$
Nonuniform field gradient	R^{-6}	Θ, E_2	A_{2245}
		Θ, E_4	$A_{42\Lambda5}, \Lambda = 4,5,6$
			$A_{24\Lambda5}, \Lambda = 4,5,6$
Back-induction	R^{-7}	$\Theta, \bar{\alpha}, \alpha_\parallel - \alpha_\perp$	A_{0001}
			A_{2021}, A_{0221}
			A_{2023}, A_{0223}
			A_{2221}
			$A_{22\Lambda3}, \Lambda = 2,3,4$
			A_{2245}
			A_{4043}, A_{0443}
		$\Theta, \alpha_\parallel - \alpha_\perp$	A_{2021}, A_{0221}
			A_{2023}, A_{0223}
			$A_{22\Lambda1}, \Lambda = 0,1,2$
			$A_{22\Lambda3}, \Lambda = 2,3,4$
			A_{2245}
			A_{4221}, A_{2421}
			$A_{42\Lambda3}, \Lambda = 2,3,4$
			$A_{24\Lambda3}, \Lambda = 2,3,4$
			$A_{42\Lambda5}, \Lambda = 4,5,6$
			$A_{24\Lambda5}, \Lambda = 4,5,6$
Dispersion	R^{-7}	$\bar{\alpha}(i\omega), B_0(0, i\omega)$	A_{0001}
		$\bar{\alpha}(i\omega), B_2(0, i\omega)$	A_{2021}, A_{0221}
			A_{2023}, A_{0223}
		$\alpha_\parallel(i\omega)-\alpha_\perp(i\omega), B_0(0, i\omega)$	A_{2021}, A_{0221}
			A_{2023}, A_{0223}
		$\bar{\alpha}(i\omega), B_4(0, i\omega)$	A_{4043}, A_{0443}
		$\alpha_\parallel(i\omega)-\alpha_\perp(i\omega), B_2(0, i\omega)$	$A_{22\Lambda1}, \Lambda = 0,1,2$
			$A_{22\Lambda3}, \Lambda = 2,3,4$
			A_{2245}
		$\alpha_\parallel(i\omega)- \alpha_\perp(i\omega), B_4(0, i\omega)$	A_{4221}, A_{2421}
			$A_{42\Lambda3}, \Lambda = 2,3,4$
			$A_{24\Lambda3}, \Lambda = 2,3,4$
			$A_{42\Lambda5}, \Lambda = 4,5,6$
			$A_{24\Lambda5}, \Lambda = 4,5,6$

larizability; $\alpha||-\alpha\perp$ is the polarizability anisotropy, which is equal to $\alpha_{zz-}\alpha_{xx}$ in the molecular axis system, where z is the symmetry axis; Φ is the hexadecapole moment; E is the dipole-octopole polarizability, which has a second-rank spherical tensor component E2 and a fourth-rank component E4. The van der Waals dispersion dipole is given by an integral over imaginary frequencies, where the integrand is a product of the polarizability at imaginary frequency $\alpha(i\omega)$ and the dipole-dipole-quadrupole hyperpolarizability $B(0,i\omega)$. The B tensor is a fourth-rank Cartesian tensor with spherical-tensor components of ranks 0, 2, and 4.

For distinct molecules 1 and 2, or for chemically identical molecules that have different bond lengths, all of the dipole coefficients listed in Table 2 are nonzero, although some of the coefficients may be quite small numerically. For chemically identical molecules, when r1 = r2, the coefficients A_{0001}, $A_{22\Lambda1}$ with $\Lambda \neq 1$, $A_{22\Lambda3}$ with $\Lambda \neq 3$, and A_{2245} vanish; the remainder are nonzero. The coefficients $A_{0\lambda\lambda L}$ and $A_{24\Lambda L}$ can be obtained from the coefficients $A_{\lambda 0\lambda L}$ and $A_{42\Lambda L}$ via the relations

$$\begin{aligned} A_{0\lambda\lambda L} &= -\mathrm{P}^{12} A_{\lambda 0\lambda L}, \\ A_{24\Lambda L} &= (-1)^{\Lambda+1}\mathrm{P}^{12} A_{42\Lambda L}, \end{aligned} \tag{4}$$

where P^{12} interchanges the labels of molecules 1 and 2. For centrosymmetric molecules such as H2, the dipole coefficients $A_{\lambda\lambda'\Lambda L}$ vanish unless λ and λ' are both even. Also, due to the Clebsch-Gordan coefficients in (2), nonvanishing contributions are found only if Λ = L–1, L, or L+1. Coefficients with higher values of λ and λ' than those listed are of higher order than R^{-7} at long-range, although they may represent significant short-range overlap effects.

From the dipole values in Table 1, we have obtained a set of A coefficients by least-squares fit (at each R value) to (2), for r1 = r2 = 1.449 a.u. From the fit, we have been able to determine the coefficients A_{2021}, A_{0221}, A_{2023}, A_{0223}, A_{2211}, A_{2233}, A_{4043}, A_{0443}, A_{4045}, A_{0445}, A_{4221}, A_{2421}, A_{4223}, A_{2423}, A_{4233}, A_{2433}, A_{4243}, A_{2443}, A_{4245}, A_{2445}, A_{4255}, A_{2455}, A_{4265}, A_{2465}, A_{4267}, and A_{2467}. We have kept all of these coefficients, as well as A_{0001} and A_{2201}, in the calculations with unequal bond lengths for molecules 1 and 2. However, for R ≥ 4.0 a.u and r1 = r2 = 1.449 a.u., the least squares fit shows that the first ten coefficients are numerically important, while the remaining coefficients are essentially negligible. At R = 4.0 a.u., the remaining coefficients do not exceed 7.0·10-5 a.u. in absolute value, and the values drop off rapidly with increasing R. Table 3 gives our results for A_{2021}, A_{2023}, A_{2211}, A_{2233}, A_{4043}, and A_{4045}; the other numerically significant coefficients are given by the relations $A_{0221} = -A_{2021}$, $A_{0223} = -A_{2023}$, $A_{0443} = -A_{4043}$, and $A_{0445} = -A_{4045}$.

Table 3. Dipole expansion coefficients $A_{\lambda\lambda'\Lambda L}$(in a.u., multiplied by 106) for H_2–H_2 with r1 = r2 = 1.449 a.u. Results from this calculation, compared with results of Meyer et al. [13] (MBF), Fu et al. [22] (FZB), long-range results [17, 18] through order R^{-7}(LR), and quadrupole-induced dipole coefficients (QID).

	R (a.u.)	4.0	5.0	6.0	7.0	8.0	9.0	10.0
A_{2021}	This work	9983	2123	407	73	13	4	2
	MBF	10401	2190	429	84	20	7	—
	FZB	10385	2184	427	83	19	6	—
	LR	279	59	16	6	2	1	0
A_{2023}	This work	−20065	−8076	−3725	−1950	−1124	−695	−455
	MBF	−19967	−7953	−3688	−1939	−1119	−692	—
	FZB	−19949	−7946	−3685	−1938	−1118	−692	—
	LR	−19687	−7652	−3603	−1921	−1118	−695	−455
	QID	−17628	−7221	−3482	−1880	−1102	−688	−451
A_{2211}	This work	402	86	18	3	0	0	0
	MBF	332	74	14	2	0	0	—
	FZB	332	74	14	2	0	0	—
	LR	−41	−9	−2	−1	0	0	0
A_{2233}	This work	2020	977	514	289	171	107	70
	MBF	1992	949	498	280	166	104	—
	FZB	1991	949	498	279	166	104	—
	LR	2588	1088	530	288	169	106	70
	QID	2726	1117	538	291	170	106	70
A_{4043}	This work	690	180	42	9	2	0	0
	LR	204	43	12	4	2	1	0
A_{4045}	This work	−845	−283	−97	−37	−16	−8	−4
	MBF	−1523	−450	−135	−47	−19	−9	—
	FZB	−1517	−447	−134	−46	−19	−9	—
	LR	−1040	−273	−91	−36	−16	−8	−4

In Table 3, the results are also compared with results from two earlier *ab initio* calculations of the H_2–H_2 dipole with r_1=r_2=1.449 a.u., reported by Meyer et al. [12], Meyer et al. [13], and Fu et al. [22]. (The signs in Table 3 follow from our choice of the positive direction of the intermolecular vector R.) Meyer et al. [12, 13] used configuration-interaction wave functions including single, double, and triple excitations from a reference Slater determinant, in a (7s 1p) basis of Gaussian primitives on each H center, contracted to [3s 1p] and augmented by a (3s, 2p, 2d) basis at the center of the H–H bond, giving a total of 31 basis functions for H_2 [11]. They performed calculations for 18 relative orientations that provided 9 nonredundant Cartesian dipole components. Fu et al. [22] employed the same basis to generate the CCSD (T) wave functions, in calculations for H_2–H_2 in 13 relative orientations, selected so that μY=0 in all cases. To find the dipoles, they used finite-field methods, with two fields that were equal in magnitude but opposite in sign. From Table 3, it is apparent that the results of Fu et al. (FZB) [22] agree well with the earlier results given by Meyer et al. (MBF) [13].

For the largest coefficients, A_{2023} and A_{0223}, our results are in excellent agreement with both of the earlier calculations: The percent differences between our results and those of Meyer et al. [13] are largest at R=5.0 a.u. (1.52%) and R=6.0 a.u. (0.99%); the remaining differences in these two coefficients average to 0.48%. We have obtained results at R=10.0 a.u., which were not given previously. The differences between our values for A_{2233} and those of Meyer et al. [13] are typically ~3% (smaller at R=4.0 a.u.). Differences in the values of A_{2021} and A_{0221} are ~5% or less at short range (R≤6.0 a.u.), where these coefficients have their largest values. At longer range, the absolute discrepancies are smaller, although the differences are larger on a relative basis. The principal differences in the dipole coefficients are attributable to the inclusion of A_{4043} and A_{0443} in our work; this affects the values of A_{4045}, A_{0445}, and A_{2211} (to a lesser extent).

In Table 3, the *ab initio* values of the coefficients are also compared with values based on the quadrupole-induced dipole model (QID) and the long range model (LR), which is complete through order R^{-7}. The LR calculations include hexadecapolar induction, back-induction, and van der Waals dispersion effects, in addition to quadrupolar induction. The QID and LR calculations are based on the value of the H_2 quadrupole computed by Poll and Wolniewicz [23], the value of Θ interpolated to r=1.449 a.u. given by Visser et al. [24], the hexadecapole computed by Karl et al. [25], the polarizabilities and E-tensor values given by Bishop and Pipin [26], and the dispersion dipoles computed from the polarizability and dipole-dipole-quadrupole polarizability at imaginary frequencies, also given by Bishop and Pipin [27].

The coefficient A_{2023} depends primarily on the quadrupole-induced dipole: the difference between the QID approximation and our result is ~12% at R=4.0 a.u., ~10.6% at R=5.0 a.u., ~6.5% for R=6.0 a.u., and smaller at larger R. The QID model gives remarkably good values for this coefficient, even when R is quite small. Agreement with the full long-range model is somewhat better, with errors of ~5.25% at R=5.0 a.u. and only 1.88% at R=4.0 a.u. Quadrupole-induced dipole effects are also present in the coefficient A_{2233}; this coefficient fits the QID and LR models quite well for R≥6.0 a.u., but the percent errors in these approximations are larger than those in A_{2023} for R=4.0 and 5.0 a.u. It should be noted that the back-induction and dispersion contributions have the same sign in A_{2023} but opposite signs in A_{2233}.

At long range the values of A_{4045} and A_{0445} depend on hexadecapolar induction, which varies as R^{-6}; there are no other contributions through order R^{-7}. We find strong agreement between the values of these coefficients and the hexadecapole-induced dipole terms (which determine LR), for R≥5.0 a.u.; short-range effects become significant when R is reduced to 4.0 a.u. In contrast, A_{2021}, A_{0221}, A_{2211}, A_{4043}, and A_{0443} seem to reflect the short-range exchange, overlap, and orbital distortion effects predominantly. For these coefficients, the leading long-range terms of back-induction and dispersion vary as R^{-7}; and they contribute with opposite signs in each case, further reducing the net effect of the long-range polarization mechanisms, in these particular dipole coefficients.

As noted above, we have carried out calculations with 28 different combinations of bond lengths in molecules 1 and 2. *Ab initio* calculations have been completed for pairs with each bond length combination, in each of the 17 relative orientations, at each of 7 separations between the centers of mass, and for at least six values of the applied field in the X, Y, or Z direction.

In the work of Meyer et al. on the absorption spectra of H2–H2 pairs in the fundamental band, results for the Cartesian components of the pair dipoles are listed for four nonredundant pairs of bond lengths, (r_o, r_o), (r_o, r_-), (r_o, r_+), and (r_-, r_+), with r_o=1.449 a.u., r_=1.111 a.u., and r_+=1.787 a.u. [13]. Fu et al. [22] augmented this set by the addition of a larger bond length, r_{++}=2.150 a.u., and reported results for all ten nonredundant pairs of configurations with the bond lengths drawn from the set {r_o,r_,r_+,r_{++}}. In the current work, we have included ro, three bond lengths smaller than ro (1.280 a.u., 1.111 a.u., and 0.942 a.u.), and four bond lengths larger than ro (1.787 a.u., 2.125 a.u., 2.463 a.u., and 2.801 a.u.), in order to examine new portions of the dipole surface, particularly those that may become significant for photon absorption at higher temperatures. The specific nonredundant length combinations used in the calculations are (r_1,r_2) = (2.801, 2.125), (2.801, 1.787), (2.801, 1.449), (2.801, 1.280), (2.801, 1.111), (2.801, 0.942), (2.463, 2.125), (2.463, 1.787), (2.463, 1.449), (2.463, 1.280), (2.463, 1.111), (2.463, 0.942), (2.125, 1.787), (2.125, 1.449), (2.125, 1.280), (2.125, 1.111), (2.125, 0.942), (1.787, 1.449), (1.787, 1.280), (1.787, 1.111), (1.787, 0.942), (1.449, 1.449), (1.449, 1.280), (1.449, 1.111), (1.449, 0.942), (1.280, 1.111), (1.280, 0.942), and (1.111, 0.942), with all bond lengths in a.u.

To illustrate the results for pairs with one or both bond lengths displaced from r_o (the averaged internuclear separation in the ground vibrational state of H_2), in Table 4 we list our values for the dipole expansion coefficients when r_1 = 1.787 a.u. and r_2 = 1.449 a.u., and we compare with the values given earlier by Fu et al. [22]. In general, we find excellent agreement. The values of A_{0001}, A_{2021}, A_{0221}, A_{2023}, A_{0223}, A_{2233}, A_{2243}, and A_{2245} agree quite closely, particularly given the extension of the basis set and the corrections for hyperpolarization effects included in the current work. A few of the coefficients show larger differences, based on differences in the fitting procedures. In the current work, we have omitted the coefficients A_{2221} and A_{2223}, which were included by Fu et al.; this contributes to the difference in the fitted values of A_{2211}. On the other hand, we have included A4043 and A0443, which were omitted by Fu et al. [22]; this probably accounts for the difference in the values of A4045 and A0445 shown in Table 4. Our inclusion of A_{4221}, A_{2421}, A_{4223}, A_{2423}, A_{4243}, A_{2443}, A_{4245}, A_{2445}, A_{4265}, A_{2465}, A_{4267}, and A_{2467} in the fitting procedure also causes slight shifts in the values of the other coefficients.

Table 4. Dipole expansion coefficients $A_{\lambda\lambda'\Lambda L}$ (in a.u., multiplied by 106) for H_2–H_2 with r1=1.787 a.u. and r2=1.449 a.u. The results from this calculation are compared with the results of Fu et al. (FZB), [22].

	R (a.u.)	4.0	5.0	6.0	7.0	8.0	9.0	10.0
A_{0001}	This work	−22960	−5786	−1241	−203	−10	13	9
	FZB	−21869	−5518	−1232	−231	−29	6	—
A_{2021}	This work	20653	4618	928	168	26	2	1
	FZB	21290	4688	963	194	45	15	—
A_{0221}	This work	−10595	−2394	−486	−95	−16	−1	−2
	FZB	−11028	−2450	−508	−108	−28	−10	—
A_{2023}	This work	−32456	−12335	−5392	−2735	−1554	−957	−624
	FZB	−32287	−12113	−5368	−2749	−1568	−966	—
A_{0223}	This work	23916	10071	4764	2525	1459	905	591
	FZB	23778	9865	4685	2488	1439	889	—
A_{2211}	This work	733	166	37	7	0	−2	0
	FZB	528	126	26	3	0	0	—
A_{2233}	This work	3079	1497	789	443	260	161	106
	FZB	2952	1415	750	423	253	158	—
A_{2243}	This work	−316	−242	−150	−88	−55	−34	−23
	FZB	−375	−263	−148	−83	−49	−30	—
A_{2245}	This work	416	180	86	39	19	10	5
	FZB	433	184	72	29	13	6	—
A_{4043}	This work	1981	529	123	29	8	3	1
A_{0443}	This work	−623	−185	−43	−9	0	1	0
A_{4045}	This work	−2079	−684	−224	−84	−35	−16	−9
	FZB	−3956	−1129	−322	−108	−43	−20	—
A_{0445}	This work	989	364	131	53	23	12	5
	FZB	1559	524	169	61	25	11	—

No previous results are available for comparison when one or both of the molecules in the pair have bond lengths of 0.942 a.u., 1.280 a.u., 2.125 a.u, 2.463 a.u., or 2.801 a.u. In Table 5, we provide results for one such combination of bond lengths, with r_1=2.463 a.u. and r_2=1.787 a.u. The coefficients listed in the top line of each set (and the corresponding coefficients for other pairs of bond lengths) were used in generating the rototranslational and vibrational spectra. These were obtained from fits that included 26 dipole coefficients all together (with A_{2211} and A_{2233}, but not A_{2221} and A_{2223}); immediately below those results in each set, we list values obtained from fits with 27 dipole coefficients (including A_{2221} and A_{2223}, but not A_{2211}). We find that the coefficients A_{0001}, A_{2021}, A_{2023}, A_{2243}, A_{2245}, A_{4043}, and A_{4045} are numerically "robust;" these coefficients are little affected by the difference in the fitting procedure. The coefficients A_{0221}, A_{0223}, A_{2233}, A_{0443}, and A_{0445} show greater sensitivity, although the agreement tends to improve as the separation between the molecular centers R increases (particularly for A_{0223} and A_{2233}). The full results for the new potential energy surface and the pair dipoles, with individual H_2 bond lengths ranging from 0.942 a.u. to 2.801 a.u., will be reported and analyzed in a subsequent paper. However, here we note that the coefficients A_{2023}, A_{0223}, A_{2233}, A_{4045}, and A_{0445} appear to be dominated by long-range induction mechanisms, specifically quadrupolar induction for A_{2023}, A_{0223}, and A_{2233}, hexadecapolar induction for A_{4045} and A_{0445}, and E-tensor induction for A_{2245}. When the logarithms of the absolute values of these coefficients are plotted versus the logarithms of the separations R between the molecular centers of mass, over the range from 8.0 a.u. to 10.0 a.u., the slopes are -4.20 for A_{2023}, -4.08 for A_{0223}, and -3.995 for A_{2233}, all close to the quadrupolar-induction value of -4. Similarly, the slopes are -6.42 for A_{4045} and -6.34 for A_{0445}, close to the value of -6 for hexadecapolar induction; and the slope is -5.84 for A_{2245}, close to the value of -6 for E-tensor induction [17].

Table 5. Dipole expansion coefficients AλλʹΛL (in a.u., multiplied by 106) for H2–H2 with r1 = 2.463 a.u. and r2 = 1.787 a.u. Results from the fit used to calculate the spectra (top line in each set) are compared with an alternate fit, which includes A2221 and A2223, but not A2211.

R (a.u.)	4.0	5.0	6.0	7.0	8.0	9.0	10.0
A_{0001}	−67727	−18429	−4357	−863	−119	8	14
	−65778	−19365	−4616	−960	−112	22	25
A_{2021}	61278	16009	3620	734	126	12	−3
	61859	15730	3543	705	128	16	0.5
A_{0221}	−23618	−6192	−1472	−322	−60	−9	−1
	−14285	−4351	−1033	−235	−47	−10	−1
A_{2023}	−77947	−28920	−11520	−5413	−2954	−1786	−1157
	−78659	−28577	−11425	−5378	−2956	−1791	−1161
A_{0223}	47485	20508	9741	5119	2948	1819	1185
	53866	22606	10265	5251	2955	1809	1178
A_{2211}	2759	674	164	37	4	−2	−1
A_{2221}	9063	937	201	10	16	8	7
A_{2223}	3803	2494	642	187	1	−19	−15
A_{2233}	6468	3571	1978	1144	678	424	278
	−2544	1369	1441	1024	666	430	282
A_{2243}	−1604	−788	−575	−362	−223	−143	−94
	−1711	−737	−561	−357	−223	−143	−95
A_{2245}	4102	1139	545	266	129	66	35
	4222	1082	530	260	129	67	36
A_{4043}	8512	2914	742	175	45	14	5
	8404	2966	756	181	44	13	4
A_{0443}	−1310	−616	−178	−41	−8	0	1
	−1531	−510	−149	−30	−8	−2	−1
A_{4045}	−7655	−3106	−984	−339	−134	−62	−32
	−7534	−3165	−1000	−345	−133	−61	−31
A_{0445}	2393	1086	415	162	70	33	17
	2640	968	382	150	71	35	19

About the Spectra

The absorption spectrum is a quasicontinuum, consisting of many thousand highly diffuse, unresolved "lines," corresponding to rotovibrational transitions from an initial state $\{v_1, j_1, v_2, j_2\}$, to a final state $\{v_1', j_1', v_2', j_2'\}$, of the binary collision complex. Under the conditions encountered in cool stellar atmospheres, vibrational quantum numbers v from 0 to about 5 occur with significant population numbers, with rotational quantum numbers j up to 20 or so, for H_2 molecules.

The isotropic potential approximation (IPA), which neglects the anisotropic terms of the intermolecular potential, is used for the calculation of the spectra [2].

Each "line" requires as input the matrix elements of the spherical dipole components [2]

$$\langle v_1 j_1 v_2 j_2 | A_{\lambda_1\lambda_2\Lambda L}(R, r_1, r_2) | v_1' j_1' v_2' j_2' \rangle, \tag{5}$$

and the isotropic component of the intermolecular potential for the initial (unprimed) state

$$\langle v_1 j_1 v_2 j_2 | V_{000}(R, r_1, r_2) | v_1 j_1 v_2 j_2 \rangle; \tag{6}$$

the potential for the final state is given by a similar expression, where all rotovibrational quantum numbers are primed. The line shape calculations proceed with these expressions as described elsewhere [2]. In (5), (6), as above, R designates the intermolecular separation and r_1, r_2 the intramolecular separations. The indices $\lambda_1\lambda_2\Lambda L$ are the expansion parameters of the spherical dipole components in (2).

Figure 1 shows the calculated absorption coefficient $\alpha(\nu; T)$, normalized by the numerical density ρ squared, at the temperature T of 297.5 K, and frequencies ν from 0 to 3000 cm^{-1} (the "rototranslational band"). Laboratory measurements [28] are shown for comparison (·). Good agreement of theory and measurements is observed.

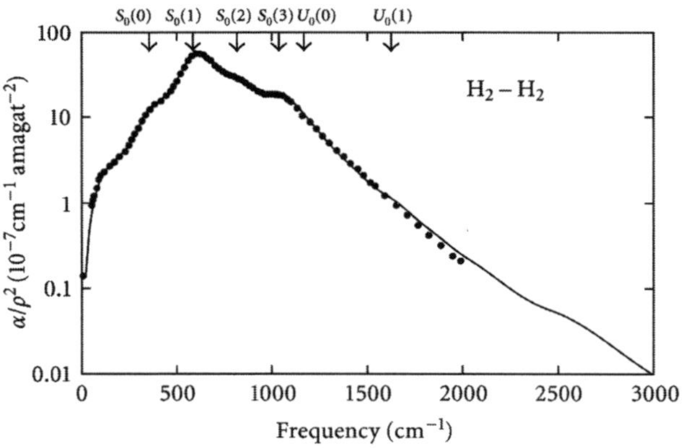

Figure 1. The calculated absorption spectrum of pairs of molecular hydrogen in the rototranslational band of H2, at the temperature of 297.5 K, and comparison with laboratory measurements (• from [28]).

We note that similarly good agreement of theory and measurement was previously observed, based on an earlier *ab initio* ID surface and a refined intermolecular potential [2, 12]. In the present work, a more complete induced dipole surface has been obtained and used, although the extension has not significantly affected the rototranslational band, shown in Figure 1. Additionally, a new potential energy surface has been obtained and used in the current work. This new potential surface (as well as the new ID surface) accounts for highly rotovibrationally excited H_2 molecules; and the new surfaces will be essential for our high-temperature opacity calculations—but again, the extensions of the potential surface are of little consequence for the rototranslational band, Figure 1, near room temperature.

The new potential surface is believed to be accurate in the repulsive region of the interaction, but it is not as extensively modeled in the well region, and at long range (dispersion part). Nevertheless, the measurements of the absorption spectra are as closely reproduced by the new *ab initio* input, Figure 1, as they are by the earlier advanced models. Apparently, the collision-induced absorption spectra arise mainly through interactions in the repulsive part of the potential, which is certainly consistent with previous observations [2].

The new opacity calculations of the fundamental and H_2 overtone bands [29] show similar agreement with measurements. Figure 2 shows the calculated normalized absorption coefficients over a frequency band ranging from the microwave region of the spectrum to the visible. In these calculations, we have used the exact equilibrium populations for the initial states, which at 2000 K consist of v=0, 1, and 2, with many different rotational states, including highly excited states. For the final states (after a photon of energy up to 2.5 eV has been absorbed), we have included much higher rotovibrational states of the molecules. We have accounted for all of these states rigorously, using the new intermolecular potential and induced dipole surfaces.

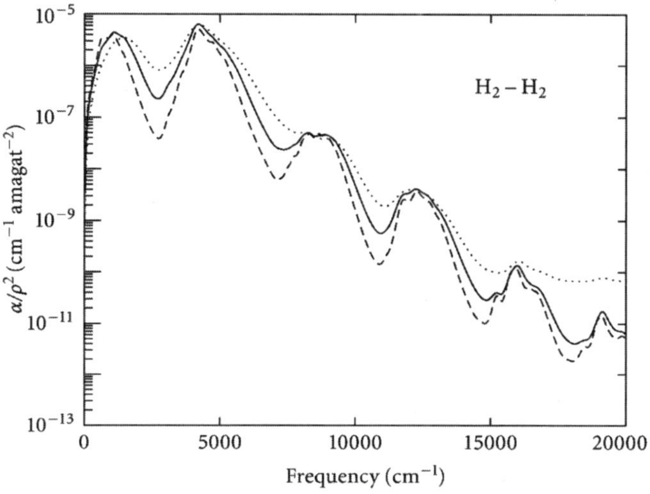

Figure 2. Calculated absorption spectrum of pairs of molecular hydrogen, from the far infrared to the visible, at the temperatures of 600 K (dashes), 1000 K (solid line), and 2000 K (dotted).

The coarse structures seen in the spectrum correspond roughly to the rototranslational band (peak near 600 cm^{-1}), the fundamental band of H_2 (peak near 4200 cm^{-1}), and the first through fourth overtone bands of H_2 (remaining peaks). Unfortunately, no measurements exist for these high-frequency data, but we feel that the results shown are of comparable reliability to the results in Figure 1.

Calculations of the type shown supplement previous estimates, especially at the highest frequencies [10, 30]. Presently, we are attempting calculations of H_2–H_2 opacities at still higher temperatures (up to 7000 K). Moreover, similar calculations are planned for H_2–He and H_2–H collisional complexes.

Conclusion

We report opacity calculations of collisional H_2–H_2 complexes for temperatures of thousands of kelvin and a frequency range from the microwave to the visible regions of the electromagnetic spectrum. The calculations are based on new *ab initio* induced dipole and potential energy surfaces of rotovibrating H_2 molecules, and are intended to facilitate modeling the atmospheres of cool stars. Agreement with earlier theoretical work and laboratory measurements, where these exist, is excellent.

Acknowledgements

This work has been supported in part by the National Science Foundation Grants AST-0709106 and AST-0708496 and by the National Natural Science Foundation of China Grant NSFC-10804008.

References

1. H. L. Welsh, "Pressure induced absorption spectra of hydrogen," in MTP International Review of Science. Physical Chemistry, Series One, Vol. III: Spectroscopy, A. D. Buckingham and D. A. Ramsay, Eds., chapter 3, pp. 33–71, Butterworths, London, UK, 1972.
2. L. Frommhold, Collision-Induced Absorption in Gases, Cambridge University Press, Cambridge, UK, 2006.
3. L. M. Trafton, "The thermal opacity in the major planets," Astrophysical Journal, vol. 140, p. 1340, 1964.
4. L. M. Trafton, "Planetary atmospheres: the role of collision-induced absorption," in Molecular Complexes in Earth's, Planetary, Cometary, and Interstellar

Atmospheres, A. A. Vigasin and Z. Slanina, Eds., pp. 177–193, World Scientific, Singapore, 1998.

5. J. Mould and J. Liebert, "Infrared photometry and atmospheric composition of cool white-dwarfs," Astrophysical Journal, vol. 226, pp. L29–L33, 1978.
6. P. Bergeron, D. Saumon, and F. Wesemael, "New model atmospheres for very cool white dwarfs with mixed H/He and pure He compositions," Astrophysical Journal, vol. 443, no. 2, pp. 764–779, 1995.
7. B. M. S. Hansen, "Old and blue white-dwarf stars as a detectable source of microlensing events," Nature, vol. 394, no. 6696, pp. 860–862, 1998.
8. D. Saumon and S. B. Jacobson, "Pure hydrogen model atmospheres for very cool white dwarfs," Astrophysical Journal, vol. 511, no. 2, pp. L107–L110, 1999.
9. P. Bergeron, S. K. Leggett, and M. T. Ruiz, "Photometric and spectroscopic analysis of cool white dwarfs with trigonometric parallax measurements," The Astrophysical Journal, Supplement Series, vol. 133, no. 2, pp. 413–449, 2001.
10. A. Borysow, U. G. Jørgensen, and Y. Fu, "High-temperature (1000–7000 K) collision-induced absorption of H2 pairs computed from the first principles, with application to cool and dense stellar atmospheres," Journal of Quantitative Spectroscopy and Radiative Transfer, vol. 68, no. 3, pp. 235–255, 2001.
11. W. Meyer and L. Frommhold, "Collision-induced rototranslational spectra of H2-He from an accurate *ab initio* dipole moment surface," Physical Review A, vol. 34, no. 4, pp. 2771–2779, 1986.
12. W. Meyer, L. Frommhold, and G. Birnbaum, "Rototranslational absorption spectra of H_2-H_2 pairs in the far infrared," Physical Review A, vol. 39, no. 5, pp. 2434–2448, 1989.
13. W. Meyer, A. Borysow, and L. Frommhold, "Absorption spectra of H2-H2 pairs in the fundamental band," Physical Review A, vol. 40, no. 12, pp. 6931–6949, 1989.
14. W. Meyer, A. Borysow, and L. Frommhold, "Collision-induced first overtone band of gaseous hydrogen from first principles," Physical Review A, vol. 47, no. 5, pp. 4065–4077, 1993.
15. H.-J. Werner, P. J. Knowles, J. Almlöf, et al., MOLPRO, Version 2000.1, Universität Stuttgart, Stuttgart, Germany and Cardiff University, Cardiff, UK, 2000.
16. X. Li, C. Ahuja, J. F. Harrison, and K. L. C. Hunt, "The collision-induced polarizability of a pair of hydrogen molecules," Journal of Chemical Physics, vol. 126, no. 21, Article ID 214302, 2007.

17. J. E. Bohr and K. L. C. Hunt, "Dipoles induced by long-range interactions between centrosymmetric linear molecules: theory and numerical results for H_2...H_2, H_2...N_2, and N_2...N_2," The Journal of Chemical Physics, vol. 87, no. 7, pp. 3821–3832, 1987.
18. X. Li and K. L. C. Hunt, "Transient, collision-induced dipoles in pairs of centrosymmetric, linear molecules at long range: Results from spherical-tensor analysis," The Journal of Chemical Physics, vol. 100, no. 12, pp. 9276–9278, 1994.
19. K. L. C. Hunt, "Long-range dipoles, quadrupoles, and hyperpolarizabilities of interacting inert-gas atoms," Chemical Physics Letters, vol. 70, no. 2, pp. 336–342, 1980.
20. L. Galatry and T. Gharbi, "The long-range dipole moment of two interacting spherical systems," Chemical Physics Letters, vol. 75, pp. 427–433, 1980.
21. K. L. C. Hunt and J. E. Bohr, "Effects of van der Waals interactions on molecular dipole moments: The role of field-induced fluctuation correlations," The Journal of Chemical Physics, vol. 83, no. 10, pp. 5198–5202, 1985.
22. Y. Fu, Ch. Zheng, and A. Borysow, "Quantum mechanical computations of collision-induced absorption in the second overtone band of hydrogen," Journal of Quantitative Spectroscopy and Radiative Transfer, vol. 67, no. 4, pp. 303–321, 2000.
23. J. D. Poll and L. Wolniewicz, "The quadrupole moment of the H_2 molecule," The Journal of Chemical Physics, vol. 68, no. 7, pp. 3053–3058, 1978.
24. F. Visser, P. E. S. Wormer, and W. P. J. H. Jacobs, "The nonempirical calculation of second-order molecular properties by means of effective states. III. Correlated dynamic polarizabilities and dispersion coefficients for He, Ne, H_2, N_2, and O_2," The Journal of Chemical Physics, vol. 82, no. 8, pp. 3753–3764, 1985.
25. G. Karl, J. D. Poll, and L. Wolniewicz, "Multipole moments of hydrogen molecule," Canadian Journal of Physics, vol. 53, pp. 1781–1790, 1975.
26. D. M. Bishop and J. Pipin, "Dipole, quadrupole, octupole, and dipole octupole polarizabilities at real and imaginary frequencies for H, He, and H2 and the dispersion-energy coefficients for interactions between them," International Journal of Quantum Chemistry, vol. 45, pp. 349–361, 1993.
27. D. M. Bishop and J. S. Pipin, "Calculation of the dispersion-dipole coefficients for interactions between H, He, and H2," The Journal of Chemical Physics, vol. 98, no. 5, pp. 4003–4008, 1993.

28. G. Bachet, E. R. Cohen, P. Dore, and G. Birnbaum, "The translational rotational absorption spectrum of hydrogen," Canadian Journal of Physics, vol. 61, no. 4, pp. 591–603, 1983.

29. M. Abel and L. Frommhold, To be published.

30. A. Borysow, U. G. Jørgensen, and Ch. Zheng, "Model atmospheres of cool, low-metallicity stars: The importance of collision-induced absorption," Astronomy and Astrophysics, vol. 324, no. 1, pp. 185–195, 1997.

Non-Linear Dielectric Spectroscopy of Microbiological Suspensions

Ernesto F. Treo and Carmelo J. Felice

ABSTRACT

Background

Non-linear dielectric spectroscopy (NLDS) of microorganism was characterized by the generation of harmonics in the polarization current when a microorganism suspension was exposed to a sinusoidal electric field. The biological nonlinear response initially described was not well verified by other authors and the results were susceptible to ambiguous interpretation. In this paper NLDS was performed to yeast suspension in tripolar and tetrapolar configuration with a recently developed analyzer.

Methods

Tripolar analysis was carried out by applying sinusoidal voltages up to 1 V at the electrode interface. Tetrapolar analysis was carried on with sinusoidal *field strengths from 0.1 V cm^{-1} to 70 V cm-1. Both analyses were performed within a frequency range from 1 Hz through 100 Hz. The harmonic amplitudes were Fourier-analyzed and expressed in dB. The third harmonic, as reported previously, was investigated. Statistical analysis (ANOVA) was used to test the effect of inhibitor an activator of the plasma membrane enzyme in the measured response.*

Results

No significant non-linearities were observed in tetrapolar analysis, and no observable changes occurred when inhibitor and activator were added to the suspension. Statistical analysis confirmed these results.

When a pure sinus voltage was applied to an electrode-yeast suspension interface, variations higher than 25 dB for the 3rd harmonic were observed. Variation higher than 20 dB in the 3rd harmonics has also been found when adding an inhibitor or activator of the membrane-bounded enzymes. These variations did not occur when the suspension was boiled.

Discussion

The lack of result in tetrapolar cells suggest that there is no, if any, harmonic generation in microbiological bulk suspension. The non-linear response observed was originated in the electrode-electrolyte interface. The frequency and voltage windows observed in previous tetrapolar analysis were repeated in the tripolar measurements, but maximum were not observed at the same values.

Conclusion

Contrary to previous assertions, no repeatable dielectric non-linearity was exhibited in the bulk suspensions tested under the field and frequency condition reported with this recently designed analyzer. Indeed, interface related harmonics were observed and monitored during biochemical stimuli. The changes were coherent with the expected biological response.

Background

An enzyme catalytic process is a cyclic reaction that responds to a periodic driving force with which the enzyme can interact. As a result of this interaction, the enzyme will oscillate between its different conformational states. For carrier enzymes, it has been assumed that this driving force is the free energy obtained from ATP hydrolysis. However, these enzymes can also couple energy conversion from transmembrane electric fields into an electrochemical gradient away from its equilibrium [1,2]. Membrane-bounded enzymes, as an electrochemical cyclic time-dependent processes, may appear as non-linear systems when exposed to a sinusoidal electric field [3]. This non-linear phenomenon was anticipated by simulations and analytical solutions [1,4-8].

Non-linear behavior of microbiological suspensions was tested with an externally applied sinusoidal electric field by the group of Woodward and co-workers [9-11]. Such non-linear system should generate harmonics on the polarization current and they can be measured with Fourier analysis [12]. Based on this analysis, Woodward designed a nonlinear spectrometer to evaluate some biological samples, mostly Saccharomyces cerevisiae suspensions [9]. They applied a sinusoidal voltage signal to a tetrapolar cell through the outer electrodes, they registered the voltage drop between the inner electrodes, and they analyzed it with Fourier series. The power spectrum obtained revealed the presence of harmonic components not included in the externally applied signal. However, each outer electrode generated an electrode-electrolyte interface (EEI) when contacted to the suspension. Within the voltage range evaluated, each EEI may be regarded as non-linear systems. The entire tetrapolar cell can be considered as a three-stage system. Two of these systems are certainly non-linear (the EEI), whereas the third one (the yeast suspension) is suspected to be so. In this scenario, the analysis of the signal measured in the inner electrodes reveals information about the biological medium and both EEI. If the measurements are carried out only in one culture cell containing yeast, there is no way to separate the EEI non-linearity from the biological one. To overcome the EEI interference, Woodward proposed to subtract the power spectrum of a similar cell with supernatant, from the power spectrum of the cell containing yeasts. This difference was attributed to the non-linear behavior of the yeasts in the suspension. The most representative finding was a prominent difference close to -25 dB of the third harmonic when the input signal was a sinusoidal electric field of $\pm$ 2 V cm^{-1} and 20 Hz.

Later, Hutchings and co-workers, with different instrumentation, reported the absence of the non-linear behavior [13] suggesting the interfacial origin of the non-linearity [14] reported by Woodward. However, the arrangement proposed by Woodward was tested with other microorganisms [15,16]. In any case, the consistency of results obtained revealed a biological contribution to the non-linear phenomenon [17].

Since then some improvements were made to the instrumentation and to the data analysis [18,19] and several solutions were tried to avoid the unstable behavior of the EEIs [20], but there was not a great deal of advance in this promissory technique. Similar results were obtained with a radical different technology when Nawarathna, Miller and co-workers used both bipolar and tetrapolar analyzers. The former was depicted with two gold electrode and the current was sensed with a magnetic device [21], and the latter was a four gold electrodes arrangement, similar to Woodward's [22]. They could partially reproduce Woodward's result using the bipolar configuration [23,24]. Certainly, the non-linear nature of electrode polarization was recognized as a significant obstacle [25] by other researchers.

We believe that in the cited literature the instrumentation has not properly been designed to eliminate the EEI contribution, and the interfaces involved have not adequately been taken into account. None of these authors have carried out a better methodological characterized of these interfaces. They have just briefly stated that the interfaces were cleaned firmed [9], or just cleaned with steel wool [13]. Neither roughness nor voltage range applied to each EEI have been measured or analyzed.

Consequently, a series of experiments were carried out to solve the controversy to either accept or reject the above authors' statements. In recent years, our group developed a non-linear dielectric spectrometer based on a commercial analyzer, which is expected to solve the EEI interference in tetrapolar cells [26,27]. In order to test the theories of these authors, tripolar and tetrapolar cells were used. The preparation and treatment of the interfaces were carefully carried out to perform the measurements.

Our findings suggest that yeast suspension does not show non-linearities within the applied voltage and frequency ranges; it rather affects the intrinsic EEI behavior. We have monitored changes in the non-linear characteristics of the EEI in presence of yeast with the addition of biochemical activators and inhibitors.

Methods

Non-Linear Spectrometer

The spectrometric system proposed is depicted in fig. 1(a) and consists of a central PC that synchronizes the instruments and collects the data, an electrochemical analyzer Solartron SI1287, a peristaltic pump, a magnetic stirrer and the measurement cell (drawn in subset b-d). The PC is connected to the SI1287 (via serial port), and generates and acquires voltage signals through digital-analog and analog-digital converters (National Instruments).

Details about the operation of the equipment have been published elsewhere [26]. Briefly, the computer generates a sinusoidal voltage signal which is applied through the SI1287 to the biological suspension contained in the cell, by means of two outer electrodes. The voltage is controlled by the SI1287 to provide a sinusoidal (with no harmonics) electric field between the inner electrodes (namely reference electrodes). The current that circulates through the cell is coherently sampled and Fourier transformed by using the periodogram method with a rectangular window [28]. The signal length is chosen to provide, at least, a five samples separation between harmonics.

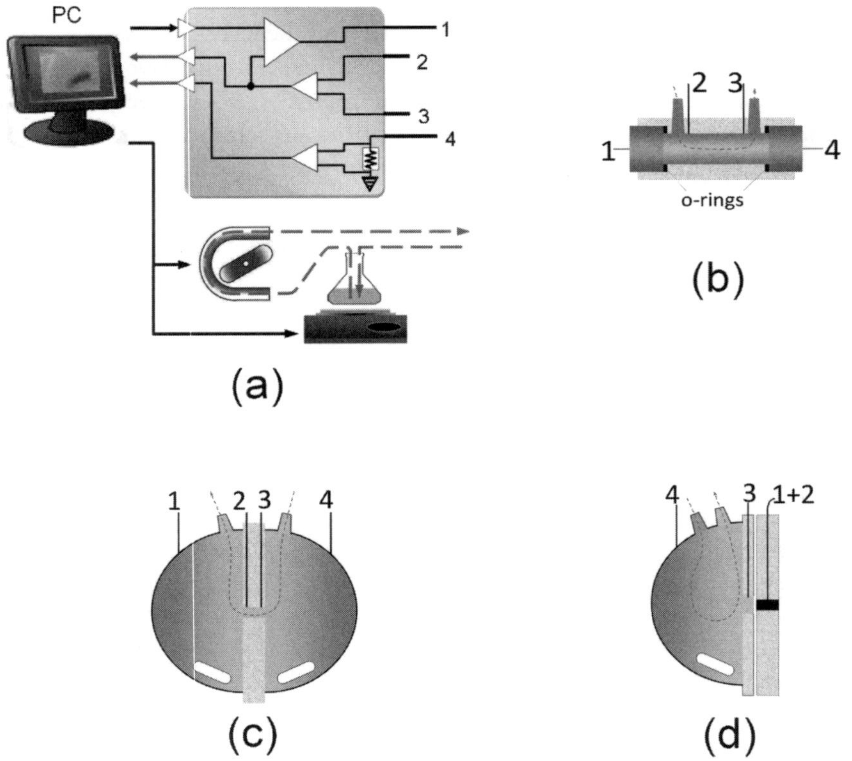

Figure 1. Non-linear dielectric spectrometer. (a) Equipment involved, (b) cylindrical flat electrodes tetrapolar cell, (c) low EEI impedance tetrapolar cell, and (d) tripolar cell. In (b) and (c) 1 and 4 correspond to outer electrodes, while 2 and 3 correspond to inner electrodes. In (d) 1 and 2 are connected together and the system acts as a tripolar analyzer, instead of a tetrapolar one. Numbers 1 to 4 correspond to the connections of the SI1287 unit depicted in (a).

The frequency estimate is obtained for the entire range, but only the amplitude of the fundamental frequency and its harmonics (2nd to 5th) are stored. For each harmonic analyzed, an amplitude grid is computed when both frequency and amplitude of the input signal are changed in discrete intervals. This grid can be used to plot a surface as function of frequency and amplitude of the input signal. All computing and operations are performed by the PC, the SI1287 only serves as electrochemical interface, but does not perform any analysis. All voltages are expressed as RMS values.

The capability of the system was tested with linear and nonlinear phantoms. It was proved that the harmonic content measured in the current signal is strictly dependent on the linear/non-linear system connected between the reference electrodes [27].

The system is completed with a reservoir with solution and a pumping system that recycles the solution through the cell. The stirrer and pump were turned on for 10 seconds after completing 90 seconds of acquisition. Whenever any product was added to the suspension (inhibitors or activator of the H+-ATPase) the mixture was stirred and pumped for 1–2 minutes to homogenize the system. Both pump and stirrer are controlled by the PC in order to ensure steady conditions during acquisition.

Tetrapolar Cells

Two different types of tetrapolar cells were developed. One of them (fig. 1(b)) was designed to provide a very uniform electric field within the biological material. It consists in a cylindrical acrylic small volume piece (length 40 mm, diameter 10 mm) with two flat stainless steel outer electrodes (AISI 304). The two inner electrodes are implemented with Dentaurum steel wire (diameter d = 1 mm), separated 30 mm in between. The electrodes and cell were designed to let the set be mounted and unmounted before each experiment. The electrodes were polished with metallographic sandpaper grit 600 (average particle size of 14.5 µm, Buehler discs, USA) at a speed of 280 RPM. Two tubing connections are provided to remove bubbles and to recycle the suspension and avoid yeast settling inside the cell.

The other tetrapolar cell (fig. 1(c)) consist of by two outer hemi spherical electrodes, each one with an area of about 150 cm2 and total cell volume of 400 cm3. An acrylic piece separates these two chambers, with a small perforation in the center which allows the communication between them. Two additional parallel flat embedded stainless steel electrodes serve as references (Dentaurum steel, d = 0.75 mm). The central piece has a hollow with a diameter d = 2.75 mm (area ≈ 6 mm2) and length l = 1 mm. The impedance between reference electrodes whit the cell filled with saline solution (NaCl 0.9%) was 830Ω, which is much greater than the EEI impedance of the outer electrodes. The central acrylic piece can be changed to achieve different dimension of the central hollow, and thus, different medium impedance.

Electric field distribution in each cell was analyzed by means of finite elements. The freely available FEMM software http://femm.foster-miller.net/ webcite was used to solve the current (AC) electromagnetic problem. Both metallic and acrylics parts of the cell were drawn, along with the suspension. Each cell was drawn and the mesh spacing was varied according to the cell's geometry. Dielectric parameters (conductivity σ and relative permittivity ε) adopted for the materials are shown in table 1.

Table 1. Dielectric parameters used in the FE analysis

Material	Conductivity σ (S/m)	Relative permittivity ε
Yeast suspension	0.65	1000
Acrylic	1×10^{-9}	3
Gold	2.2×10^{6}	1
AISI 304	1.3×10^{6}	1

The EEIs were not modeled, and voltage was applied to the boundaries of the injection electrodes in all cells. The field frequency was fixed at 20 Hz, and its magnitude was adjusted to match an estimated 1 V/cm electric field between the reference electrodes. In the longitudinal cells (fig. 1(b) and 1(c)) it was simply calculated as 1 V/cm multiplied by the separation between the outer electrodes. In the spherical cell (fig. 1(d)), the voltage applied to the external electrodes was manually adjusted to match an electric field of 1 V/cm between reference electrodes. Woodward's cell was configured as planar and the other two were modeled as axisymetric relative to the y-axis.

Tripolar Cells

The tripolar cell (fig. 1(d)) has a central working electrode (circular with diameter d = 8 mm) bounded in an acrylic disc (external diameter d = 100 mm), while a second thin central-hollowed disc (external diameter = 100 mm, diameter of the central hollow d = 10 mm) with a stainless steel wire acts as a reference electrode (Dentaurum steel, d = 1 mm). The working electrode is hand-polished before each experiment with diamond past and aluminum powder, up to a final roughness of 1 μm. We tested a golden electrode (24 carat) and a stainless steel (AISI 304) working electrode. We have also tested the stainless steel electrode polished with sandpaper under the same conditions of the tetrapolar cell electrodes. The system is completed with a large-area hemi-spherical stainless steel counter electrode. Thus the counter electrode-electrolyte-interface impedance (Z_{ce}) is much smaller than the working electrode-electrolyte-interface impedance (Z_{we}). The counter electrode is hermetically secured with rubber seals over the acrylic disc, and is provided with tubes to allow the suspension to flow-in and out. A magnetic stirring bar is introduced inside the cell.

In the tripolar cell the voltage is controlled between working and reference electrodes, that is, the voltage drop in the working electrode-electrolyte-interface.

The loss in the medium or counter electrode interface is negligible, due to the low medium impedance Z_{medium} ($Z_{Medium} + Z_{ce} \ll Z_{we}$).

Initially, two opposite tripolar (ideally equal) cells were used, one as reference and the other one as test cell. The responses were subtracted to correct the non-linear behavior of the interface. However, this became almost unpractical because these two interfaces were hardly similar (in their impedance response) even they were made with identical material and equally treated [26]. As an alternative, we decided to use the same interface repeatedly and test the biological response by using an inhibitor or activator of the enzyme, as it was also carried out by Woodward in his latest publications [18,19]. After some reference measurements, the activator was added and some test measurements were performed.

Experimental procedures

Experiments were initially divided in two groups: bulk or electrode-interface measurements carried out either with tetrapolar or tripolar cells, respectively. In each case all harmonics were analyzed, but statistics and graphics correspond only to the third harmonic because it has been showed to be the most representative of the biological response (see Woodward's and Nawarathna's references).

The tetrapolar experiments were performed by using a single cell and consecutive measurements up to four hours after yeast hydrating. These procedures were divided in two stages. During the first two hours, measurements were carried out on the hydrated yeast, subsequently, a biochemical stimulus was added and measurements were continued for two hours. Measurements before and after the stimulus are used as reference and test, respectively. The stimuli were selected upon the previous references and they are sodium metavanadate (SMV, 1 mM, an inhibitor of the enzyme [29]) and glucose (100 mM, substrate of the enzyme, the activator [30,31]). Both affect the H+-ATPase, which was recognized as the source of the non-linear biological response.

A total of 12 experiments were carried for tetrapolar cells with several ranges of electric field. The lowest electric field range tested varied from 0.1 V cm^{-1} to 1.5 V cm^{-1}, whereas the maximum ranged from 7 V cm^{-1} to 70 V cm^{-1}. This high field range was attained with the cell depicted in fig 1(c). The field range was divided in 11 or 21 logarithmic steps. The frequency range was always the same, between 1 Hz and 100 Hz, divided both into 11 or 21 logarithmic steps. Most of previous harmonics generation reports in yeast suspension were accounted in the 1 Hz–100 Hz range.

The amount of measurements depends on the length of both frequency and voltage ranges. When both ranges were built with 11 steps, each complete measurement (11×11 single measurement) took about 10 minutes to be completed. These experiments usually covered 10 measurements previous and 10 measurements posterior to the stimuli. When the ranges were divided into 21 steps, each complete measurement took over 40 minutes. In this case, each experiment consisted of 2~3 measurements previous and posterior to the stimuli.

Since electrode corrosion does not influence tetrapolar measurements, and yeast cell count does not significantly vary during the experiments, repeated measurement can be considered independent observations. Tetrapolar data were analyzed with the ANOVA model for repeated measurements with no interaction between factors. Frequency, electric field and type of measurement (reference or test measurement) are treated as independent predictors. Amplitude of the third harmonic, expressed in dB, is used as dependent variable of the model. Values of p less than 0.01 are considered significant.

Interface measurements were carried out likewise, with reference and test measurements and stimulus in between. However, as tripolar measurements are influenced by electrode electrochemical corrosion, they were divided in two groups, single and repeated (> 5) measurements prior and after the stimulus. Single measurements are intended to avoid as much as possible electrode corrosion. Consequently, only two measurements are carried out with the stimuli in between. In long term experiments, it is expected to observe a change in the harmonic amplitude coincident with the biochemical stimuli, along with a drift due to electrode corrosion. In such cases, any response should be observed as a time depending process. It was also tested the difference (subtraction) of the average measurement previous and posterior to the stimulus. No statistics analysis was performed over the tripolar measurements, and the data provided is analyzed qualitatively. These experiments were carried out with both gold and stainless steel electrodes with a voltage range of 150 mV~1.5 V between 1 Hz and 100 Hz applied to the working-electrode-electrolyte interface.

Samples preparation

In order to facilitate comparison with results reported by Woodward, we used the same biological material (e.g. S. cerevisiae). The microorganism was obtained locally as a freeze-dried powder, and resuspended up to a concentration of ca. 50 mg dry wt ml-1, in the same saline solution. All chemicals used were of analytical grade and the water was glass-distilled (final conductivity less than 5 µS cm^{-1}). The suspension concentration (109 CFU ml^{-1}) was determined by plate count.

Results

Electric Field Distribution in Tetrapolar Cells

The electric field distribution obtained with the software simulation is presented in fig. 2. The first cell depicted corresponds to Woodward's design (fig. 2(a)). The voltage difference between the boundary of the injection electrodes was 0.75 V (their center are separated by 7.5 mm). The geometry and the field distribution are highly non-uniform in this cell. There are peak values as high as 3 V/cm close to the injection electrodes, and a large peripheral area with much lower values (< 0.5 V/cm). A histogram analysis performed to fig. 2(a) (data not shown) revealed that ~60% of the electric field is minor than 0. 5 V/cm and only 11.3% of the field ranges into the 0.85 V/cm~1.05 V/cm interval.

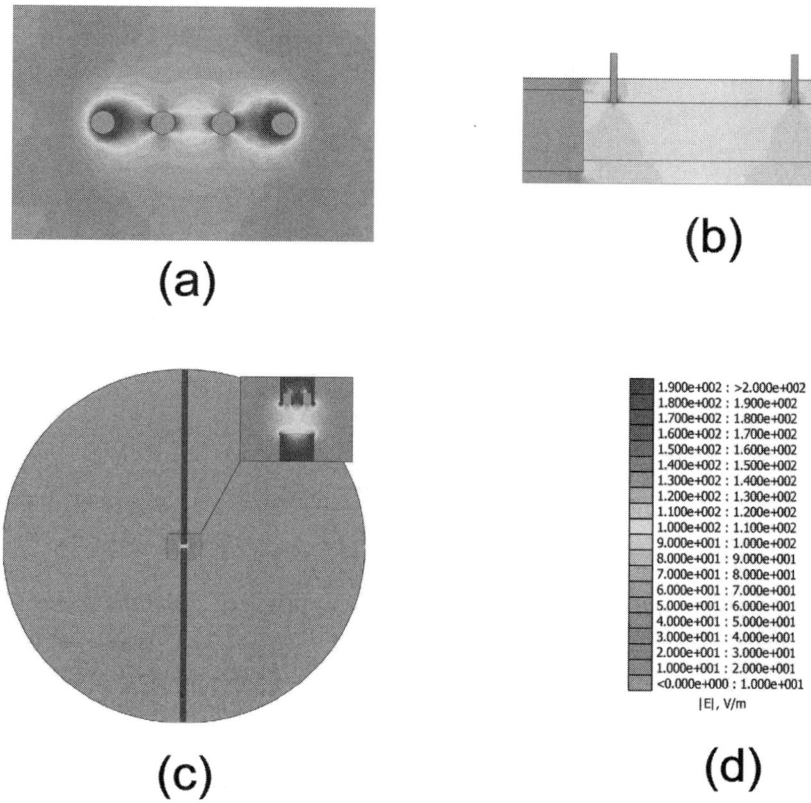

Figure 2. Electric field distribution in the tetrapolar cells. Woodward's cell, (b) tetrapolar cell with flat electrodes, and (c) low EEI cell. All figures have been color-coded according to the scale showed in (d). For a better visualization, the field scale was clipped to a maximum of 2 V/cm. However higher values (~3.5 V/cm) were observed close to the outer electrodes in (a) and close to the acrylic corners in (c).

The analysis of the flat electrodes cylindrical cell (fig. 2(b)) revealed a very uniform electric field but still a field gradient (between 0.2 V/cm and 2 V/cm) was observed in the region close to the inner electrodes. However, 95% of the computed electric field was close to 1 V/cm.

The high volume cell (fig. 2(c)) revealed a wider range of electric field measured in the inner hole. This is mainly due to the gradient generated at the boundaries of the inner hole and also close to the inner electrodes. The field values in this inner hole are normally distributed around the mean value of 1 V/cm.

Experimental Procedures

The first five harmonics were obtained, however, the third harmonic was considered to be the most important and we shall refer to it from now on. We first analyze the measurements performed with tetrapolar cells. Results of the statistical analysis along with some experimental parameters are presented in table 2.

Table 2. Results of the tetrapolar analysis

#	Cell	Electric field [V/cm]	Frequency [Hz]	Inhibitor or activator	# of measurements before-after	*p*
1	Cylindrical	0.16~1.33	1~100	MVS	9 -- 10	0.72
2	Cylindrical	0.1~1.5	1~100	MVS	12 -- 18	0.09
3	Cylindrical	0.1~1.5 *	1~100 *	MVS	3 -- 5	0.89
4	Cylindrical	0.16~1.67	5~50	MVS	12 -- 17	0.96
5	Hemispheric	0.5~15	1~100	MVS	9 -- 10	0.38
6	Hemispheric	7~70 *	1~1000 *	MVS	2 -- 2	0.39
7	Hemispheric	0.1~5	1~100	MVS	6 -- 9	0.14
8	Hemispheric	0.5~30	1~100	Glucose	6 -- 5	0.60
9	Hemispheric	0.5~30 *	1~100 *	MVS	5 -- 3	0.65
10	Hemispheric	0.3~30 *	1~100 *	Glucose	3 -- 5	0.28
11	Hemispheric	0.3~30	1~100	Glucose	6 -- 6	0.27
12	Hemispheric	0.3~30	1~100	Glucose	6 -- 7	0.07

In every tetrapolar measurement the relative amplitude of the third harmonic (as expressed to the fundamental amplitude) was lower than -60 dB. Neither electric field nor frequency seemed to affect the harmonic content observed. Furthermore, addition of SMV or glucose did not either modify the harmonics. Fig. 3 shows the difference between averaged measurements previous and posterior to a single application of SMV, which corresponds to the first experiment described in table 2. The algebraic operations were performed upon single surfaces expressed in dB.

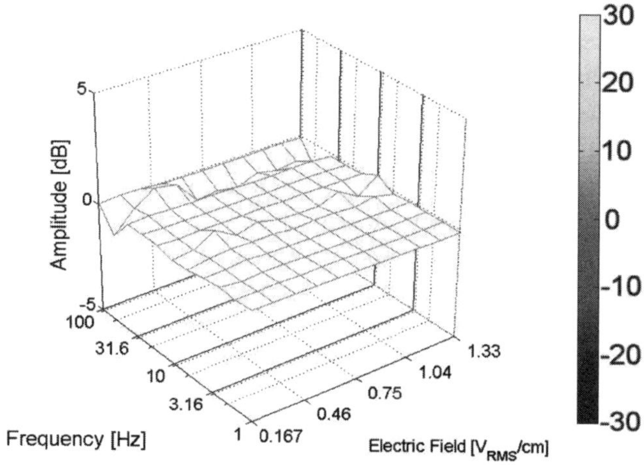

Figure 3. Difference between averaged measurements before and after addition of SMV in tetrapolar cells. The horizontal axes stand for the frequency and amplitude of the applied electric field. The vertical axis stands for the amplitude of the third harmonic measured at that combination of frequency and amplitude. The experiment was performed with the flat parallel electrodes cylindrical cell showed in fig. 1(b). Frequency range was 1 Hz–100 Hz (logarithmic scale, 11 steps), and field range was: (a) 0.167 V cm^{-1} –1.33 V cm^{-1} (logarithmic scale, 11 steps).

Surface Is Color-Coded According to the Color Bar Presented

The surface obtained is almost flat, with its amplitude very close to 0 dB, with no response attributable to biological source. The same result was obtained when the stimulus was glucose and when other harmonics were analyzed. The same response was obtained in other experiments using even higher voltages (up to 1.67 V cm^{-1}). A total of four experiments are reported using the same cell and moderated electric fields, all tested with SMV (row 1 to 4, in table 2).

The low EEI impedance hemispheric tetrapolar cell was then evaluated with even higher electric fields, but the results obtained were almost identical to the former cell. The response was a noisy surface, with its amplitude comprised between -5 dB and 5 dB, but no reproducible biological response was observed. It was also tested the effect of glucose, and also all harmonic were analyzed, and the same results were obtained. All these experiments are detailed in table 2 (rows 5 to 12).

There was no evidence of harmonic generation or biological interaction within the electric field and frequency ranges tested. Statistical analysis revealed no difference between test and reference measurements ($p > 0.01$) for all the tetrapolar experiments performed and this result was independent from the type of stimuli and cell used.

After performing tetrapolar measurements without noticeable results, measurements at the EEI in the tripolar configuration were carried out as showed in fig. 1(d). Earliest experiments were conducted with only two measurements with the stimulus in between, to reduce the electrode corrosion process.

When SMV was added, a reproducible change in the third harmonic amplitude was observed, as presented in fig. 4. Subset (a) shows the difference between the measurements before and after the SMV injection. Two distinctive peaks were observed at the surface, and their amplitude was greater than 25 dB. For each peak, the corresponding frequency spectra were plotted at the right of the image (subsets (b) and (c)). From top to bottom, each spectrum corresponds to test (hydrated yeast), reference (hydrated yeast plus SMV) and its difference, respectively. Both peaks are generated because the third harmonic was decreased or abolished after the addition of SMV.

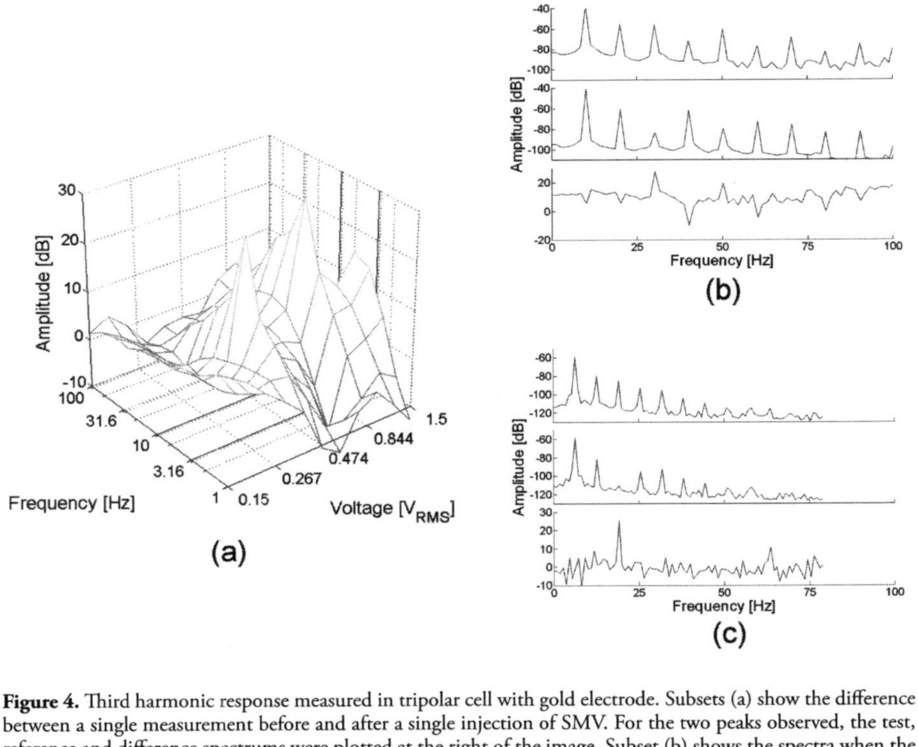

Figure 4. Third harmonic response measured in tripolar cell with gold electrode. Subsets (a) show the difference between a single measurement before and after a single injection of SMV. For the two peaks observed, the test, reference and difference spectrums were plotted at the right of the image. Subset (b) shows the spectra when the signal strength was 946 mV at 10 Hz. The difference measured for the third harmonic is 27.4 dB peak. Subset (c) shows the spectra when the signal strength was 376 mV at 6.3 Hz. The difference measured for the third harmonic is 26.3 dB. The surface in (a) is color-coded according to the color bar presented in fig. 2.

The experiment was repeated, but mostly only one peak was observed. However, in every case the third harmonic was abolished after the addition of SMV, and there is a difference close to 25 dB between the test and reference spectrum. The frequency and voltage values for every peak detected are detailed in table 3.

Table 3. Results of tripolar analysis

#	Material	Frequency [Hz]	Voltage [V_{RMS}]	Amplitude [dB]
1	Gold	10.0	0.94	27.4
2	Gold	6.31	0.37	26.3
3	Gold	8.57	0.69	42.4
4	Gold	6.31	0.94	26.3
5	AISI 304/1 μm	6.31	0.94	-20.6
6	AISI 304/papersand 600	3.98	0.94	-25.1
7	Gold	10.0	0.46	14.5
8	Gold	25.1	0.46	10.0
9	Gold/glucose	15.9	0.94	11.8

When using the polished (1 μm diamond past) stainless steel electrode (fig. 5), the surface showed a valley (instead of a peak) for very low frequencies (~4 Hz). However, for this material, the third harmonic was not present in the test spectrum, and it appeared in the reference spectrum. Similar result was observed when the electrode was polished with sandpaper grit 600 (table 3, experiments 5 and 6 respectively).

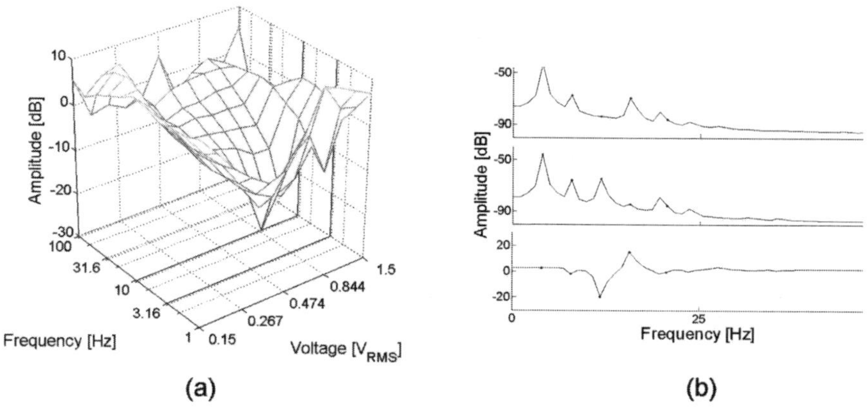

Figure 5. Third harmonic response measured in tripolar cell with stainless steel electrode. Subsets (a) show the difference between a single measurement before and after a single injection of SMV. For the valley observed, the test, reference and difference spectrums were plotted at the right of the image, in subset (b). The surface in (a) is color-coded according to the color bar presented in fig. 2.

Thereafter, several measurements were carried out, the stimulus was applied and then measured again. The measurements were performed throughout the entire voltage and frequency range stated in materials and methods. After that, the amplitude measured for a single combination of voltage and frequency was extracted from each surface and plotted against the time of measurement. Time zero corresponds to the time when hydration and mixture of the dried yeast with the saline solution started.

Fig. 6 shows the temporal evolution of the third harmonic in two different experiments. Both lines correspond to a voltage of 460 mV, but the frequencies were 25 Hz (-■-) and 10 Hz (-o-) for each experiment. The arrows indicate the time when SMV was added. There is a shift of the harmonic baseline and its amplitude decreased after adding SMV. These two plots corresponds to experiments #7 and #8 of table 3. Subset (b) shows the time evolution of the third harmonic at 946 mV and 15.9 Hz, when glucose (100 mM) was added, as indicated by the arrow. A decrease of 30 dB was observed after ~30 minutes, and it was reestablished one hour after the glucose injection (experiment #9 in table 3).

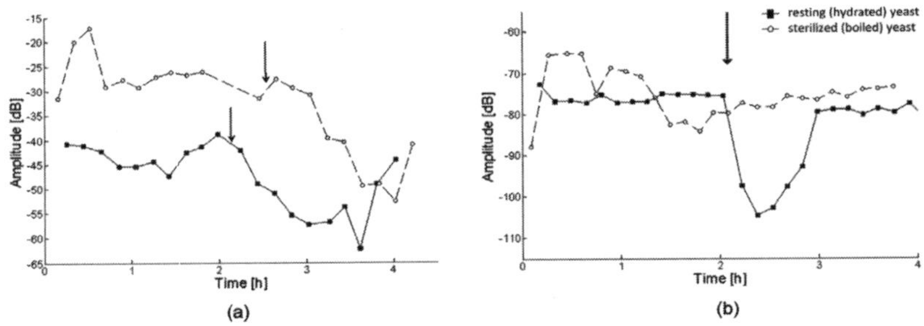

Figure 6. Time evolution of the third harmonic when measured repeatedly on gold electrode adding SMV and glucose. (a) two different experiments where 1 mM of SMV was added at the time indicated by the arrow. (b) Effect of glucose on resting and sterilized yeast. Glucose (100 mM) was added at the time indicated by the arrow in both experiments.

When measurements were repeated with the stainless steel electrodes, the electrochemical corrosion of the electrode produced a high drift of the baseline, previous to the stimulus injection. We did not observe reproducible responses on steel electrode, despite of the electrode polishing or previous experimental treatment. The magnitude of the drift was close to 20 dB/hour and did not stabilize during the four hours of experiments (data not shown).

Discussion

Both bulk medium and electrode-interfaces were tested with the same treatment and analysis in order to obtain biological non-linear interaction. In all cases, changes were analyzed up to the fifth harmonic, and the relevant ones were documented here. In order to confirm true bulk non-linear response, tetrapolar cells were first used. No relevant response were found for any harmonics, independently of the cell arrangement (low EEI or flat parallel electrodes), despite that high electric fields (> 10 V cm^{-1}) were applied. Statistical analysis confirmed these statements. All measurements showed linear response, and no variations were observed even when SMV or glucose was added.

Repeatable results were found when the analysis was focused on the current through the EEI, while sustaining a sinusoidal voltage. Changes have been observed in the 3rd harmonic when SMV was added, and the spectral changes observed, close to 30 dB, were similar to the previously published. However, the voltage and frequency of appearance of each peak were not the same in all experiments; there was a day-by-day variation in the exact magnitude and location of the 3rd harmonic.

Time evolution changes were observed when both SMV and glucose were added. These changes are also consistent with previous results of other authors [18,23], nevertheless changes did not occur always at the same voltage-frequency values.

Due to the non-linear nature of the EEI, tripolar measurements must be corrected to "subtract" the interface contribution. Consequently, stable interface are still required with this arrangement. The cells were intended to facilitate the polishing procedure, but this did not guarantee repeatable interfaces. We have tested other materials for electrodes, including copper and graphite. The graphite electrode was the most stable one, but no detectable changes were found when the stimulus was applied. We have also tested several polishing degrees for stainless steel electrodes, but it was very unstable and suffered rapid and severe corrosion due to the high voltage applied. The effect of electrode instability and material was extensively analyzed by Woodward and co-workers [18]. Gold electrodes were chosen among several metal to be the most stable one, but the surface treatment of the material had not been emphasized. Our results were highly dependent on the polishing degree, and reproducible results were only observed for polished gold electrodes.

A brief analysis of all instrumentation of cited groups reveals that there are neither two identical spectrometers nor two identical measurement cells. All reported spectrometers used sinusoidal generators applied to outer electrodes of bi and tetrapolar cells, and none of them controlled the waveform of the electric field measured applied to the bulk. This required further correction, as performed by Woodward and Claycomb [9-11,15,16,18,23,24]. Other chose to use a single cell and traced the change in the harmonics amplitude [25,32]. Our design compensated well enough the distortion of the EEI to ensure a true sine electric field applied to the bulk without need for correction. Medium-related harmonics in the voltage signal were sensed and corrected by the driver electrodes, and harmonics were only present in the sensed current.

The cell geometry should also be analyzed as a possible source or harmonics generation. The electric field presented in fig. 2(a) shows how the cell geometry affects the potential measured by the sensing electrodes. The electric field distribution, governed by differential equations, is nothing but homogenous. It seems feasible to observe harmonics in the voltage sensed by the inner electrodes, even if in a true tetrapolar sinusoidal signal was applied to the injection (outer) electrodes. Similar results would be obtained with the pin-type cell used by other groups [23,24,32].

The tetrapolar cells developed and presented in this paper were designed taking into consideration the electric field distribution. The difference between Woodward's results and ours could be related not only to the waveform of the electric field, but also the magnitude and dispersion of the electric field. Even though there is no enough evidence to relate harmonic generation to the cell geometry, the geometry should be considered in future analysis and simulations. On the contrary, it has been pointed out experimentally that linear homogeneous electric fields do not generate harmonics in the sensed current.

Our findings are partially consistent with the hypothesis of Blake-Coleman and coworkers, [13,14] where the interface region is responsible for the generation of harmonics, and it can be altered by the presence of biological cells. It has been proved that cell presence can modify the linear and non-linear properties of the interface region [32-34]. The proposed mechanism of linear interaction, however, must be extended to the non-liner domain to explain these finding. Certainly, the interface width and the electrode roughness are several decades lower than the cell size [17]. The interface will probably be better affected by ionic modifications due to the presence of the yeast cell, rather than a direct modification of the double-layer itself.

Recently, the patch-clamp technique was employed successfully to synchronize and drive the energy conversion of the Na+/K+ ATPase [35-37]. This procedure simplifies the dielectric suspension model (there is a direct measurement of the transmembrane voltage and ionic current). It also resolves the random orientation of the enzymes to the electric field and the membrane surface conductivity of the cell. The information about enzymatic rates, field amplitudes, and signal symmetry should be integrated into a dielectric model of the suspension to provide an equivalent whole-cell equivalent model. This information would contribute to understand the macroscopic dielectric phenomenon of the non-linear interaction.

Conclusion

The biological phenomenon interaction well reported by many authors (as proved with enzyme inhibitors, variable cell concentration, glucose addition, yeast sterilization and genetic alterations) could not be reproduced in tetrapolar analysis. There was no significant harmonics generation in the bulk suspensions of the organisms tested so far, when applying a pure sinusoidal electric field up to 70 VRMS cm^{-1}.

Our results emphasize the hypothesis that there is no electric distortion in a biological suspension when a strictly sinusoidal electric field is applied. The harmonics observed were due principally to electrode-generated non-linearities which affected the bulk current, and thus modulated the sensed voltage. The cell-induced modulation of the harmonics produced at an electrode-suspension interface may be modified by the microorganism response to inhibition or activation.

Competing Interests

The authors declare that they have no competing interests.

Authors' Contributions

EFT carried out the spectrometer development and the experimental procedures, and drafted the manuscript. EFT and CJF participated in the design of the study and performed the statistical analysis. CJF conceived of the study, and participated in its design and coordination and helped to draft the manuscript. All authors read and approved the final manuscript.

Acknowledgements

This work was supported by grants from the Agencia Nacional de Promoción Científica y Tecnológica, the Consejo Nacional de Investigaciones Científicas y Técnicas (CONICET), the Consejo de Investigaciones de la Universidad Nacional de Tucumán (CIUNT), and Institutional funds from INSIBIO (Instituto Superior de Investigaciones Biológicas). We also thank Professor Santiago Caminos for his valuable help in the translation of this paper.

References

1. Tsong TY, Liu DS, Chauvin F, Gaigalas A, Astumian RD: Electroconformational coupling (ECC): An electric field induced enzyme oscillation for cellular energy and signal transductions. Bioelectrochemistry and Bioenergetics 1989, 21:319–331.
2. Blank M, Soo L: Ion activation of the Na, K-ATPase in alternating currents. Bioelectrochemistry and Bioenergetics 1980, 24:51–61.
3. Astumian RD, Robertson B: Nonlinear effect of an oscillating electric field on membrane proteins. The Journal of Chemical Physics 1989, 91:4891–4901.
4. Tsong TY, Astumian RD: 863 – Absorption and conversion of electric field energy by membrane bound atpases. Bioelectrochemistry and Bioenergetics 1986, 15:457–476.
5. Westerhoff HV, Tsong TY, Chock PB, Chen YD, Astumian RD: How Enzymes Can Capture and Transmit Free Energy from an Oscillating Electric Field. PNAS 1986, 83:4734–4738.
6. Astumian RD, Chock PB, Tsong TY, Chen YD, Westerhoff HV: Can Free Energy Be Transduced from Electric Noise? PNAS 1987, 84:434–438.
7. Tsong TY, Astumian RD: Electroconformational coupling and membrane protein function. Progress in Biophysics and Molecular Biology 1987, 50:1–45.
8. Tian YT, Dao-Sheng L, Francoise C, Astumian RD: Resonance electroconformational coupling: A proposed mechanism for energy and signal transductions by membrane proteins. Biosci Rep 1989, 9(1):13–26.
9. Woodward AM, Kell DB: On the nonlinear dielectric properties of biological systems: Saccharomyces cerevisiae. Bioelectrochemistry and Bioenergetics 1990, 24:83–100.
10. Woodward AM, Kell DB: Confirmation by using mutant strains that the membrane-bound H+-ATPase is the major source of non-linear dielectricity in Saccharomyces cerevisiae. FEMS Microbiology Letters 1991, 84:91–95.
11. Woodward AM, Kell DB: Dual-frequency excitation: a novel method for probing the nonlinear dielectric properties of biological systems, and its application to suspensions of S. cerevisiae. Journal of Electroanalytical Chemistry 1991, 320:395–413.
12. Lang ZQ, Billings SA: Output frequency characteristics of nonlinear systems. International Journal of Control 1996, 64:1049–1067.
13. Hutchings MJ, Blake-Coleman BC, Silley P: Harmonic generation in "nonlinear" biological systems. Biosensors and Bioelectronics 1994, 9:91–103.

14. Blake-Coleman BC, Hutchings MJ, Silley P: Harmonic 'signatures' of micro-organisms. Biosensors and Bioelectronics 1994, 9:231–242.

15. Woodward AM, Kell DB: On the relationship between the nonlinear dielectric properties and respiratory activity of the obligately aerobic bacterium Micrococcus luteus. Journal of Electroanalytical Chemistry 1991, 321:423–439.

16. McShea A, Woodward AM, Kell DB: Non-linear dielectric properties of Rhodobacter capsulatus. Bioelectrochemistry and Bioenergetics 1992, 29:205–214.

17. Woodward AM, Kell DB: On harmonic generation in nonlinear biological systems. Biosensors and Bioelectronics 1995, 10:639–641.

18. Woodward AM, Jones A, Zhang Xz, Rowland J, Kell DB: Rapid and non-invasive quantification of metabolic substrates in biological cell suspensions using non-linear dielectric spectroscopy with multivariate calibration and artificial neural networks. Principles and applications. Bioelectrochemistry and Bioenergetics 1996, 40:99–132.

19. Woodward AM, Gilbert RJ, Kell DB: Genetic programming as an analytical tool for non-linear dielectric spectroscopy. Bioelectrochemistry and Bioenergetics 1999, 48:389–396.

20. Woodward AM, Davies EA, Denyer S, Olliff C, Kell DB: Non-linear dielectric spectroscopy: antifouling and stabilisation of electrodes by a polymer coating. Bioelectrochemistry 2000, 51:13–20.

21. Nawarathna D, Claycomb JR, Miller J, Benedik MJ: Nonlinear dielectric spectroscopy of live cells using superconducting quantum interference devices. Applied Physics Letters 2005, 86:23902–23903.

22. Miller J, Nawarathna D, Warmflash D, Pereira F, Brownell W: Dielectric Properties of Yeast Cells Expressed With the Motor Protein Prestin. Journal of Biological Physics 2005, 31:465–475.

23. Nawarathna D, Miller J, Claycomb JR, Cardenas G, Warmflash D: Harmonic Response of Cellular Membrane Pumps to Low Frequency Electric Fields. Physical Review Letters 2005, 95:158103–158104.

24. Nawarathna D, Claycomb JR, Cardenas G, Gardner J, Warmflash D, Miller J, et al.: Harmonic generation by yeast cells in response to low-frequency electric fields. Physical Review E (Statistical, Nonlinear, and Soft Matter Physics) 2006, 73:51914–51916.

25. McLellan CJ, Chan ADC, Goubran RA: Aspects of Nonlinear Dielectric Spectroscopy of Biological Cell Suspensions. 28th Annual International Conference

of the Engineering in Medicine and Biology Society IEEE/EMBS; 30 August - 3 September 2006455–458.

26. Treo EF, Felice CJ, Madrid RM: Non linear dielectric properties of microbiological suspensions at electrode-electrolyte interfaces. IEEE-EMBS 27th Annual International Conference of the Engineering in Medicine and Biology Society; 1-4 September 2005; Shangai4588–4591.
27. Treo EF: Estudio de los Espectros de Impedancia Dieléctrica No-Lineal de suspensiones Biológicas y su Aplicación en el Monitoreo de Proteínas de Membrana. PhD Thesis. Universidad Nacional de Tucumán; 2009.
28. Welch P: The use of fast Fourier transform for the estimation of power spectra: A method based on time averaging over short, modified periodograms. IEEE Transactions on Audio and Electroacustics 1967, 15:70–73.
29. Wach A, Graber P: The plasma membrane H+ -ATPase from yeast. Effects of pH, vanadate and erythrosine B on ATP hydrolysis and ATP binding. European Journal of Biochemistry 1991, 201:91–97.
30. Serrano R: In vivo glucose activation of the yeast plasma membrane ATPase. FEBS Letters 1983, 156:11–14.
31. Alexis NC: Activation of the plasma membrane H+-ATPase of Saccharomyces cerevisiae by glucose is mediated by dissociation of the H+-ATPase - acetylated tubulin complex. Eur J Biochem 2005, 272:5742–5752.
32. Yositake H, Muraji M, Tsujimoto H, Tatebe W: The estimation of the yeast growth phase by nonlinear dielectric properties of the measured electric current. Journal of Electroanalytical Chemistry 2001, 496:148–152.
33. Ehret R, Baumann W, Brischwein M, Schwinde A, Stegbauer K, Wolf B: Monitoring of cellular behaviour by impedance measurements on interdigitated electrode structures. Biosensors and Bioelectronics 1997, 12:29–41.
34. Felice CJ, Valentinuzzi ME, Vercellone MI, Madrid RE: Impedance bacteriometry: medium and interface contributions during bacterial growth. IEEE Trans Biomed Eng 1992, 39:1310–1313.
35. Chen W, Dando R: Electrical Activation of Na/K Pumps Can Increase Ionic Concentration Gradient and Membrane Resting Potential. Journal of Membrane Biology 2006, 214:147–155.
36. Chen W, Dando R: Synchronization modulation of Na/K pump molecules can hyperpolarize the membrane resting potential in intact fibers. Journal of Bioenergetics and Biomembranes 2007, 39:117–126.

37. Chen W, Dando R: Membrane Potential Hyperpolarization in Mammalian Cardiac Cells by Synchronization Modulation of Na/K Pumps. Journal of Membrane Biology 2008, 221:165–173.

38. Raicu V, Raicu G, Turcu G: Dielectric properties of yeast cells as simulated by the two-shell model. Biochimica et Biophysica Ac ta (BBA) - Bioenergetics 1996, 1274:143–148.

Near-Infrared Spectroscopy-Derived Tissue Oxygen Saturation in Battlefield Injuries: A Case Series Report

Greg J. Beilman and Juan J. Blondet

ABSTRACT

Background

Near-infrared spectroscopy technology has been utilized to monitor perfusion status in animal models of hemorrhagic shock and in human traumatic injury. To observe the effectiveness of such a device in a combat setting, an FDA-approved device was used in conjunction with standard resuscitation and therapy of wounded patients presenting to the 228th Combat Support Hospital (CSH), Company B, over a three-month period.

Materials and Methods

These observations were performed on patients presenting to the 228th CSH, Co B, at Forward Operating Base Speicher, outside of Tikrit, Iraq, between *the dates of June 15 and September 11, 2005. We utilized the Inspectra™ 325 tissue oxygen saturation (StO_2) monitor (Hutchinson Technology, Inc; Hutchinson, MN, USA) with the probe placed on the thenar eminence or on another appropriate muscle bed, and used to monitor StO_2 during early resuscitation and stabilization of patients.*

Results

During the above time period, 161 patients were evaluated at the CSH as a result of traumatic injury and the device was placed on approximately 40 patients. In most patients, StO_2 readings of greater than 70% were noted during the initial evaluation. No further information was collected from these patients. In 8 patients, convenience samples of StO_2 data were collected along with pertinent physiologic data. In these patients, StO_2 levels of below 70% tracked with hypotension, tachycardia, and clinical shock resulted in increases in StO_2 after resuscitation maneuvers.

Conclusion

Near-infrared spectroscopy-derived StO_2 reflected and tracked the resuscitation status of our patients with battlefield injuries. StO_2 has significant potential for use in resuscitation and care of patients with battlefield injuries.

Background

Optimal treatment for early hemorrhagic shock includes adequate control of bleeding followed by restoration of tissue oxygen delivery with appropriate resuscitation. Unfortunately, from a military perspective, this optimal strategy may not be available for many patients due to field situations that preclude prompt transport to the appropriate treatment facility [1]. Therefore, determination of the magnitude of shock using a rapid, non-invasive method may be useful at the point of care in the field in both military and urban trauma settings. Such a method has the potential to be of use for appropriate triage depending on availability of medical resources.

Near-infrared (NIR) spectroscopy utilizes fiber-optic light to non-invasively determine the percentage of oxygen saturation of chromophores (e.g. hemoglobin) based on spectrophotometric principles [2]. This technology has been utilized to experimentally determine regional tissue oxygen saturation (StO_2) [3-5] by monitoring the differential tissue optical absorbance of near-infrared light. Unlike pulse oximetry, NIR spectroscopy measures not only arterial, but also venous oxyhemoglobin saturation at the microcirculatory level (Figure 1). This measurement therefore is a reflection of both oxygen delivery (DO_2) and oxygen consumption (VO_2) of the tissue bed sampled [6,7]. Non-invasive determination of these parameters using NIR spectroscopy has been described as has its correlation with DO_2 and mixed venous oxygen saturation (SvO_2) [3-7]. NIR-derived StO_2 has been demonstrated to be predictive of severity of shock states in an animal model of hemorrhagic shock [8].

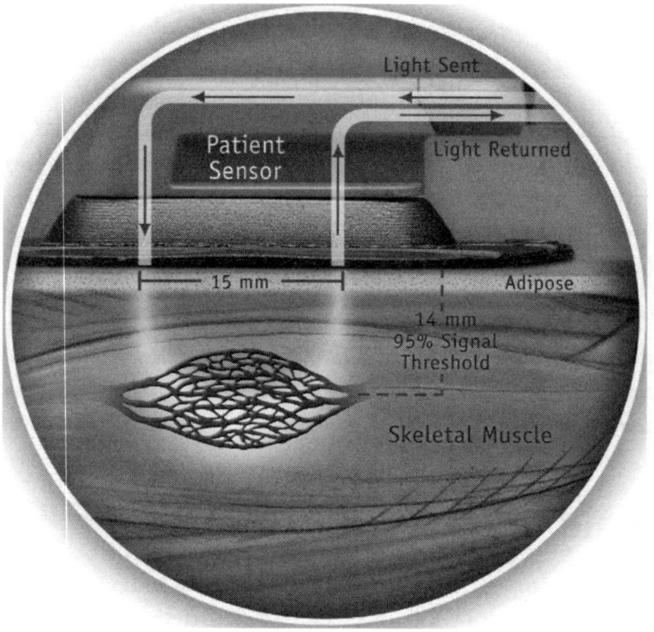

Figure 1. StO_2 is derived from measurement of the near-infrared spectra of the tissue bed sampled. A near-infrared light source shines light into the tissue bed. A spectrum, measured using reflectance of near-infrared light, is used to measure the percentage of hemoglobin saturation.

To observe the effectiveness of such a device in a combat setting, an FDA-approved device was used in conjunction with standard resuscitation and therapy of wounded patients presenting to the 228th Combat Support Hospital (CSH), Company B, over a three-month period.

Materials and Methods

These observations were performed on patients presenting to the 228th Combat Support Hospital (CSH), Company B, at Forward Operating Base Speicher, outside of Tikrit, Iraq, between the dates of June 15 and September 11, 2005. These observations were performed during use of the Inspectra™ 325 as a clinical monitor (Figure 2). The Brooke Army Medical Center Institutional Review Board waived the need for informed consent. The Inspectra™ StO_2 tissue oxygenation monitor (Hutchinson Technology, Inc; Hutchinson, MN, USA) is currently FDA-approved for use in monitoring patients continuously during circulatory or perfusion examinations of skeletal muscle, or when there is a suspicion of compromised circulation. A recent large observational and descriptive study found a mean thenar StO_2 of 87 ± 6% in 707 normal human volunteers [9]. In the present observations, a 70% cutoff value of StO_2 was selected to screen for patients to be followed in time because data obtained from severely injured trauma patients has verified that a StO_2 value of less than 75% is predictive of multiple organ failure and mortality [10].

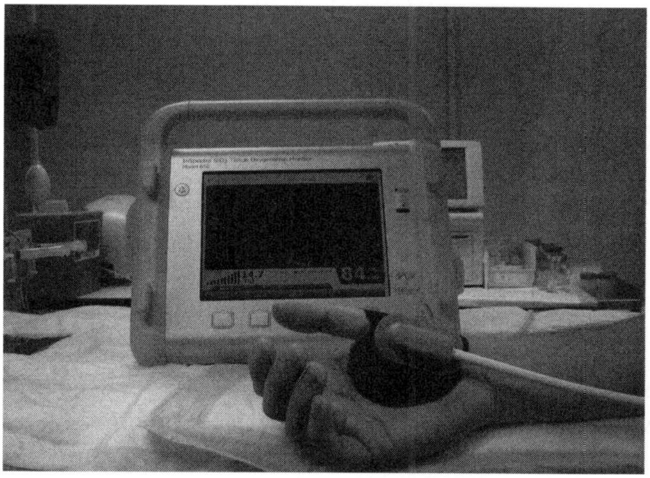

Figure 2. The non-invasive StO_2 probe is placed directly over the thenar eminence of the patient. The device will continuously generate StO_2 readings every 4 seconds.

Patients were brought to the 228th CSH via ground ambulance or helicopter after traumatic injury. Patients were evaluated by a team of physicians and health care providers using a standardized ATLS protocol and after stabilization taken as appropriate to the operating room and/or prepared for transfer to a higher level of care. Patients were monitored during resuscitation and early evaluation using clinical parameters, continuous EKG and pulse oximetry, and other monitors (e.g. bladder catheterization) as appropriate. In situations where more than one patient was evaluated concurrently, an attempt was made to place the StO_2 monitor on the most severely injured patient. Convenience samples of demographic data, vital signs, laboratory data, and StO_2 data were collected on patients as patient care permitted.

Case Presentations

Between June 15 and September 11, 2005, there were 161 patients evaluated at the 228th CSH, Co B as a result of traumatic injury. The StO_2 monitor was placed on approximately 40 patients during this period of time. In most patients, StO_2 readings of greater than 70% were noted during the initial evaluation. No further information was collected from these patients. In 8 patients, convenience samples of StO_2 data were collected along with pertinent physiologic data. In these patients, StO_2 levels of below 70% tracked with hypotension, tachycardia, and clinical shock resulted in increases in StO_2 after resuscitation maneuvers (Table 1). Four cases are presented in greater detail to illustrate the use of the device as correlated with patient status.

Table 1. Comparison of StO_2 levels at presentation and after resuscitation maneuvers.

Injury	Initial StO_2	Resuscitation Maneuver	Post resuscitation StO_2
Bilateral lower extremity IED	60	2 LR, 2 PRBCs	78
IED blast, right leg, left flank	51	2 LR, 1 PRBCs	71
GSW left thigh	54	1 LR	88
Abdominal compartment syndrome	62	Open abdomen	91
Bilateral lower extremity IED	51	1 LR	76
GSW abdomen	50	1 LR	82
GSW right arm	55	0.5 LR (9 y/o)	76
Blast injury	1	CPR	1

Case 1

A 36-year-old male was injured from an improvised explosive device (IED) and presented with near amputations of both lower extremities. He arrived at the emergency medical treatment area (EMT) with blood pressure (BP) of 110/70 mm Hg and heart rate (HR) of 120/min. His initial StO_2 reading was 51% from the right thenar eminence. He received 1 liter of lactated ringers (LR) with an increase in StO_2 to 76% and was taken to the operating room (OR) where he underwent a right below the knee amputation and debridement and external fixator placement for a complex left tibia fracture.

The next morning, the patient's StO_2 was noted to be low at 40%. His BP was 105/72 mm Hg and HR was 130/min with hemoglobin of 8.9 g/dl. Over the next 2 hours, the patient received 300 cc of 25% albumin, 1 liter of LR, and 1 unit of packed red blood cells (PRBCs) with HR decreasing to 110/min, and BP increasing to 130/70 mm Hg, and urine output of 150 cc over the previous hour. StO_2 increased to 73%.

This patient's post-injury course was long and complicated. After multiple operations including debridements and skin grafting, the patient was discharged from the hospital approximately 2.5 months after his initial injury.

Case 2

A 24-year-old male was seen in the EMT after a gunshot wound (GSW) to the abdomen. His initial vital signs included a BP of 90/60 mm Hg and HR of 120/min. His initial StO_2 from the thenar eminence was 50%. He received 1 liter of LR with an increase of his BP to 110/70 mm Hg and StO_2 to 82%. He was taken to the OR where he was found to have a tangential transverse colon injury. He underwent a primary repair and recovered and was discharged from the hospital approximately 2 weeks post-injury.

Case 3

A 20-year-old male presented to the EMT after a high-velocity GSW to the left hip. At the time of presentation, two peripheral intravenous (IV) lines, which had been placed in the field, were infiltrated. One wound was noted in the left lateral hip and the patient had a distended, tense, and tender abdomen. His initial BP was 56/30 mm Hg and HR was 150/min. Arterial oxygen saturation (SaO_2) was 100% and thenar StO_2 was 54%. A left subclavian line was placed and patient received 1 liter of crystalloid with a response of BP to 110/70 mm Hg and HR to 120/min. His StO_2 increased to 88%.

He was taken to the OR where exploratory laparotomy and repair of small bowel enterotomies was carried out. Proctoscopy was negative. He received 4 units of PRBCs and 2500 cc of crystalloid in the OR. His postoperative vitals were BP of 110/68 mm Hg, HR of 100/min, SaO_2 of 100% and StO_2 of 89%. Two hours later, he became hypotensive and oliguric and StO_2 decreased to 65%. He received 2 liters of crystalloid, 2 units of fresh frozen plasma (FFP), and 1 unit of PRBCs with an improvement of BP, urine output, and StO_2 (82%).

Approximately 8 hours after the patient's initial presentation he developed recurrent oliguria, increased airway pressures (Peak pressures of 50 cm H_2O with tidal volumes of 6 cc/Kg). His BP was 100/60 mm Hg and HR of 150/min with a base deficit of 12 mEq/L. StO_2 had dropped to 62%. The patient was taken to the OR where his abdomen was opened and a Bogota bag was placed with immediate improvement of all parameters (StO_2 increased to 91%). (Initial hospital course: Figure 3)

His post-injury course was complicated and included development of necrotizing muscle infection, internal iliac arterial bleed, and ureteral fistula requiring left nephrectomy. He was eventually discharged from the hospital 3 months after his injury.

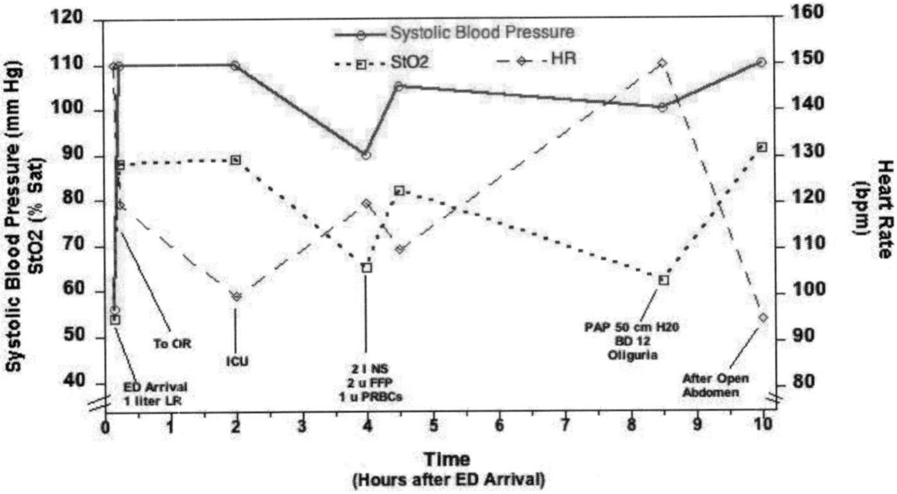

Figure 3. Graphic representation of systolic blood pressure, heart rate, and StO_2 of patient described in case 3 during the first 10 hours of hospital course.

Case 4

A 36-year-old male suffered an IED injury resulting in a massive injury to the right lower extremity. He was hypotensive in the field with a systolic BP (SBP) of 77 mm Hg. A tourniquet was placed and the patient was transferred via air to our facility. He arrived at the EMT with a SBP of 69 mm Hg, HR of 150/min, SaO_2 of 91%, and StO_2 of 51%. In the ED he received 2 liters of LR and 1 unit of O negative PRBCs with an improvement of his vital signs and StO_2 (SBP 110 mm Hg, HR 125/min, StO_2 71%). Initial injuries noted included left pulmonary contusion, open right femur fracture, large soft tissue injury in left buttocks, and laceration of the right radial artery.

He was taken to the OR where the tourniquet was removed and injuries to the profunda femoral artery and vein were noted. Multiple branches were ligated and oversewed. The sciatic nerve and superficial femoral artery were both intact. The patient had massive soft tissue injury that was widely debrided. The shrapnel in his left buttocks was removed (proctoscopy was negative). He developed coagulopathy, an external fixator was placed, and the patient was returned to the intensive care unit (ICU) for further resuscitation (INR: 10, platelets: 33,000, and hemoglobin: 3.9 g/dl). During his OR course the patient's StO_2 dropped to 51% just prior to transfer to the ICU. His final OR temperature was 36.6°C. OR fluids included 13 liters of crystalloid, 4 units of FFP, and 9 units of PRBCs.

On arrival in the ICU, the patient's initial SBP was 82 mm Hg, HR 130/min, and StO_2 50%. Initial hemoglobin was 7.9 g/dl and base deficit was 16 mEq/L.

Over the next 4 hours the patient received 9 units of FFP, 10 mg of vitamin K, 2 units of fresh whole blood, 4 units of PRBCs, 200 cc of 25% albumin, 2 liters of LR, and 6500 mcg of Factor VIIa. Two hours into the resuscitation 2 plateletpheresis packs arrived via helicopter and were given. With this therapy the patients' vital signs and urine output improved gradually (BP 100/70 mm Hg, HR 90/min, and urine output 150 cc/hour) and his laboratory parameters likewise showed improvement with a normal INR, hemoglobin of 8.6 g/dl, platelets of 70,000/ml, and base deficit of 7 mEq/L. StO_2 likewise slowly improved (65%).

The next morning the patient was weaned and extubated. His platelet count and INR were normal. His StO_2 was 82% (initial hospital course: Figure 4). He received debridement and progressive closure of his wound every other day and 10 days post-injury received intramedullary femoral rod for stabilization of his femur fracture. He was discharged from the hospital 24 days post-injury.

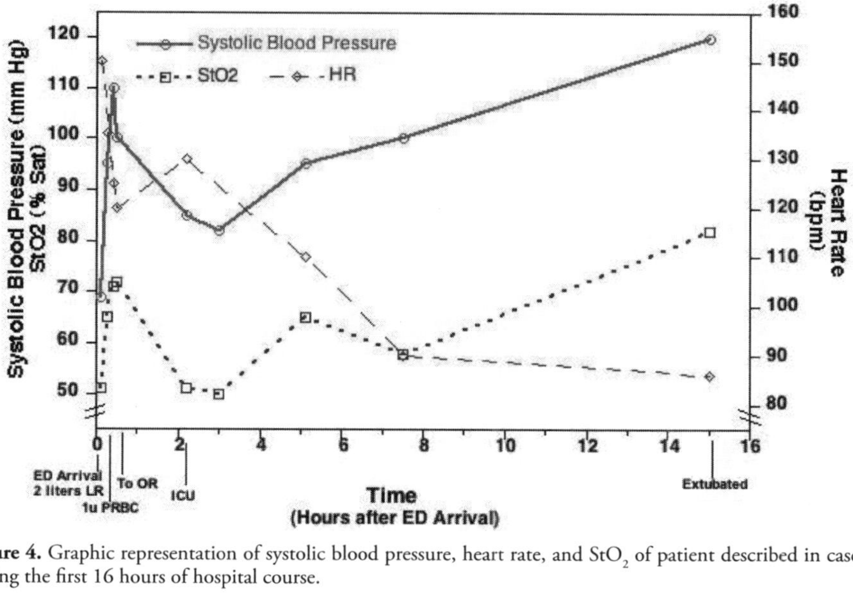

Figure 4. Graphic representation of systolic blood pressure, heart rate, and StO_2 of patient described in case 4 during the first 16 hours of hospital course.

Discussion

Care of patients in the austere environment of the battlefield presents challenges to the clinician, including limited access to invasive monitoring techniques readily available in the care of civilian trauma patient. Equipment utilized in a field situation must be readily transportable, rugged, reliable, and easy to use. Over the years, many technologies originally developed for civilian use have found their way into the armamentarium of battlefield care, including bedside ultrasound and computed tomography. Near-infrared spectroscopy has a similar promise for field use.

The patient experiences described above suggest that NIR spectroscopy-derived StO_2 is able to serve as a non-invasive tool for early identification and treatment of hypoperfusion in the severely injured trauma patient. Nevertheless, in the present case series, the small number of patients described and the observational nature of this report preclude any generalization or formal recommendation.

A recent study of 383 trauma patients at 7 civilian trauma centers has identified the association of a low StO_2 with both multiple organ failure and mortality [10]. There are currently no prospective studies examining its use as an endpoint for therapy in hemorrhagic shock. In the 8 patients described, StO_2 followed the clinical course of the patient and in the 7 surviving patients tracked resuscitation status, suggesting that this measure may be potentially useful as such an endpoint. In addition, the readings from the StO_2 monitor were more rapidly available and easier to categorize than other invasive or non-invasive hemodynamic tools used for determining need for additional resuscitation, and also had the advantage of easy interpretation regardless of the level or experience of the care provider. While there were no instances in this small series of abnormally low StO_2 before clinical symptoms of shock were present, there is also the potential for such a device to be useful in early identification of "sub-clinical" shock.

Equally appealing is the possible use of StO_2 in a triage setting in either civilian or military trauma. Such a use has the added benefit of giving a number to confirm the presence of tissue hypoperfusion for less experienced care providers. These potential benefits have led to the incorporation of StO_2 as another tool for early evaluation of trauma patients at several civilian trauma centers.

Previous work from our lab in a porcine model of severe hemorrhagic shock identified StO_2 as a significant predictor of eventual mortality in this setting [8], with StO_2 significantly lower in the cohort of animals that were unsuccessfully resuscitated.

Conclusion

Near-infrared spectroscopy-derived StO_2 reflected and tracked the resuscitation status in the observed severely injured patients suffering battlefield injuries. StO_2 has significant potential for use in resuscitation and care of patients with battlefield injuries.

Competing Interests

GJB has served on an Advisory Board and is the recipient of grant support from Hutchinson Technology, Inc. He is funded by the Office of Naval Research (#N00014-05-1-0344).

Authors' Contributions

GJB collected data from patients, collated data, and drafted the manuscript. JJB performed statistical analysis and coordinated manuscript preparation. All authors read and approved the final manuscript.

About the Authors

GJB serves as a Colonel in the United States Army Reserve. He's also Professor of Surgery and Anesthesia, Chief of the Division of Surgical Critical Care/Trauma, Vice Chair of Perioperative Services and Quality Improvement in the Department of Surgery at the University of Minnesota, and a Fellow of the American College of Surgeons.

JJB served as a postdoctoral research associate at the Division of Surgical Critical Care/Trauma and currently is a general surgery resident in the Department of Surgery at the University of Minnesota.

Acknowledgements

The authors would like to acknowledge the contributions of the staff of the 228th Combat Support Hospital, Company B.

References

1. Holcomb JB: Fluid resuscitation in modern combat casualty care: lessons learned from Somalia. J Trauma. 2003, 54(5 Suppl):S46–S51.
2. Myers DE, Anderson LD, Seifert RP, Ortner JP, Cooper CE, Beilman GJ, Mowlem JD: Noninvasive method for measuring local hemoglobin oxygen saturation in tissue using wide gap second derivative near-infrared spectroscopy. J Biomed Opt 2005, 10(3):034017.

3. Mancini DM, Bolinger L, Li H, Kendrick K, Chance B, Wilson JR: Validation of near-infrared spectroscopy in humans. J Appl Physiol 1994, 77(6):2740–2747.
4. Beilman GJ, Groehler KE, Lazaron V, Ortner JP: Near-infrared spectroscopy measurement of regional tissue oxyhemoglobin saturation during hemorrhagic shock. Shock 1999, 12(3):196–200.
5. Cohn SM, Varela JE, Giannotti G, Dolich MO, Brown M, Feinstein A, McKenney MG, Spalding P: Splanchnic perfusion evaluation during hemorrhage and resuscitation with gastric near-infrared spectroscopy. J Trauma 2001, 50(4):629–634.
6. Rhee P, Langdale L, Mock C, Gentilello LM: Near-infrared spectroscopy: continuous measurement of cytochrome oxidation during hemorrhagic shock. Crit Care Med 1997, 25(1):166–170.
7. Simonson SG, Welty-Wolf K, Huang YT, Griebel JA, Caplan MS, Fracica PJ, Piantadosi CA: Altered mitochondrial redox responses in gram negative septic shock in primates. Circ Shock 1994, 43(1):34–43.
8. Taylor JH, Mulier KE, Myers DE, Beilman GJ: Use of near-infrared spectroscopy in early determination of irreversible hemorrhagic shock. J Trauma 2005, 58(6):1119–1125.
9. Crookes BA, Cohn SM, Bloch S, Amortegui J, Manning R, Li P, Proctor MS, Hallal A, Blackbourne LH, Benjamin R, Soffer D, Habib F, Schulman CI, Duncan R, Proctor KG: Can near-infrared spectroscopy identify the severity of shock in trauma patients? J Trauma 2005, 58(4):806–813.
10. Cohn SM, Nathens AB, Moore FA, Rhee P, Puyana JC, Moore EE, Beilman GJ, the StO_2 in Trauma Patients Trial Investigators: Tissue oxygen saturation predicts the development of organ dysfunction during traumatic shock resuscitation. J Trauma 2007, 62(1):44–54.

Rapid Etiological Classification of Meningitis by NMR Spectroscopy Based on Metabolite Profiles and Host Response

Uwe Himmelreich, Richard Malik, Till Kühn, Heide-Marie Daniel, Ray L. Somorjai, Brion Dolenko and Tania C. Sorrell

ABSTRACT

Bacterial meningitis is an acute disease with high mortality that is reduced by early treatment. Identification of the causative microorganism by culture is sensitive but slow. Large volumes of cerebrospinal fluid (CSF) are required to maximise sensitivity and establish a provisional diagnosis.

We have utilised nuclear magnetic resonance (NMR) spectroscopy to rapidly characterise the biochemical profile of CSF from normal rats and animals with pneumococcal or cryptococcal meningitis. Use of a miniaturised capillary NMR system overcame limitations caused by small CSF volumes and low metabolite concentrations. The analysis of the complex NMR spectroscopic data by a supervised statistical classification strategy included major, minor and unidentified metabolites.

Reproducible spectral profiles were generated within less than three minutes, and revealed differences in the relative amounts of glucose, lactate, citrate, amino acid residues, acetate and polyols in the three groups. Contributions from microbial metabolism and inflammatory cells were evident. The computerised statistical classification strategy is based on both major metabolites and minor, partially unidentified metabolites. This data analysis proved highly specific for diagnosis (100% specificity in the final validation set), provided those with visible blood contamination were excluded from analysis; 6–8% of samples were classified as indeterminate.

This proof of principle study suggests that a rapid etiologic diagnosis of meningitis is possible without prior culture. The method can be fully automated and avoids delays due to processing and selective identification of specific pathogens that are inherent in DNA-based techniques.

Introduction

Bacterial meningitis is an acute disease with a high mortality [1], [2]. Severe neurological sequelae have been reported in 25% of cases [3], [4]. Outcomes are directly related to the speed with which the diagnosis is established and therapy initiated. Conventional diagnosis relies on screening examination of cerebrospinal fluid (CSF) for non-specific markers such as inflammatory cells, proteins and glucose, and the more specific but insensitive Gram stain for micro-organisms, to distinguish rapidly between bacterial and non-bacterial (most commonly viral and fungal) infection. Ultimate confirmation of the diagnosis and identification of the specific microbial pathogen by culture is sensitive and specific, but slow. This might delay the most appropriate treatment. Relatively large volumes of CSF are recommended to maximise sensitivity and provide sufficient material for completion of the various diagnostic tests.

Alternative approaches based on chemicals in CSF due to production by the infection causing microorganism or due to the immune response of the host are potentially useful for the diagnosis of a variety of neurological diseases, including meningitis [5]. Metabolic changes that reflect infection such as low glucose or disproportionately elevated lactate are characteristic of bacterial meningitis but have been poor discriminators in some studies [6]–[8]. A disadvantage of these biochemical methods is that only particular compounds are targeted by individual tests. High-throughput technologies such as Nuclear Magnetic Resonance (NMR), Infrared (IR) and Raman spectroscopy, chromatographic methods and mass spectrometry, generate complex data ("fingerprints") based on chemical composition and metabolite profiles of micro-organisms (metabolome) [9]–[13]. Such metabolomic methods can detect genotypic and phenotypic differences even in closely related microorganisms and also in genetically modified strains [13]–[17], with greater discriminatory power than transcriptomics and proteomics [17], [18]. Spectroscopic techniques characterise rapidly and simultaneously multiple chemical compounds in biological fluids [10], [19]–[28]. In addition, NMR spectroscopy detects low molecular weight metabolites, requires no time-consuming sample preparation and is non-destructive.

The metabolite composition of normal CSF has been studied intensively by NMR spectroscopy [22], [29], [30]. CSF contains a relatively limited repertoire of metabolites, compared with serum or urine, that are relatively stable with diet and medication, and on storage at room temperature for a short time [22], [29], [31]. CSF is therefore a potentially useful biofluid for identifying metabolic profiles of pathogens and the host response.

Computerised methods utilize the whole NMR spectrum and include unnoticed (or unknown) compounds. In a pilot study of acute meningitis, CSF spectra from small numbers of controls, patients with viral meningitis and bacterial/fungal meningitis were distinguished using an unsupervised cluster analysis method [8], suggesting that classification according to etiology is possible using larger data sets. Since human cases of meningitis caused by different bacterial or fungal species are relatively uncommon in developed countries and acquisition of sufficient spectral data would require several years, a proof of principle study was performed using rat models of meningitis. Repeated sampling of small volumes of CSF from individual animals was made possible by the use of a miniaturised micro NMR system [32], [33].

In clinical practice, rapid identification of a specific bacterial or fungal pathogen is important as it enables immediate initiation of appropriate antimicrobial therapy based on predictable patterns of antibiotic susceptibility and hence improves clinical outcomes. We have utilised a metabolomic approach based on NMR spectra from CSF of animal models to evaluate if metabolites produced by an infection-causing microorganism or a specific immune response can be utilised for diagnosis. This could provide the basis for further development of the method on human CSF specimens. The two disease models used were meningitis caused by Cryptococcus neoformans and Streptococcus pneumoniae.

Results

Meningitis Models and Metabolite Profiles

Meningitis was induced in Fisher 344 rats by injection of Cryptococcus neoformans (N = 31) or Streptococcus pneumoniae (N = 30) into the cisterna magna. CSF samples were collected from these animals after first signs of meningitis and at several time points thereafter. Control CSF samples were repeatedly collected from sham injected animals (phosphate buffered saline (PBS), N = 5) or prior infection.

Metabolite composition of the CSF samples was studied by NMR spectroscopy using either an NMR system equipped with an 1 mm micro probe or a conventional 5 mm probe. Figure 1 shows typical 1D NMR spectra of CSF samples from control animals (injected with PBS) and collected from animals with

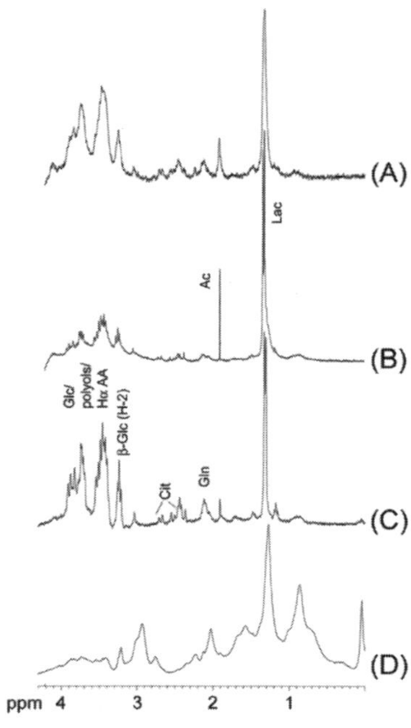

Figure 1. 1H NMR spectra acquired with a 400 MHz spectrometer equipped with probe for 1 mm sample tubes.
NMR spectra were acquired from the following samples: (A) CSF from a control animal (before initiation of the infection), (B) CSF from an animal five days after injection of S. pneumoniae (animal showed symptoms of meningitis; heavy growth of S. pneumoniae from CSF confirmed streptococcal meningitis), (C) CSF from an animal four days after injection of C. neoformans (animal showed symptoms of meningitis, heavy growth of C. neoformans from CSF confirmed cryptococcal meningitis), (D) CSF heavily contaminated with blood (same animal as C but collected three days after initiation of the infection). Abbreviations refer to 1H NMR signal of the following metabolites: Ac—acetate, Ala—alanine, Cit—citrate, β-glc (H-2)—H-2 resonance of β-glucose, Glc—glucose, Gln—glutamine, Hα AA—resonances of H-α from amino acid residues, Lac—lactate.

histologically confirmed meningitis, four days after injection of 104 cfu of C. neoformans and S. pneumoniae, respectively. Resonances in the 1D NMR spectra from each pathology were assigned to respective metabolites using 2D NMR correlation spectra from at least five independent samples per category.

Although the major metabolites in CSF samples from all three categories were qualitatively similar for control and post-infection CSF samples from individual animals, quantitative differences were noted. CSF from infected animals contained increased amounts of lactate (5–90%, using the resonance at 1.31 ppm), decreased glucose (5–30%, resonance at 3.25 ppm) and marginally increased citrate (0–15% resonance at 2.45–2.75 ppm) and amino acid residues (0–15% for alanine (1.49 ppm) and glutamine/glutamate (2.2–2.4 ppm)). For most samples, only glutamine but no glutamate was detected. Some infected CSF samples showed increased acetate (for 55% of S. pneumoniae infections and for 30% of C. neoformans infections) and polyol signals (mannitol and glycerol resonances (3.90–3.95 ppm) for 20% of C. neoformans infections).

The quantitative changes that occurred in individual animals during the course of infection overlapped with variations in samples collected from different animals of the same class (control, S. pneumoniae and C. neoformans), precluding the use of metabolite ratios for diagnosis of streptococcal and cryptococcal meningitis.

Statistical Classification of NMR Data

A statistical classification strategy (SCS) [42] (an earlier version of which was previously used for metabolomic studies in pathogenic yeasts [16]) was utilised to improve predictability of class assignment based on NMR spectra, and also to identify metabolite differences occurring during pathogenesis in the meningitis models. A genetic-algorithm-based optimal region selection (GA-ORS) algorithm was used to identify discriminatory regions in the NMR spectra [34]. The three most discriminatory spectral regions identified by the GA-ORS for each pair-wise comparison (S. pneumoniae versus C. neoformans, S. pneumoniae versus control and C. neoformans versus control) are summarised in Table 1 (B). Metabolites within these spectral regions that were identified by 2D NMR techniques include the amino acid residues glutamine (traces of glutamate were found in 10% of the CSF samples independent of the pathology), valine, leucine, isoleucine; acetate, lactate, citrate, polyols and carbohydrate residues and α-hydroxybutyrate. Notably, polyols (mannitol/glycerol) were determined to contribute to the distinction between C. neoformans and S. pneumoniae meningitis but not to that between C. neoformans meningitis and controls, presumably because polyols overlap with glucose resonances, and glucose levels are higher in normal CSF than in that from animals with cryptococcal meningitis.

Table 1. Statistical Classification Strategy using NMR spectra of CSF samples.

(A) SCS analysis with and without blood contaminated CSF samples.			
	N	**correct [%]**	**crisp [%]**
1. Training set (all incl.)			
Control	52	97	73
C.neoformans	25	91	88
S.pneumoniae	24	95	88
1. Validation set (all incl.)			
Control	6	50	67
C.neoformans	15	64	93
S.pneumoniae	15	100	87
Final Training set (excl. blood contaminations)			
Control	49	96	94
C.neoformans	34	100	94
S.pneumoniae	39	97	92
Final Validation set (excl. blood contaminations)			
Control	12	100	92
C.neoformans	6	100	100
S.pneumoniae	6	100	84
Inclusion of an additional clinical isolate			
C.neoformans (WM1128)	5	80	100
S.pneumoniae (99-235-2193)	5	100	80
(B) Spectral regions used for the final SCS-based classifiers and potential metabolites of these regions that were also identified in 2D NMR spectra.			
	Method	**Regions [ppm]**	**Potential metabolites**
C. neoformans vs. *S. pneumoniae*	Rank-ordered 1st derivatives	3.91–3.95	polyols (mannitol/glycerol) and carbohydrate residues
		2.66–2.70	citrate
		2.25–2.27	glutamine
C. neoformans vs. control	1st derivatives	2.25–2.31	glutamine
		1.89–1.92	acetate, glutamine
		0.90–0.95	α-hydroxybutyrate, valine, leucine, isoleucine
S. pneumoniae vs. control	Rank-ordered 1st derivatives	1.84–1.92	acetate, glutamine
		1.15–1.29	lactate
		0.77–0.88	valine, leucine, isoleucine

After identification of the most discriminatory regions in the NMR spectra of CSF, these regions were utilised to distinguish between the two pathologies and the uninfected rats. The final classification accuracy was achieved after repeated re-development and validation of the pair-wise classifiers. The first set of classifiers were based on all CSF samples (N = 101) and was still relatively inaccurate (Table 1). Although 91–97% of the spectra were assigned correctly to one of the classes, 12–27% of these assignments were 'fuzzy' (e.g., with low assignment confidence). These classifiers performed poorly on an independent validation sample set. It was noted that a high percentage of those spectra that were fuzzy or incorrect, were visibly contaminated with blood. Classifier redevelopment following exclusion of samples contaminated with blood resulted in final assignments with only 6–8% of samples classified fuzzy and 4% or less as incorrect (misclassification). Testing against a newly acquired, independent validation set (N = 34) from animals that were not part of the data set used for classifier development, resulted in 8–16% unreliable classification and 100% correct class assignment (Table 1). Animals were infected with an additional clinical isolate of C. neoformans (WM1128, n

= 5) or S. pneumoniae (99-235-2193, n = 5). Both new isolates were classified mostly correctly (one isolate was misclassified and one was fuzzily assigned).

Repeated CSF collections from the same infected animal (N = 27) were mainly used for classifier validation. The time of collection of CSF after onset of meningitis did not influence class assignment. One possible explanation is that quantitative changes in metabolite composition between day 4 and 11 after infection in an individual animal were similar to differences in CSF composition between different animals collected at the same time point.

Discussion

Micro-NMR spectroscopy of CSF when analysed by a supervised classification method distinguished rapidly and reproducibly between controls, rats with pneumococcal meningitis and rats with cryptococcal meningitis. This proof of principle study in experimental rats indicates that the etiology of meningitis can potentially be established rapidly without prior culture of the infection-causing microorganism using a very small amount of CSF. If able to be translated into clinical practice and extended to other pathogens, this is potentially a major diagnostic advance that would assist in the rapid selection of appropriate antimicrobial therapy and hence improve patient outcomes.

Human CSF is rich in NMR detectable metabolites. More than hundred have been identified based on NMR spectroscopy in CSF from patients with various neurological, metabolic and other non-neurological diseases [29], the most abundant being glucose and lactate, with lesser amounts of acetate, citrate, formate, 3-hydroxyl-butyrate, alanine, valine and glutamine [30]. The major metabolites in CSF samples from humans without infections or other apparent diseases are similar to those in healthy rats as observed in the present study (Figure 2). Rat models have previously proven to be suitable for NMR spectroscopic studies of staphylococcal and cryptococcal brain abscesses [35], [36].

Semi-quantitative analysis revealed that elevation of lactate and reduction in glucose levels were the dominant CSF metabolite changes in rats with meningitis. These changes are also documented in human cases of bacterial and fungal meningitis, but are not specific [8]. Changes in metabolite concentrations in CSF samples are due to a combined effect of metabolites from infecting microorganisms, inflammatory cells and changes due to the effect of meningitis on brain cell metabolism. For example, increased lactate levels in CSF may result from anaerobic cerebral metabolism (glycolysis) which occurs in response to brain cell ischemia and is a feature of both infectious and non-infectious neurological disorders [7], [37] but it is also a dominant metabolite of some bacteria under those conditions [38]. Notably, lactate is a prominent end product of metabolism in S. pneumoniae and levels in pneumococcal meningitis exceed those produced in individual human cases of staphylococcal and cryptococcal meningitis [8], suggesting a contribution from the organisms themselves. Further evidence of low concentrations of microbial metabolites in the CSF comes from mannitol identified in 20% of the CSF samples from animals infected with C. neoformans [39]. Low levels of glucose in bacterial meningitis might be due to reduced membrane carrier-facilitated glucose transport, microbial metabolism and increased cerebral glycolysis secondary to ischemia and cytokine release [7], but are also seen in subarachnoid hemorrhage [37].

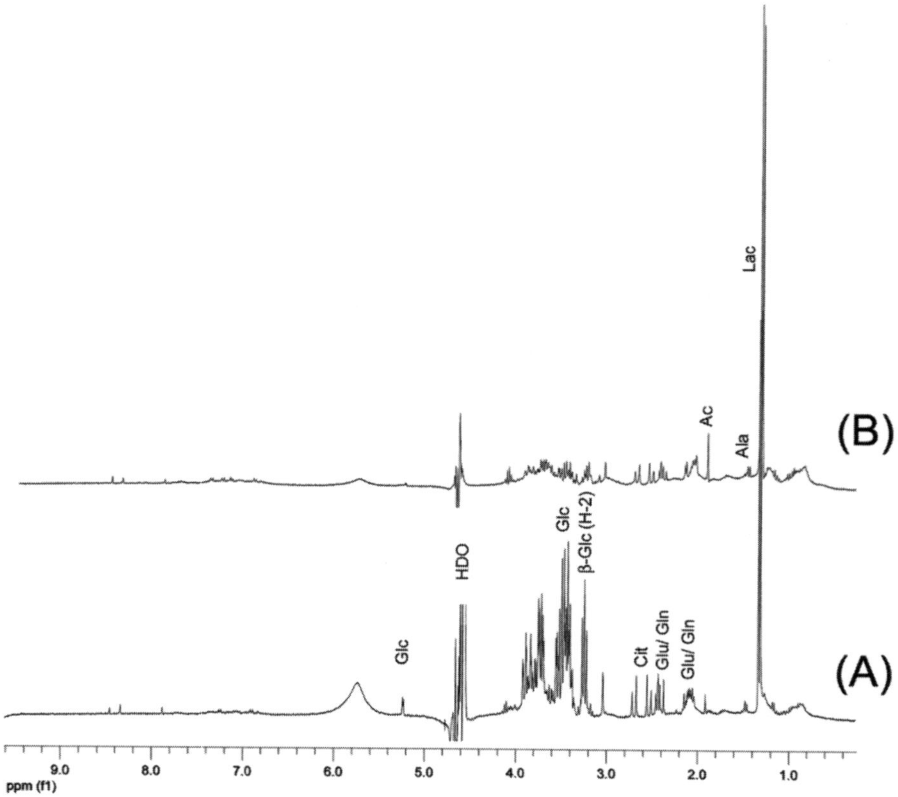

Figure 2. 1H NMR spectra of 100 µl CSF samples from (A) a healthy rat and (B) a human not suffering from meningitis or other infections.

The spectra acquired at 360 MHz using a susceptibility-matched 5 mm NMR tube. Abbreviations refer to 1H NMR signal of the following metabolites: Ac—acetate, Ala—alanine, β-glc (H-2)—H-2 resonance of β-glucose, Cit—citrate, HDO—remaining, partly deuterated water resonance, Glc—glucose, Gln—glutamine, Lac—lactate.

Although some metabolic differences were detected by analysing subjectively chosen spectral regions, discrimination between the two pathologies and controls would not have been possible if based solely on these major metabolites. In the present study, the selection of the most discriminatory regions of the NMR spectra by GA-ORS suggests that glutamine (2.25–2.27/2.3 ppm) may have contributed most to the distinction between the three pathologies (control CSF, cryptococcal and pneumococcal meningitis specimen), which was in principle confirmed by the comparison of individual integral regions. However, other unidentified minor metabolites may have been more discriminatory, underlining the advantages of spectral analysis by a statistical classification strategy compared to assignment of individual, preselected key metabolites. Elevated glutamine, acetate and the amino acids valine, leucine and/or isoleucine as well as polyols may have contributed to the classification between pneumococcal or cryptococcal meningitis and the respective controls as indicated in the spectral regions of Table 1. For example, increased glutamine has been observed in bacterial meningitis [40] and acute subarachnoid haemorrhage [37] and reflects decreased neuronal activity.

Differences in the relative abundance of recognisable metabolites, exemplified by the integral ratios of lactate, acetate, glutamine, citrate and polyols relative to glucose in the present study, were insufficient for classification based on changes in relative concentrations of those individual compounds. Although the analysis of NMR spectra by GA-ORS identified most discriminatory regions that might indicate certain key metabolites, the complex nature of the spectra with composite resonances makes it difficult to identify all, even minor contributors to the successful classification. Such low concentration metabolites are most likely not recognised by operator-based assignment of the simultaneously detected resonances. This is particularly true for the spectral region between 3.5 to 4.0 ppm where many metabolites resonate, including those of importance for classification (for example carbohydrates, polyols like mannitol and all Hα resonances of amino acid residues). Those metabolites can often only be identified by time consuming 2D correlation spectra.

Technical Considerations

Accurate classification was achieved in the present study without any pre-processing of the CSF samples. As also reported by Maillot [29], contamination of CSF with blood did influence classification accuracy and resulted in exclusion of 10% of samples. Blood contamination is a potentially greater problem with repeated cisternal taps in small animals than it is with a single spinal fluid collection in humans.

A desirable precondition for the incorporation of NMR spectroscopy into clinical diagnostics is automation of sample handling and data analysis, and reduction of the sample volume to a minimum. Even more important is the utilization of small volumes for repeated monitoring in animal models. We therefore chose a 1 mm micro-NMR probe for acquisition of NMR spectra. Low volumes of CSF samples (5 to 35 μl) were directly collected into 1 mm micro NMR tubes. Subsequent snap-freezing and later, automated acquisition of NMR spectra resulted in a satisfactory signal-to-noise ratio and spectral quality. Comparison between data acquisition using the 1 mm micro NMR probe with those in conventional 5 mm tubes confirmed previous findings of dramatically improved signal-to-noise ratios and equal or better spectral resolution and water suppression by using 1 mm micro tubes [33]. No additional sample preparation was performed apart from addition of deuterated water (D2O) to some samples (N = 30) for evaluation of spectral quality (resolution) after shimming. No significant differences were found between these samples and sample acquisition performed without shimming. Thus the micro-volume NMR system represents a powerful advance. The stability of CSF samples at room temperature for some hours makes the combination of the NMR system with automatic sample changers feasible, allowing rapid, high-throughput data acquisition and analysis.

Concluding Points

We have demonstrated for C. neoformans and S. pneumoniae infections that diagnosis of meningitis according to etiology of the infection causing microorganisms is possible without prior culture of the microorganism using a metabolomics based approach. The method can be fully automated and is suitable for incorporation in clinical diagnosis. This would potentially accelerate and hence greatly improve turn-around-times compared with current diagnostic procedures once it has been extended to other meningitis-causing pathogens and validated on clinical samples.

Materials and Methods

Ethics Statement

Animal experimentation was carried out according to the Australian National Health and Medical Research Council Guidelines and with ethical approval from the University of Sydney Animal Ethics Committee (approval number K14/12-97/3/2668).

Clinical samples were studied in compliance with ethical approval granted by the Human Ethics Review Committee of Sydney Western Area Health Service.

Animal Studies

A total number of 175 CSF samples were collected from 76 animals (Fisher 344 rat strain, 150–250 g, Animal Research Council, Perth, WA, Australia) in this study. Rats were anaesthetized by inhalation of 2% halothane (May & Baker, Degenham, UK) in oxygen delivered from a precision out-of-circuit vaporiser using a tight-fitting facemask. Following induction of deep anaesthesia, a 25-gauge needle was inserted into the cisterna magna and 15–35 μl of CSF was removed by suctioning the hub of the needle using an insulin syringe. These CSF samples were used as control (sterile) specimens. Meningitis was induced by the slow injection of 15–35 μl PBS containing 104 cfu of clinical isolates of Cryptococcus neoformans (isolate WM628, 31 rats infected) or Streptococcus pneumoniae (isolate 99-241-1187, 30 rats infected), followed by rapid withdraw of the needle. Sham injection was performed for five animals by injection of sterile PBS. For validation of the classification five animals per pathogen were injected using an additional clinical isolates of C. neoformans (isolate WM1128) and S. pneumoniae (isolate 99-235-2193).

CSF samples (15–35 μl) from infected rats were harvested when animals first showed signs of meningitis (such as head tilt, seizures, loss of appetite; typically four to eight days following inoculation) or after day eight if the animals remained asymptomatic. Additional collection of CSF was performed three days thereafter for all animals that had not been euthanized (N = 27). CSF was collected as before from the cisterna magna. Part of the CSF sample from all animals was used for confirmation of the presence of microorganisms using standard microbiological tests (10–15 μl CSF). Only CSF samples from which the respective microorganisms had been cultured were included in the NMR study. The presence of cryptococcal organisms in CSF was confirmed by thin smears of CSF (10 μl), which were stained with a rapid Romanowsky stain (DiffQuik; Lab Aids, Australia). Typically, large numbers of organisms were evident in stained smears. Plate counts were performed on CSF samples from thirteen animals (six C. neoformans and seven S. pneumoniae) by culture of five different dilutions on horse blood agar plates for 24 hours (S. pneumoniae) and on Sabouraud dextrose agar plates for 48 hours (C. neoformans) at 35°C. Three inoculated animals were excluded from the study due to an absence of microorganisms in CSF samples.

Animals were given free access to food and water ad libitum and maintained in a 12 hours light/dark cycle at 25°C throughout the study. Animals were monitored daily for any signs of distress. Rats were sacrificed after the experiments. Euthanasia was generally performed with CO_2.

Surplus CSF from clinical samples was studied in compliance with ethical approval granted by the Human Ethics Review Committee of Sydney Western Area Health Service. CSF from patients with cryptococcal meningitis were stored at –80°C for up to three months before NMR analysis. The diagnosis of cryptococcal meningitis was made independently by attending clinicians and reviewed by TCS for consistent clinical features. Leukocyte counts, biochemistry, microbial strains and cultures were performed for all samples. Cryptococcal meningitis was defined by a positive India Ink stain and/or CSF cryptococcal antigen titre >8 and/or culture of C. neoformans from CSF.

NMR Studies

Five to fifteen microliters of CSF were directly collected into micro NMR tubes (1 mm outer diameter, 0.8 mm inner diameter, Bruker BioSpin AG, Fällanden, Switzerland). The samples were transferred within 15 minutes into liquid nitrogen and stored at –70°C for up to 60 days for NMR experiments. Samples were carefully thawed before NMR experiments and centrifuged for 30 seconds using a manual centrifuge to avoid air bubbles in the micro tubes. Deuterated water (2–4 µl) was added to 30 samples to adjust for low filling heights.

NMR experiments were carried out as described [33]. In brief, 1D 1H NMR spectra for SCS analysis were acquired using a Bruker Biospin Avance 400 spectrometer. Two-dimensional (2D) correlation NMR spectra for resonance assignment were acquired using a Bruker Biospin Avance 600 spectrometer. An appropriate TXI 1 mm MicroProbe with z-gradients was used on either NMR spectrometer. In addition, NMR spectra were also acquired from CSF samples collected at the terminal time point (volume 50–150 µl) using a Bruker Biospin Avance 600 spectrometer equipped with a 5 mm {1H, 13C} inverse-detection dual-frequency probe. All measurements were carried out without spinning at 37°C. 1D 1H NMR spectra were acquired with 16 k data points, a spectral width of 10 ppm, a repetition time of three seconds and accumulation of 64 averages. Residual water was suppressed using an 1D NOESY presaturation sequence with a mixing time of 100 ms [41]. For the suppression of contributions from macromolecules and other compounds with short T2 values, the Carr-Purcell-Meiboom-Gill (CPMG) experiment was performed on a small set of CSF samples. Experimental parameters were: number of repeated cycles 200, echo time 10 ms.

Resonance Assignment

2D homo- and heteronuclear correlation spectra were acquired for six to eight CSF samples per class to assign 1H NMR resonances to respective metabolites. Standard {1H, 1H} COSY and {1H, 1H} TOCSY experiments were acquired with the following parameters: spectral width in t2 10 ppm, t2 time domain 2 K, 256 increments of 8 or 16 acquisitions each, relaxation delay 1 s. TOCSY spectra with mixing times of 40 ms were acquired with 256 increments of 2 K data points and 16 acquisitions. Standard sensitivity-enhanced gradient inverse-detection HSQC spectra were acquired with the following parameters: optimisation for one-bond coupling of 125 and 145 Hz, total of 256 increments with 32 transients, 4 K complex data points, and 13C decoupling using GARP-1. HMBC spectra were optimised for one-bond coupling of 125 Hz and long range coupling constants of 6 Hz.

Relative quantification of metabolites was achieved by integration of resonances in the chemical shift region between 0.0–4.0 ppm, following polynomial baseline correction.

Statistical Classification Strategy

1H NMR spectra of CSF from controls and animals with confirmed S. pneumoniae and C. neoformans meningitis (obtained three to eight days after infection) were used to develop three pair-wise classifiers (for S. pneumoniae versus C. neoformans; C. neoformans versus control and S. pneumoniae versus control) as described previously [16], [42], [43]. In brief: magnitude NMR spectra were normalized to the total integral between 0.35 to 4.0 ppm, which contains 1500 data points. Two to three maximally discriminatory regions of these spectra were identified by a genetic-algorithm-based Optimal Region Selector (GA-ORS) [34]. These regions are summarised in Table 1. Using the first derivatives or the rank ordered first derivatives of the spectral regions, pair-wise Linear Discriminant Analysis based classifiers were developed. The robustness of the LDA classifiers was tested using a bootstrap-based crossvalidation by randomly selecting half of the spectra to develop the classifiers and the remaining half to validate the classifiers [44]. This process was repeated 1000 times with random replacements. The classifiers yielded probabilities of class assignments for the individual spectra. Class assignment was called crisp if class assignment probabilities were >66%. Software developed in-house was used for all steps of the statistical classification (IBD, NRC Canada, Winnipeg) as described before [34], [42], [43].

Acknowledgements

The authors wish to thank Ms Susan Dowd for assistance with the animal experiments and Mrs Ok Cha Lee for assistance with yeast identification.

Authors' Contributions

Conceived and designed the experiments: UH RM TCS. Performed the experiments: UH RM TK HMD. Analyzed the data: UH RLS BD. Contributed reagents/materials/analysis tools: RM TK HMD RLS BD. Wrote the paper: UH RLS TCS.

References

1. Aronin SI, Peduzzi P, Quagliarello VJ (1998) Community-acquired meningitis. Risk stratification for adverse clinical outcome and effect of antibiotic timing. Ann Intern Med 129: 862–869.
2. Lu CH, Huang CR, Chang WN (2002) Community-acquired bacterial meningitis in adults: the epidemiology, timing of appropriate antimicrobial therapy, and prognostic factors. Clin Neurol Neurosurg 104: 352–358.
3. Tunkel AR, Scheld WM (2000) Acute meningitis. In: Mandell GL, Bennett JE, Dolin R, editors. Mandell, Douglas and Bennett's Principles and Practice of Infectious Diseases. 5th ed. Philadelphia: Churchill Livingstone. pp. 959–997.
4. De Gans J, Van de Beek D (2002) Dexamethasone in adults with bacterial meningitis. N Engl J Med 347: 1549–1556.
5. Fishman R (1992) Cerebrospinal fluid in diseases of the nervous system. Philadelphia: WB Saunders.
6. Van de Beek D, De Gans J, Spanjaard L, Weisfelt M, Reitsma JB, et al. (2004) Clinical features and prognostic factors in adults with bacterial meningitis. N Engl J Med 351: 1849–1859.
7. Wood M, Anderson M (1998) Cerebrospinal fluid and infections of the central nervous system. In: Walton Sir J, editor. Major problems in Neurology. London: W.B. Saunders.
8. Coen M, O'Sullivan M, Bubb WA, Kuchel PW, Sorrell T (2005) Proton nuclear magnetic resonance - based metabonomics for rapid diagnosis of meningitis and ventriculitis. Clin Infect Dis 41: 1582–1590.
9. Mariey L, Signolle JP, Amiel C, Travert J (2001) Discrimination, classification, identification of microorganisms using FTIR spectroscopy and chemometrics. Vibrational Spectroscopy 26: 151–159.
10. Himmelreich U, Mountford CE, Sorrell TC (2004) NMR spectroscopic determination of microbiological profiles in infectious diseases. Trends Appl Spectrosc 5: 269–283.

11. Maquelin K, Kirschner C, Choo-Smith LP, van den Braak N, Endtz HP, et al. (2002) Identification of medically relevant microorganisms by vibrational spectroscopy. Journal of Microbiological Methods 51: 255–271.
12. Fenselau C, Demirev PA (2001) Characterization of intact microorganisms by MALDI mass spectrometry. Mass Spectrometry Reviews 20: 157–171.
13. Allen JK, Davey HM, Broadhurst D, Heald JK, Rowland JJ, et al. (2003) High-throughput characterisation of yeast mutants for functional genomics using metabolic footprinting. Nature Biotechnology 21: 692–696.
14. Raamsdonk LM, Teusink B, Broadhurst D, Zhang N, Hayes A, et al. (2001) A functional genomics strategy that uses metabolome data to reveal the phenotype of silent mutations. Nature Biotechnology 19: 45–50.
15. Dunn WB, Bailey NJC, Johnson HE (2005) Measuring the metabolome: current analytical technologies. Analyst 130: 606–625.
16. Himmelreich U, Somorjai RL, Dolenko B, Lee OC, Daniel HM, et al. (2003) Rapid identification of Candida species by using nuclear magnetic resonance spectroscopy and a statistical classification strategy. Appl Environ Microbiol 69: 4566–4674.
17. Pope GA, MacKenzie DA, Defernez M, Aroso MAMM, Fuller LJ, et al. (2007) Metabolic footprinting as a tool for discriminating between brewing yeasts. Yeast 24: 667–679.
18. Urbanczyk-Wochniak E, Luedemann A, Kopka J (2003) Parallel analysis of transcript and metabolic profiles: a new approach in systems biology. EMBO Report 4: 989–993.
19. Nicholson JK, Wilson ID (1989) High resolution proton magnetic resonance spectroscopy of biological fluids. In: Emsley JW, Feeney J, editors. Progress in Nuclear Magnetic Resonance Spectroscopy. Oxford: Pergamon Press. pp. 449–501.
20. Hiraoka A, Miura I, Hattori M, Tominaga I, Kushida K, et al. (1994) Proton magnetic resonance spectroscopy of cerebrospinal fluid as an aid in neurological diagnosis. Biol Pharm Bull 17: 1–4.
21. Nicoli F, Vion-Dury J, Maloteaux JM, Delwaide C, Confort-Gouny S, et al. (1993) CSF and serum metabolic profile of patients with Hungington's chorea: a study by high resolution proton NMR spectroscopy and HPLC. Neurosci Lett 154: 47–51.
22. Wevers RA, Engelke U, Wendel U, De Jong JGN, Gabreels FJM, et al. (1995) Standardized method for high resolution 1H-NMR of cerebrospinal fluid. Clin Chem 41: 744–751.

23. Hu Y, Malone JP, A.M. F, Townsend RR, Holtzman DM (2005) Comparative proteomic analysis of intra- and interindividual variation in human cerebrospinal fluid. Mol Cell Proteomics 4: 2000–2009.
24. Selle H, Lamerz J, Buerger K, Dessauer A, Hager K, et al. (2005) Identification of novel biomarker candidates by differential peptidomics analysis of cerebrospinal fluid in Alzheimer's disease. Comb Chem High Troughput Screen 8: 801–806.
25. Dekker LJ, Boogerd W, Stockhammer G, Dalebout JC, Siccama I, et al. (2005) MALDI-TOF mass spectrometry analysis of cerebrospinal fluid tryptic peptide profiles to diagnose leptomeningeal metastases in patients with breast cancer. Mol Cell Proteomics 4: 1341–1349.
26. Pfyffer GE, Kissling P, Jahn EMI, Welscher HM, Salfinger M, et al. (1996) Diagnostic performance of amplified Mycobacterium tuberculosis direct test with cerebrospinal fluid, other nonrespiratory, and respiratory specimens. J Clin Microbiol 34: 834–841.
27. Lewczuk P, Esselmann H, Meyer M, Wollscheid V, Neumann M, et al. (2003) The amyloid-beta (A beta) peptide pattern in cerebrospinal fluid in Alzheimer's disease: evidence of a novel carboxyterminally elongated A beta peptide. Journal of Rapid Communications in Mass Spectrometry 17: 1291–1296.
28. Yuan X, Desiderio DM (2005) Human cerebrospinal fluid peptidomics. J Mass Spectrom 40: 176–181.
29. Maillet S, Vion-Dury J, Confort-Gouny S, Nicoli F, Lutz NW, et al. (1998) Experimental protocol for clinical analysis of cerebrospinal fluid by high resolution proton magnetic resonance spectroscopy. Brain Res Protoc 3: 123–134.
30. Bell JD, Brown JC, Sadler PJ, Macleod AF, Sonksen PH, et al. (1987) High resolution proton nuclear magnetic resonance studies of human cerebrospinal fluid. Clin Sci 72: 563–570.
31. Levine J, Panchalingam K, McClure RJ, Gershon S, Pettegrew JW (2000) Stability of CSF metabolites measured by proton NMR. J Transm 107: 843–848.
32. Griffin JL, Nicholls AW, Keun HC, Mortishire-Smith RJ, Nicholson JK, et al. (2002) Metabolic profiling of rodent biological fluids via H-1 NMR spectroscopy using a 1 mm microlitre probe. Analyst 127: 582–584.
33. Schlotterbeck G, Ross A, Hochstrasser R, Senn H, Kuehn T, et al. (2002) High-resolution capillary tube NMR. A miniaturized 5 ul high-sensitivity TXI probe for mass-limited samples, off-line LC NMR, and HT NMR. Anal Chem 74: 4464–4471.

34. Nikulin AE, Dolenko B, Bezabeh T, Somorjai RL (1998) Near-optimal region selection for feature space reduction: novel preprocessing methods for classifying MR spectra. NMR Biomed 11: 209–216.

35. Himmelreich U, Accurso R, Malik R, Dolenko B, Somorjai RL, et al. (2005) Identification of Staphylococcus aureus brain abscesses: rat and human studies with 1 H MR spectroscopy. Radiology 236: 261–270.

36. Himmelreich U, Dzendrowskyj T, Allen C, Dowd S, Malik R, et al. (2001) Cryptococcomas distinguished from gliomas with magnetic resonance spectroscopy: an experimental rat and cell culture study. Radiology 220: 122–128.

37. Dunne VG, Bhattachayya S, Besser M, Rae C, Griffin JL (2005) Metabolites from cerebrospinal fluid in aneurysmal subarachnoid haemorrhage correlate with vasospasm and clinical outcome: a pattern-recognition 1H NMR study. NMR Biomed 18: 24–33.

38. Schlegel HG (1992) Allgemeine Mikrobiologie. Stuttgart: Georg Thieme Verlag.

39. Himmelreich U, Allen C, Dowd S, Malik R, Shehan BP, et al. (2003) Magnetic Resonance Spectroscopy of rat lung cryptococcomas identifies compounds of importance in pathogenesis. Microbes Infect 5: 285–290.

40. Subramanian A, Gupta A, Saxena S, Kumar R, Nigam A, et al. (2005) Proton MR CSF analysis and a new software as predictors for the differentiation of meningitis in children. NMR Biomed 18: 213–225.

41. Palmer AG III, Cevanagh J, Wright PE, Rance M (1991) J Magn Reson 93: 151–170.

42. Somorjai RL, Baumgartner R, Booth S, Bowman C, Demko A, et al. (2004) A data-driven, flexible machine learning strategy for the classification of biomedical data. In: Dubitzky W, Azuaje F, editors. Artificial intelligence methods and tools for systems biology. Berlin: Springer. pp. 67–85.

43. Somorjai RL, Dolenko B, Nikulin AE, Nickerson P, Rush D, et al. (2002) Distinguishing normal from rejecting renal allografts: Application of a three stage classification strategy to MR and IR spectra of urine. Vib Spectrosc 28: 97–102.

44. Efron B, Tibshirani R (1993) An introduction to the bootstrap. New York: Chapman & Hall.

Traditional Biomolecular Structure Determination by NMR Spectroscopy Allows for Major Errors

Sander B. Nabuurs, Chris A. E. M. Spronk, Geerten W. Vuister and Gert Vriend

ABSTRACT

One of the major goals of structural genomics projects is to determine the three-dimensional structure of representative members of as many different fold families as possible. Comparative modeling is expected to fill the remaining gaps by providing structural models of homologs of the experimentally determined proteins. However, for such an approach to be successful it is essential that the quality of the experimentally determined structures is adequate. In an attempt to build a homology model for the protein dynein light chain 2A (DLC2A) we found two potential templates, both experimentally determined nuclear magnetic resonance (NMR) structures originating from structural genomics efforts. Despite their high sequence identity (96%), the folds of the two structures are markedly different. This urged us to perform in-depth analyses of both structure ensembles and the deposited experimental data, the results of which clearly identify one of the two models as largely incorrect. Next, we analyzed the quality of a large set of recent NMR-derived structure ensembles originating from both structural genomics projects and individual structure determination groups. Unfortunately, a visual inspection of structures exhibiting lower quality scores than DLC2A reveals that the seriously flawed DLC2A structure is not an isolated incident. Overall, our results illustrate that the quality of NMR structures cannot be reliably evaluated using only traditional experimental input data and overall quality indicators as a reference and clearly demonstrate the urgent need for a tight integration of more sophisticated structure validation tools in NMR structure determination projects. In contrast to common methodologies where structures are typically evaluated as a whole, such tools should preferentially operate on a per-residue basis.

Synopsis

Three-dimensional biomolecular structures provide an invaluable source of biologically relevant information. To be able to learn the most of the wealth of information that these structures can provide us, it is of great importance that the quality and accuracy of the protein structure models deposited in the Protein Data Bank are as high as possible. In this work, the authors describe an analysis that illustrates that this is unfortunately not the case for many protein structures solved using nuclear magnetic resonance spectroscopy. They present an example in which two strikingly different models describing the same protein are analyzed using commonly available structure validation tools, and the results of this analysis show one of the two models to be incorrect. Subsequently, using a large set of recently determined structures, the authors demonstrate that unfortunately this example does not stand on its own. The analyses and examples clearly illustrate that relying solely on the experimental data to evaluate structural quality can provide a false sense of correctness and the combination of multiple sophisticated structure validation tools is required to detect the presence of errors in protein nuclear magnetic resonance structures.

Introduction

Experimentally determined three-dimensional structures of biomolecules form the foundation of structural bioinformatics, and any structural analysis would be impossible without them. Two main techniques are available for biomolecular structure determination: x-ray crystallography and nuclear magnetic resonance (NMR) spectroscopy. It is important to realize that all resulting structure models are derived from their underlying experimental data. Unfortunately, any experiment and thus any structure model will have errors associated with it. Random errors depend on the precision of the experimental measurements and are propagated to the precision of the final models. Systematic errors and mistakes often result from errors in the interpretation of the experimental data and relate directly to the accuracy of the final structure models. For example, in NMR spectroscopy errors can be introduced by misassignment of the spectral signals; in X-ray crystallography errors are most likely made when the protein structure is positioned in the electron density [1,2].

Several studies have shown that not all experimentally determined biomolecular structure models are of equally high quality [3–6]. Many different types of errors can be identified in protein structures, ranging from too tightly restrained bond lengths and angles, to molecules exhibiting a completely incorrect fold. Where the former type of errors often does not have large consequences for the analysis of the structure and typically can be easily remedied by refinement in a proper force field [7,8], the latter renders a structure model completely useless for all practical purposes. Throughout the years several such errors have been uncovered in the Protein Data Bank (PDB) [9], which often resulted in the replacement of the incorrect models with improved ones.

A typical example of an incorrectly folded structure model is the first crystal structure of photoactive yellow protein. The structure was solved initially in 1989 [10] and deposited under the now obsolete PDB entry 1PHY. An updated model released 6 y later showed that in the original model the electron density had been misinterpreted [11] (PDB entry 2PHY). Similar chain tracing problems led to an incorrect model for a DD-peptidase [12] (the now obsolete PDB entry 1PTE), which was corrected 10 y later when the structure was solved again but now at higher resolution [13] (PDB entry 3PTE).

Also, for structures determined using NMR spectroscopy, cases are known where reevaluation of the experimental data, often prompted by publication of a corresponding structure, has resulted in the replacement of structures in the PDB. A well-known example is the original NMR structure of the oligomerization domain of p53 [14]. In this dimer of dimers, a difference in the orientation of the two dimers was observed between the NMR and crystal structure, the latter published shortly after the NMR structure [15] (PDB entry 1C26). Reexamination of the nuclear Overhauser enhancement (NOE) data led to the identification of three misinterpreted peaks in the original p53 NOE assignments and the inclusion of several new NOEs, resulting in a revision of the original PDB entry [16] (PDB entry 1OLH). A similar low number of misinterpreted NOE signals (17 in total) resulted in a largely incorrect fold for the anti–σ factor AsiA [17] (the now obsolete PDB entry 1KA3). In this case, it was not until a second solution structure of AsiA was published [18] (PDB entry 1JR5) that the experimental data of the original AsiA structure were reexamined and the assignment errors were discovered [19] (updated PDB entry 1TKV).

In this paper, we describe a detailed analysis of two recently released NMR structures of the protein dynein light chain 2A (DLC2A), one from human (PDB entry 1TGQ) and one from mouse (PDB entry 1Y4O). Both structures originate from large structural genomics initiatives: the structure of human DLC2A (hDLC2A) was determined by the Northeast Structural Genomics Consortium (NESGC, http://www.nesg.org), and the mouse variant (mDLC2A) was determined by the Center for Eukaryotic Structural Genomics (CESG, http://www.uwstructuralgenomics.org). Despite 96% sequence identity, large structural differences are observed between the two ensembles; an unexpected and extremely unlikely result. Using the deposited experimental data we show that only the 1Y4O structure ensemble is correct. Subsequently, we analyze both ensembles using various structure and data validation methods to show that the erroneous structure ensemble could have been identified prior to deposition. Finally, we validate a large set of protein NMR structures that were released from the PDB in the period 2003 to 2005 and show that the DLC2A example does not stand on its own, but that more errors of this magnitude can be found. We conclude with some suggestions on how, in the future, such large errors can be identified during the structure determination process using readily available validation software.

Results/Discussion

Our interest in DLC2A originated from a request by one of our collaborators to build a homology model for this protein. A BLAST search in February 2005 [20] against the PDB revealed that construction of a homology model should be straightforward: two NMR structures of DLC2A (PDB entries 1Y4O and 1TGQ), both with more than 95% sequence identity to the target sequence, had been released in the months prior to our query. Surprisingly, a first visual inspection of both structures revealed striking differences, as shown in Figure 1.

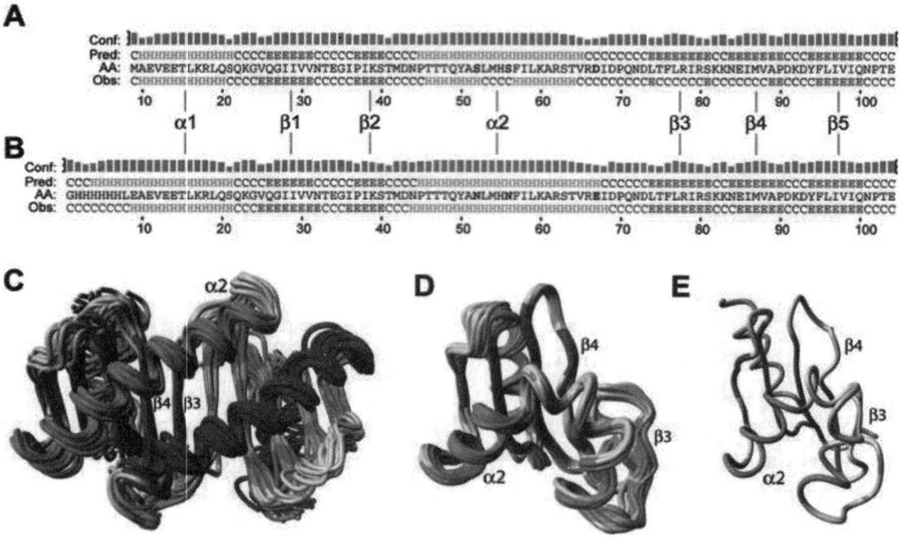

Figure 1. Sequence and Structure Ensembles of Two DLC2A Structures
(A) The sequence of human DLC2A (hDLC2A) (AA).
(B) The sequence of mouse DLC2A (mDLC2A) proceeded by an eight-residue His-tag (AA). The secondary structure as predicted using PSIPRED [33,50] (Pred) and the confidence of this prediction (Conf) are shown above the sequences. The secondary structure as observed in the ensembles (Obs) is indicated below the sequences. Except for the His-Tag, the mouse and human sequences differ at three positions (indicated in bold).
(C) Ribbon diagram of the structure ensemble of mDLC2A (PDB entry 1Y4O). The residues of the His-tag have been omitted for clarity.
(D) Ribbon diagram of the structure ensemble of hDLC2A (PDB entry 1TGQ).
(E) The refined average structure of the ensemble calculated using the reconstructed 1TGQ dataset, as discussed in the text. Secondary structure is indicated using colors: helices are shown in blue and purple, strands are shown in red and orange. A numbering scheme for the secondary structure elements is indicated between the two sequences.

It is immediately obvious that DLC2A forms a dimer in the 1Y4O structure models (Figure 1C), whereas the 1TGQ ensemble contains DLC2A in monomeric form (Figure 1D). Additionally, the DLC2A models feature remarkably different folds. The central α-helix (α2 in Figure 1A and 1B), which extends from Asn44 to Ile68 in the 1Y4O ensemble, consists in the 1TGQ ensemble of two separate, almost antiparallel, α-helices (Thr46-Ser52 and Phe57-Thr64) connected by a turn-like region (Leu53-Ser56). Beta strands β3 (Leu71-Ser80) and β4 (Glu85-Pro90) pack tightly against each other in the 1Y4O structure models. In the 1TGQ structures, the β3 region forms a hairpin-like structure, and the β4 strand is much less tightly packed against the core of the protein.

During evolution, protein structure has always been more stable and has changed much slower than the associated sequence [21]. As a result, similar sequences fold into practically identical structures and remotely related sequences still adopt similar folds [22]. An accurate limit for this rule was recently derived by

Rost [23], who found that two sequences that share over 30% sequence identity in 100 aligned residues are practically guaranteed to have the same fold. Given this knowledge, it is extremely unlikely for mouse and human DLC2A, which share 96% sequence identity, to fold into the different structures shown in Figure 1C and 1D.

Visual inspection of the two ensembles made us realize quickly that the large differences probably originate from the oligomeric state of the two structures. Using NMR spectroscopy (and in most structural genomics initiatives [24]), the presence of tertiary structure in a soluble protein is typically assessed using a proton-nitrogen correlation (15N-HSQC) spectrum [25]. The observed pattern of dispersed signals, ideally one for each amino acid, provides a "fingerprint" of the protein. However, the formation of a symmetric dimer, as shown in Figure 1A, does not result in a doubling of the number of observed NMR signals. Consequently, it is not straightforward to determine the oligomeric state of a protein from its 15N-HSQC NMR spectra alone, and typically assessments have to be made from estimates of the protein's relaxation rates [26]. Therefore, if the oligomeric state of a protein is not known or is incorrectly known, the NMR spectra of a dimeric protein could be easily interpreted as originating from a monomer. Below, we present evidence that such a misinterpretation is the root-cause of the observed differences between the human and mouse DLC2A structure ensembles.

Figure 1C shows that the two α2-helices in the dimer interface are oriented in an antiparallel fashion. As a result, intermolecular signals arising from, for example, contacts between the N-terminal and C-terminal sides of these respective helices are to be expected. When it is a priori known that the protein under investigation is a dimer, specific experiments can be performed to distinguish such intermolecular contacts from the intramolecular ones [27]. However, if the intermolecular contacts are wrongly interpreted as intramolecular, the residues involved would appear to be close to each other also in the monomeric structure, something that is indeed observed in the structure models shown in Figure 1D.

To further test this hypothesis, we used the experimental NMR restraints from the 1Y4O structure ensemble (as those for the 1TGQ ensemble were not available) and changed all 72 intermolecular NOEs into 36 intramolecular distance restraints. With this simulated subset of 36 erroneous intramolecular NOEs (hereafter referred to as the 1TGQsim dataset) and the experimentally observed intramolecular restraints, structure calculations were performed. An ensemble of 20 structures without any distance violations larger than 0.5 Å was readily obtained. The refined geometric average of this ensemble is shown in Figure 1E, and it exhibits a fold very similar to that observed for the 1TGQ ensemble. These results provide a strong indication that the NMR spectra of hDLC2A were indeed interpreted as those of a monomer, while the protein, like its mouse homolog, is actually a dimer in solution. Conclusive evidence that the human DLC2A protein does indeed form a dimer was obtained from the NESGC Web site, where the aggregation screening records associated with hDLC2A clearly show that this protein forms dimers in solution (http://spine.nesg.org/buffer_exchange.pl?id=HR2106). During the reviewing process of this paper, one of the referees pointed us to the publication of an independent structure determination of the human homolog in August 2005 (PDB entry 1Z09) [28], which was indeed also solved as a dimer. Subsequently, in November 2005, 1.5 y after its original deposition, the monomeric PDB entry 1TGQ was replaced by a correct dimeric structure (PDB entry 2B95).

Data and Structure Analyses

Having established the origin of the errors present in the 1TGQ ensemble, we can now ask the most important question: Could these errors have been discovered during the structure determination and validation process? To investigate this issue, the deposited structure ensembles were evaluated using common structure validation tools. In addition, both structure ensembles were refined in explicit solvent [7,8] and subsequently also included in the structure validation process. The DLC2A models of the 1Y4O ensemble were refined against the deposited NOE distance restraints and dihedral angle restraints. As mentioned before, for the 1TGQ ensemble no experimental restraints had been deposited, and therefore the intramolecular restraints as obtained from the 1Y4O dataset were used. In addition, the restraints from the 1TGQsim dataset were also included in the refinement of the 1TGQ structures. The structure validation results for the two original and the two re-refined structural ensembles are shown in Table 1.

Table 1. Average Quality Indicators of the 1Y4O and 1TGQ Structure Ensembles before and after

Criteria	Characteristic	1Y4O (Original)	1Y4O (Refined)	1TGQ (Original)	1TGQ (Refined)
Agreement with experimental data	RMS violation 1Y4O distance restraints (Å)	0.0129	0.0097	0.607	0.0284
	Violations >0.5 Å 1Y4O distance restraints	0	0	63	0
	RMS violation $1TGQ_{sim}$ restraints (Å)	12.8	12.6	0.521	0.0231
	Violations >0.5 Å $1TGQ_{sim}$ restraints	32	32	4	0
	RMS violation 1Y4O dihedral restraints (°)	0.497	0.336	25.0	1.59
	Violations >5° 1Y4O dihedral restraints	0	0	34	4
PROCHECK validation results[a]	Most favored regions	91.2	90.5	67.7	85.8
	Additionally allowed regions	8.4	9.0	27.3	12.8
	Generously allowed regions	0.2	0.2	4.7	0.5
	Disallowed regions	0.2	0.3	0.2	0.9
WHAT IF structure Z-scores[b]	Packing quality	−0.4	0.1	−2.1	−1.5
	Ramachandran plot appearance	−3.6	−3.3	−6.6	−4.6
	χ_1/χ_2 rotamer normality	−0.3	−0.7	−5.8	−3.0
	Backbone conformation	−0.8	−1.1	−5.4	−5.4

[a]Percentage of residues present in the four different regions of the Ramachandran plot.
[b]A *Z*-score [31,32] is defined as the deviation from the average value for this indicator observed in a database of high-resolution crystal structures, expressed in units of the standard deviation of this database-derived average. Typically, *Z*-scores below a value of −3 are considered poor, those below −4 are considered bad.

Refinement in Explicit Solvent

The 1Y4O ensemble demonstrates a good agreement with the experimentally deposited restraints. For the distance restraints, no violations larger than 0.5 Å are observed, for the dihedral angle restraints, we find no violations larger than 5°. Both these thresholds are widely considered as compatible with and representative for a good structure within the NMR community. As expected, the 1TGQsim dataset of erroneous intramolecular restraints exhibits very large violations for the 1Y4O ensembles. The validation scores, as determined by the programs PROCHECK [29] and WHAT IF [30], all fall within acceptable ranges; only the Ramachandran plot Z-score [31] of –3.3 might be considered poor [32]. Still, this score is substantially better than that of a typical NMR structure taken from the PDB [8]. The refinement in explicit solvent slightly improves the quality indicators of the 1Y4O ensemble and the agreement of the structures with the experimental data. For comparison, we also evaluated of the quality of the recently released and the updated DLC2A entries in the PDB (entries 1Z09 and 2B95, respectively). Both exhibit quality scores much comparable to those of the 1Y4O ensemble, with again only the Ramachandran plot score being somewhat poor (data not shown).

The quality indicators for the deposited 1TGQ ensemble are, however, considerably worse when compared to those of the 1Y4O structure models: the majority of the quality Z-scores identify this structure as an outlier (Z-score < –4). The agreement of the original 1TGQ ensemble with the experimental restraints from 1Y4O is quite poor, but this is to be expected as these restraints were not used in the actual 1TGQ structure determination. The agreement of the 1TGQsim dataset with the 1TGQ ensemble is much better than for the 1Y4O ensemble. After a refinement in explicit solvent, the 1TGQ ensemble has accommodated to all distance restraints and does not show any violations larger than 0.5 Å. It is, however, unable to completely fulfill the experimental dihedral angle restraints of the 1Y4O dataset. On average four dihedral angle restraints per structure are violated by more than 5° in the refined 1TGQ ensemble, but none of these violate more than 15°. The refinement results in a considerable improvement of the PROCHECK validation scores and the percentage of residues in the most favored regions of the Ramachandran plot increases to a commonly considered acceptable score of 85.8%. Most of the WHAT IF quality Z-scores improve, but both the Ramachandran plot and the backbone normality scores remain at a very worrisome level (below –4). Also the $\chi 1/\chi 2$ rotamer normality does not reach the level of quality typically observed for this quality indicator after a refinement in explicit solvent [8].

All in all, our results show that an incorrectly folded NMR structure is easily refined to a good agreement with the experimental input data and acceptable PROCHECK Ramachandran plot statistics. The overall WHAT IF quality indicators identify the structure as problematic, but only the $\chi 1/\chi 2$ rotamer normality score is significantly worse than the 100 refined structures present in the DRESS database [8]. When judged by its overall quality parameters, it is understandable, but nevertheless worrisome, that the erroneous 1TGQ ensemble went unnoticed through the structure determination and validation pipeline at the NESGC. However, a more detailed inspection of the validation results shows that the problematic regions of this ensemble of structures could have been identified.

Structure Validation on a Per-Residue Basis

One of the first and very straightforward indicators that something might be wrong with the 1TGQ structure ensemble is the large discrepancy between the predicted and observed secondary structure, as shown in Figure 1A. Modern secondary structure prediction algorithms, such as the PSIPRED algorithm [33] applied here, typically yield predictions with an accuracy of 75% to 80%. The large deviations between predicted and observed secondary structure for the $\alpha 2$, $\beta 3$, and $\beta 4$ regions justify a further detailed inspection of these parts of the protein.

Figure 2 shows the per-residue scores of the two refined ensembles for four different WHAT IF quality indicators. The refined 1TGQ ensemble exhibits lower values for the packing quality [34] (see Figure 2A) compared to the refined 1Y4O ensemble, most notably in the $\alpha 2$, $\beta 4$, and $\beta 5$ regions. When the packing quality scores of 1TGQ are evaluated by themselves, the problematic regions do not particularly stand out. The same notion holds for the rotamer normality Z-scores (see Figure 2C), although the continuous stretch of residues from Pro45 to Arg80 with relatively low-quality scores should be considered suspicious. This is also expressed in the lower overall rotamer normality score, as already shown in Table 1. A nearly identical stretch of low scoring residues (from Met55 to Ile85) is observed when evaluating the Ramachandran plot quality scores (see Figure 2B). The finding that similar regions of consecutively low scoring residues are highlighted by different quality indicators provides more circumstantial evidence of the underlying problems, but again, no exceptional outliers are found.

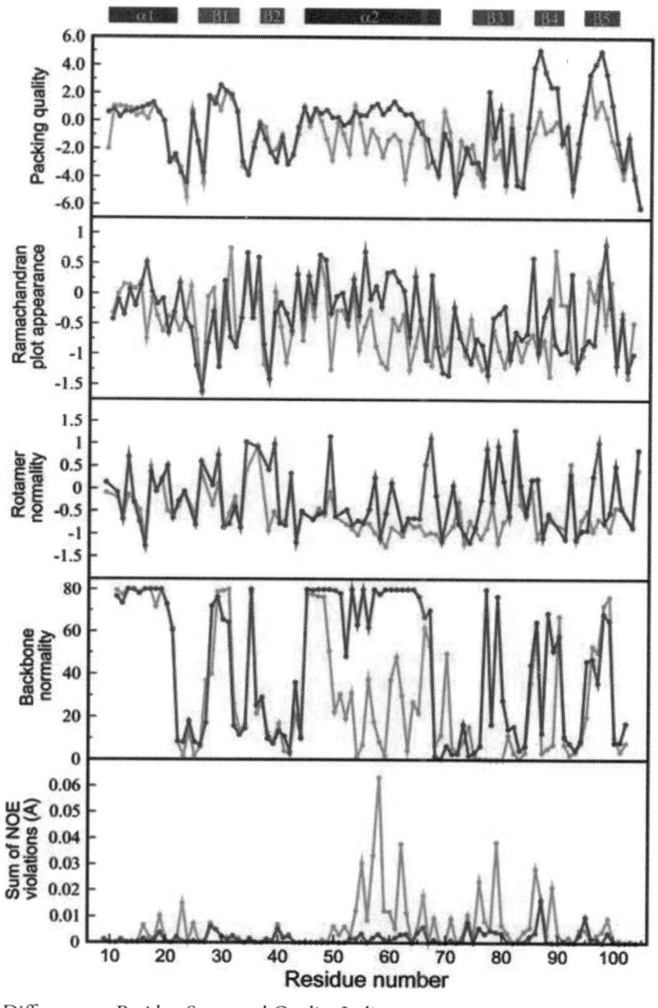

Figure 2. Five Different per-Residue Structural Quality Indicators

(A) Packing quality Z-score.

(B) Ramachandran plot appearance Z-score.

(C) Rotamer normality Z-score.

(D) Backbone normality score. The values listed on the y-axis indicate the number of times the local backbone (defined by the current residue plus or minus two residues) was found in WHAT IF's internal database (with a cut-off on the number of hits at 80).

(E) Sum of the NOE violations. Scores for the refined 1Y4O ensemble are shown in green; those for the refined 1TGQ ensemble are shown in orange. Secondary structure of the 1Y4O ensemble is indicated using colored boxes: α-helices are shown in blue, β-strands are shown in red.

Our analysis shows that only the backbone normality score unambiguously identifies the erroneous regions in the 1TGQ structure ensemble. Figure 2D shows the number of occurrences of the local backbone conformation of each residue in WHAT IF's nonredundant internal database. For NMR structures, it is quite common to find low backbone normality scores in loops and other flexible regions, as evidenced by the validation results of the 1Y4O ensemble where most low scoring regions are found between the different secondary structure elements. These low scoring loops do, however, not influence the overall backbone normality score, which for the 1Y4O structures falls well within the normal range (Table 1).

Regular secondary structure elements, such as α-helices, typically score very well on the backbone normality check (e.g., the α1 region in both ensembles and the α2 region of 1Y4O). In the 1TGQ ensemble, however, unusually low backbone normality scores are observed for most residues in the α2 region. A near-zero number of hits is obtained for several residues (e.g., Met54, His55, Leu59, and Ser63), most of which are involved in bending the α2-helix. Alarming are the successive residues Thr75-Arg80, which all have a backbone occurrence score of 0, indicating that no similar backbone conformations are observed in the WHAT IF internal database of high-quality crystal structures [35]. This is not uncommon for occasional residues in loops but highly unlikely for consecutive residues in a well-defined region of the structure and is indicative of either a very unique or a very wrong backbone conformation. In either case, these results indisputably warrant an in-depth investigation of these regions of the structure and the experimental data that define them.

To assess if the experimental data also indicate the same regions as problematic, the sum of the NOE violations per residue is shown in Figure 2E. The found violations are small and would under normal circumstances not be considered problematic, but again they are clustered in the α2/β3 region. To further investigate this finding, we also analyzed the dataset constructed for the 1TGQ ensemble using the QUEEN program [36]. Using a representation of the structure in distance space and concepts derived from information theory, QUEEN can quantify the information contained in both individual restraints and sets of restraints. For the 1TGQ dataset, the total information content (Itotal) and, for each of the individual restraints, the unique information content (Iuni) and the average information content (Iave) were determined. We previously showed that combining the unique and average information content can be very useful in the identification of problematic restraints in an experimental dataset [36]. The [Iuni,Iave] plot shown in Figure 3 clearly illustrates the varying information content of the different restraints in the 1TGQ dataset. Similar to previous work [37], we evaluated the 30 most important and most informative restraints, all located above the dashed line in Figure 3. In total, 13 of the 30 most crucial restraints (indicated by the black squares in Figure 3) are located in regions of the structure ensemble that score low on the backbone normality check. As such, an analysis of the 1TGQ dataset using QUEEN would also have highlighted the α2 and β3 regions as parts of the molecule deserving further investigation.

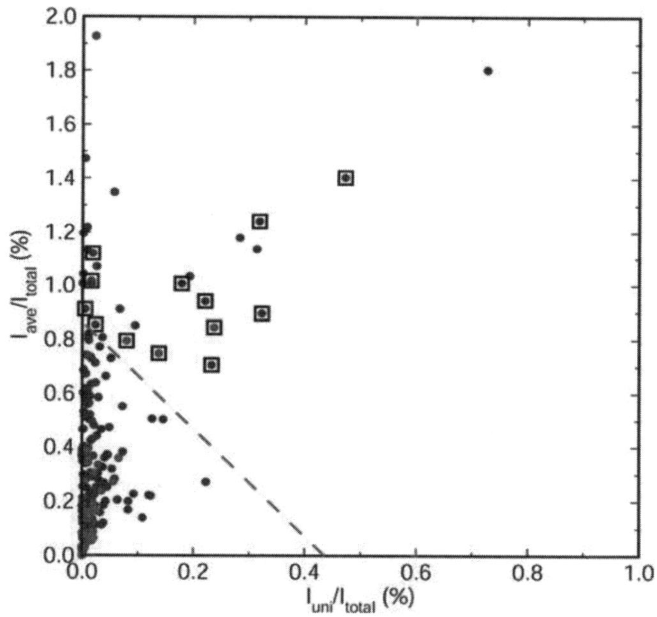

Figure 3. [Iuni, Iave] Plot for 1TGQ Calculated Using the QUEEN Program [36]
Long-range restraints (blue filled circles) and the 1TGQsim restraints (red filled circles) are indicated. Restraints that are among the 30 most unique and most important (those above the dashed gray line) and that involve residues in either the α2 or β3 region (cf. Figure 1A) are indicated by black boxes.

In summary, our analyses of both the structure ensemble and the supposedly observed experimental data of PDB entry 1TGQ clearly reveal the erroneous regions present in this set of structural models. Such a severe error therefore should not have gone undiscovered in any structure determination project.

Evaluation of a Large Set of Recent NMR Structures

The fact that the erroneous 1TGQ ensemble made it into the PDB inevitably raises the question if more comparatively large errors might have gone unnoticed. To answer this question, we performed a quality analysis of a large set of protein NMR structures, the results of which are shown in Figure 4. The presented dataset was constructed by selecting from the PDB all NMR structures that were deposited after January 2003, consisted of at least 45 amino acids, and had more than 40% of their amino acids involved in secondary structure elements. The latter criterion was imposed to remove the models of largely unfolded structures that might bias our analysis. From this set all structural genomics target were filtered (310 in total), their quality scores are shown in orange in Figure 4. From the remaining NMR structures, originating from individual structure determination laboratories, an equally sized random selection of structures was made, whose quality scores are shown in green in Figure 4. For comparison, the average quality scores of the 1TGQ ensemble, both before and after refinement, are also indicated.

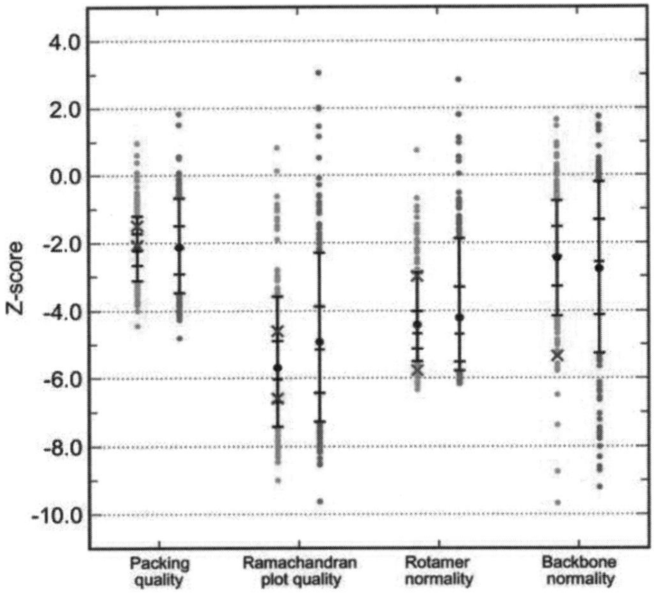

Figure 4. Structure Quality Z-Scores for a Large Set of Recent NMR Structures
The quality scores of 620 NMR ensembles released from the PDB after January 1, 2003, are shown. For comparison, the dataset is separated in structures solved as part of structural genomics projects (orange) and structures originating from individual research groups (green). For each quality indicator, the average Z-score is indicated with a filled black circle. The black horizontal markers indicate (from top to bottom) the 90th, 75th, 50th (the median), 25th, and 10th percentiles of the data points for each quality indicator. The distribution of the outliers outside the markers is indicated using colored data points. The quality scores of the original and refined 1TGQ ensemble (cf. Table 1) are indicated by red and blue crosses, respectively. The backbone normality score of 1TGQ is identical for the original and refined ensemble.

The data show no significant difference between the distributions of the quality indicators of structural genomics structures compared to those structures originating from individual research groups. In general, the distribution of the quality scores appears to be somewhat narrower for the structural genomics structures, but the average scores are similar, a result in-line with recent other studies [38]. Surprisingly, for both the packing and Ramachandran plot quality scores, the 1TGQ ensembles score comparable to the majority of the NMR structures. The rotamer normality score initially places the 1TGQ ensemble among the 10% worst scoring structural genomics structures, but after refinement it is amidst the top 10%. As before, the backbone normality score consistently identifies the erroneous 1TGQ structures as one of the outliers. Given the serious errors present in the 1TGQ ensemble, one might consider the fact that several NMR structures solved over the past years demonstrate backbone normality scores lower than those of 1TGQ rather worrisome.

Visual inspection of the structural ensembles exhibiting lower backbone normality scores than 1TGQ revealed that in some instances these low scores resulted from the corresponding proteins exhibiting unusual folds or dynamic behavior. For others, however, we noted some striking structural abnormalities of which we will discuss two examples. First, our attention was drawn to the NMR structure with the lowest backbone normality Z-score ($Z = -9.8$). It corresponds to an alternatively spliced PDZ domain of PTP-Bas [39] (PDZ-Bas, PDB entry 1Q7X), which was determined in the context of the Structural Proteomics In Europe project (SPINE, http://www.spineurope.org). In this structure ensemble, an arginine side chain deeply penetrates the hydrophobic core (cf. Figure 5A). Arginine, however, is a very hydrophilic residue and is typically not observed in hydrophobic environments. In the highly identical alternative spliced second PDZ domain of PTP-BL [37] (PDZ-BL, PDB entry 1OZI, sequence identity 95% with PDB entry 1Q7X) and, to the best of our knowledge, in all other homologous PDZ domains, the corresponding arginine is indeed solvent exposed (cf. Figure 5B), rendering it very unlikely for the 1Q7X ensemble to be correct. This finding is corroborated by the backbone residual dipolar coupling (RDC) data [40] measured for the PDZ-BL protein [37]. To allow for a fair comparison, an ensemble of 20 PDZ-BL structures was calculated and refined using only the experimental distance and dihedral data and the procedures described above, as the deposited structures [37] were refined against the RDC restraints. The RDC R-factor [41] obtained for the newly calculated PDZ-BL ensemble is 43%, whereas the RDC R-factor of 69% for the PDZ-BAS ensemble is significantly higher. This clearly demonstrates the ability of RDC-derived orientational restraints to also distinguish incorrect backbone orientations, but unfortunately these data are typically not acquired in structural genomics pipelines.

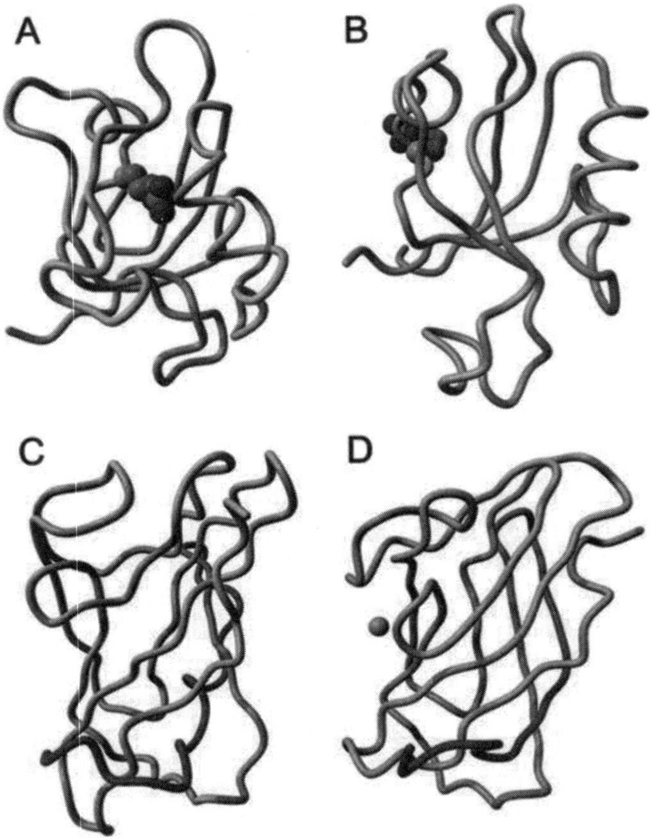

Figure 5. Examples of Observed Structural Anomalies
(A) An arginine side chain protruding the hydrophobic core of the second PDZ domain of PTP-Bas [39].
(B) The corresponding arginine in the highly homologous second PDZ domain of PTP-BL [37] is solvent exposed.
(C) The C-terminal region of DR1885 [42] (residues 120 to 149 are color-coded from yellow to red) forms a knot-like structure in the apo-form of DR1885.
(D) In the copper bound form of DR1885, the C-terminus wraps around the protein, instead of traversing through it. For each of the four structure ensembles, only the first, and presumably best, model is shown.

As a second example, we noticed striking differences between the apo- and copper bound forms of the protein DR1885 [42] (PDB entries 1X7L and 1X9L), also originating from the SPINE project. Most notable are the differences in the conformation of the C-terminal region of the protein (residues 120 to 149, Figure 5C and 5D). In the apo-form these residues are in a very unusual knot-like conformation, with the C-terminus passing through a loop consisting of residues 118 to 125. In the copper bound structures, the backbone of the C-terminal residues assumes a much more normal conformation and wraps around the DR1885 protein, instead of traversing through it. Given that there are no significant changes in the chemical shifts of the residues involved upon binding of copper to DR1885 (see Figure 2C in [42]), one of the two structure ensembles is almost certain to be incorrect.

In the publications describing the DR1885 protein [42] and the alternatively spliced PDZ domain from PTP-Bas [39], structural quality is foremost assessed by the number and size of the restraint violations and PROCHECK Ramachandran plot statistics. Our findings for the DLC2A protein already illustrated that these quality indicators are relatively insensitive to large structural errors, a result corroborated by the relatively acceptable scores found for these two datasets. Therefore, these examples clearly illustrate that the fact that no distance or dihedral angle violations are observed above a given threshold and that majority of the residues are found in allowed regions of the Ramachandran can be indicative of a good structure but does not provide any guarantees. It is interesting to note here that the three erroneous structures described in this paper stem from premier protein NMR groups, all involved in the development of structure validation and refinement methodologies [43–46], and that these methodologies either failed or were not or incorrectly applied in identifying the serious errors present in these structure ensembles.

To hopefully prevent such large errors from reoccurring in the future, we strongly suggest that validation results from normality checks, such as those implemented in the WHAT IF program [4,30], should be evaluated (and reported on) in any structure determination project. For high-throughput structural genomics projects, the application of multiple and sophisticated validation tools is even more critical, as much effort is geared towards minimizing the amount of expert time required for the determination and refinement of NMR structures [47]. Since this amount is deliberately continuously reduced, we expect structural genomics projects to become increasingly dependent on data and structure validation software to direct the spectroscopist to the regions that warrant his or her expert assessment.

Conclusions

We have shown that, when using only distance and dihedral restraints, even a largely incorrect structure is readily refined to seemingly acceptable levels of quality. As a result, the quality of biomolecular NMR structures cannot be safely assessed by the size and number of residual restraints violations, the precision of the structure ensemble, or even the fact that most residues are located in the allowed regions of the Ramachandran plot. Relying solely on these indicators to evaluate an ensemble of NMR structures therefore provides a false sense of correctness.

The fundamentally different nature of residual dipolar couplings renders them complementary to traditional NMR data and a powerful tool to identify large errors in NMR structures. Unfortunately, in many instances, such as in most structural genomics efforts, they are not routinely acquired and proper use of structure validation tools then becomes crucial. Furthermore, our results show that also more sophisticated quality indicators, e.g., the overall WHAT IF backbone normality score, do not unambiguously identify problematic structures. In contrast, we showed that only the simultaneous evaluation of multiple quality indicators on a per-residue basis, however, combined with a careful evaluation of the experimental data (e.g., using QUEEN), does allow for the well-supported identification erroneous regions in biomolecular NMR structures, thereby avoiding errors as those reported here.

Materials and Methods

NMR Structures and Data

For both mDLC2A and hDLC2A, the structure ensembles were obtained from the PDB (PDB entries 1Y4O and 1TGQ, respectively). The residue numbering of the 1TGQ ensemble was adjusted to match to that of the 1Y4O ensemble, as shown in Figure 1. The coordinates describing the His-tag in the 1Y4O ensemble (residues 1 to 8) were removed so that all DLC2A models contained an equal number of residues.

The experimental restraints for the 1Y4O ensemble, solved as a dimer, were obtained from the PDB, for the 1TGQ ensemble no experimental restraints were available at the time of writing. All stereospecifically assigned NOEs were deassigned for the violation analyses, structure calculations, and refinements. To be able to apply the same dataset to both structures, all restraints involving unique atoms of the three amino acids that are different in both sequences (cf. Figure 1A and 1B) were removed from the dataset. The final dataset contained 1,395 distance restraints, consisting of 553 intraresidue, 341 sequential, 278 medium-range, 187 long-range, and 72 intermolecular restraints. In addition, 146 dihedral angle restraints were included in all refinements. The deposited dataset also contained 96 hydrogen bond restraints, but as it is not clear how these were derived, and as they showed considerable violations in the deposited 1Y4O ensemble, these restraints were excluded from all analyses.

Structure Calculation and Refinement Protocols

All structure calculations were performed using CNS [48] and the default simulated annealing protocol, as provided with the software package. All refinements in explicit solvent [7] were performed using XPLOR-NIH [49] using the refinement procedure as described before [8]. Both the deposited and newly generated structure ensembles were validated using PROCHECK [29] and WHAT IF [30]. The deposited and constructed datasets were evaluated using the QUEEN program [36].

Authors' Contributions

SBN, CAEMS, GWV, and GV conceived and designed the experiments. SBN performed the experiments. SBN and CAEMS analyzed the data. SBN wrote the paper.

References

1. Kleywegt GJ (2000) Validation of protein crystal structures. Acta Crystallogr D 56: 249–265.
2. DePristo MA, de Bakker PI, Blundell TL (2004) Heterogeneity and inaccuracy in protein structures solved by X-ray crystallography. Structure (Camb) 12: 831–838.
3. Branden CI, Jones TA (1990) Between objectivity and subjectivity. Nature 343: 687–689.
4. Hooft RW, Vriend G, Sander C, Abola EE (1996) Errors in protein structures. Nature 381: 272.
5. Doreleijers JF, Rullmann JA, Kaptein R (1998) Quality assessment of NMR structures: A statistical survey. J Mol Biol 281: 149–164.
6. Spronk CA, Linge JP, Hilbers CW, Vuister GW (2002) Improving the quality of protein structures derived by NMR spectroscopy. J Biomol NMR 22: 281–289.
7. Linge JP, Williams MA, Spronk CA, Bonvin AM, Nilges M (2003) Refinement of protein structures in explicit solvent. Proteins 50: 496–506.
8. Nabuurs SB, Nederveen AJ, Vranken W, Doreleijers JF, Bonvin AM, et al. (2004) DRESS: A Database of REfined Solution nmr Structures. Proteins 55: 483–486.
9. Berman HM, Westbrook J, Feng Z, Gilliland G, Bhat TN, et al. (2000) The Protein Data Bank. Nucleic Acids Res 28: 235–242.
10. McRee DE, Tainer JA, Meyer TE, Van Beeumen J, Cusanovich MA, et al. (1989) Crystallographic structure of a photoreceptor protein at 2.4 A resolution. Proc Natl Acad Sci USA 86: 6533–6537.

11. Borgstahl GE, Williams DR, Getzoff ED (1995) 1.4-A Structure of photoactive yellow protein, a cytosolic photoreceptor: Unusual fold, active site, and chromophore. Biochemistry 34: 6278–6287.
12. Kelly JA, Knox JR, Moews PC, Hite GJ, Bartolone JB, et al. (1985) 2.8-A Structure of penicillin-sensitive D-alanyl carboxypeptidase-transpeptidase from Streptomyces R61 and complexes with beta-lactams. J Biol Chem 260: 6449–6458.
13. Kelly JA, Kuzin AP (1995) The refined crystallographic structure of a DD-peptidase penicillin-target enzyme at 1.6 A resolution. J Mol Biol 254: 223–236.
14. Clore GM, Omichinski JG, Sakaguchi K, Zambrano N, Sakamoto H, et al. (1994) High-resolution structure of the oligomerization domain of p53 by multidimensional NMR. Science 265: 386–391.
15. Jeffrey PD, Gorina S, Pavletich NP (1995) Crystal structure of the tetramerization domain of the p53 tumor suppressor at 1.7 angstroms. Science 267: 1498–1502.
16. Clore GM, Omichinski JG, Sakaguchi K, Zambrano N, Sakamoto H, et al. (1995) Interhelical angles in the solution structure of the oligomerization domain of p53: Correction. Science 267: 1515–1516.
17. Lambert LJ, Schirf V, Demeler B, Cadene M, Werner MH (2001) Flipping a genetic switch by subunit exchange. EMBO J 20: 7149–7159.
18. Urbauer JL, Simeonov MF, Urbauer RJ, Adelman K, Gilmore JM, et al. (2002) Solution structure and stability of the anti-sigma factor AsiA: Implications for novel functions. Proc Natl Acad Sci USA 99: 1831–1835.
19. Lambert LJ, Schirf V, Demeler B, Cadene M, Werner MH (2004) Flipping a genetic switch by subunit exchange. EMBO J 23: 3186.
20. Altschul SF, Madden TL, Schaffer AA, Zhang J, Zhang Z, et al. (1997) Gapped BLAST and PSI-BLAST: A new generation of protein database search programs. Nucleic Acids Res 25: 3389–3402.
21. Chothia C, Lesk AM (1986) The relation between the divergence of sequence and structure in proteins. EMBO J 5: 823–826.
22. Sander C, Schneider R (1991) Database of homology-derived protein structures and the structural meaning of sequence alignment. Proteins 9: 56–68.
23. Rost B (1999) Twilight zone of protein sequence alignments. Prot Eng 12: 85–94.
24. Montelione GT, Zheng D, Huang YJ, Gunsalus KC, Szyperski T (2000) Protein NMR spectroscopy in structural genomics. Nat Struct Biol 7(Suppl): 982–985.

25. Bodenhausen G, Ruben DJ (1980) Natural abundance N-15 NMR by enhanced heteronuclear spectroscopy. Chem Phys Lett 69: 185–189.
26. Anglister J, Grzesiek S, Ren H, Klee CB, Bax A (1993) Isotope-edited multidimensional NMR of calcineurin B in the presence of the non-deuterated detergent CHAPS. J Biomol NMR 3: 121–126.
27. Burgering MJ, Boelens R, Caffrey M, Breg JN, Kaptein R (1993) Observation of inter-subunit nuclear Overhauser effects in a dimeric protein. Application to the Arc repressor. FEBS Lett 330: 105–109.
28. Ilangovan U, Ding W, Zhong Y, Wilson CL, Groppe JC, et al. (2005) Structure and dynamics of the homodimeric dynein light chain km23. J Mol Biol 352: 338–354.
29. Laskowski RA, MacArthur MW, Moss DS, Thornton JM (1993) PROCHECK: A program to check the stereochemical quality of protein structures. J Appl Cryst 26: 283–291.
30. Vriend G (1990) Vriend G (1990) WHAT IF: A molecular modeling and drug design program. J Mol Graph 8: 52–56, 29.
31. Hooft RW, Sander C, Vriend G (1997) Objectively judging the quality of a protein structure from a Ramachandran plot. Comp Appl Biosci 13: 425–430.
32. Spronk CA, Nabuurs SB, Krieger E, Vriend G, Vuister GW (2004) Validation of protein structures derived by NMR spectroscopy. Prog Nucl Magn Reson Spectrosc 45: 315–337.
33. Jones DT (1999) Protein secondary structure prediction based on position-specific scoring matrices. J Mol Biol 292: 195–202.
34. Vriend G, Sander C (1993) Quality control of protein models: Directional atomic contact analysis. J Appl Cryst 26: 47–60.
35. Hooft RWW, Sander C, Vriend G (1996) Verification of protein structures: Side-chain planarity. J Appl Cryst 29: 714–716.
36. Nabuurs SB, Spronk CA, Krieger E, Maassen H, Vriend G, et al. (2003) Quantitative evaluation of experimental NMR restraints. J Am Chem Soc 125: 12026–12034.
37. Walma T, Aelen J, Nabuurs SB, Oostendorp M, van den Berk L, et al. (2004) A closed binding pocket and global destabilization modify the binding properties of an alternatively spliced form of the second PDZ domain of PTP-BL. Structure (Camb) 12: 11–20.
38. Snyder DA, Bhattacharya A, Huang YJ, Montelione GT (2005) Assessing precision and accuracy of protein structures derived from NMR data. Proteins 59: 655–661.

39. Kachel N, Erdmann KS, Kremer W, Wolff P, Gronwald W, et al. (2003) Structure determination and ligand interactions of the PDZ2b domain of PTP-Bas (hPTP1E): splicing-induced modulation of ligand specificity. J Mol Biol 334: 143–155.
40. Tjandra N, Bax A (1997) Direct measurement of distances and angles in biomolecules by NMR in a dilute liquid crystalline medium. Science 278: 1111–1114.
41. Clore GM, Garrett DS (1999) R-factor, free R, and complete cross-validation for dipolar coupling refinement of NMR structures. J Am Chem Soc 121: 9008–9012.
42. Banci L, Bertini I, Ciofi-Baffoni S, Katsari E, Katsaros N, et al. (2005) A copper(I) protein possibly involved in the assembly of CuA center of bacterial cytochrome c oxidase. Proc Natl Acad Sci USA 102: 3994–3999.
43. Gronwald W, Kirchhofer R, Gorler A, Kremer W, Ganslmeier B, et al. (2000) RFAC, a program for automated NMR R-factor estimation. J Biomol NMR 17: 137–151.
44. Bertini I, Cavallaro G, Luchinat C, Poli I (2003) A use of Ramachandran potentials in protein solution structure determinations. J Biomol NMR 26: 355–366.
45. Huang YJ, Moseley HN, Baran MC, Arrowsmith C, Powers R, et al. (2005) An integrated platform for automated analysis of protein NMR structures. Methods Enzymol 394: 111–141.
46. Huang YJ, Powers R, Montelione GT (2005) Protein NMR recall, precision, and F-measure scores (RPF scores): Structure quality assessment measures based on information retrieval statistics. J Am Chem Soc 127: 1665–1674.
47. Liu G, Shen Y, Atreya HS, Parish D, Shao Y, et al. (2005) NMR data collection and analysis protocol for high-throughput protein structure determination. Proc Natl Acad Sci USA 102: 10487–10492.
48. Brünger AT, Adams PD, Clore GM, Delano WL, Gros P, et al. (1998) Crystallography and NMR system (CNS): A new software system for macromolecular structure determination. Acta Cryst D54: 905–921.
49. Schwieters CD, Kuszewski JJ, Tjandra N, Marius Clore G (2003) The Xplor-NIH NMR molecular structure determination package. J Magn Reson 160: 65–73.
50. McGuffin LJ, Bryson K, Jones DT (2000) The PSIPRED protein structure prediction server. Bioinformatics 16: 404–405.

A New On-Axis Multimode Spectrometer for the Macromolecular Crystallography Beamlines of the Swiss Light Source

Robin L. Owen, Arwen R. Pearson, Alke Meents, Pirmin Boehler, Vincent Thominet and Clemens Schulze-Briesea

ABSTRACT

X-ray crystallography at third-generation synchrotron sources permits tremendous insight into the three-dimensional structure of macromolecules. Additional information is, however, often required to aid the transition from structure to function. In situ spectroscopic methods such as UV–Vis absorption and (resonance) Raman can provide this, and can also provide a means of detecting X-ray-induced changes. Here, preliminary results are introduced from an on-axis UV–Vis absorption and Raman multimode spectrometer currently being integrated into the beamline environment at X10SA of the Swiss Light Source. The continuing development of the spectrometer is also outlined.

Keywords: single-crystal microspectrophotometry, kinetic crystallography, structural enzymology, radiation damage, macromolecular crystallography, complementary techniques

Introduction

In order to fully understand biological processes, a knowledge of the atomic structure of the macromolecules involved is essential. The most common method of structure determination is X-ray crystallography, and the many protein and DNA structures already determined have provided incredible insights into the molecular underpinnings of life. Most X-ray crystal structures, however, are static average snapshots of a molecule and only yield information about a single state of a complex reaction. In addition, the deleterious effects of radiation damage can cast uncertainty on the validity of the model itself. More information about function can be obtained via kinetic crystallography, and one form this can take is soaking experiments. These take advantage of the fact that, as long as the active site is not blocked by crystal packing, and turnover does not require any major structural rearrangements that could disrupt the crystalline lattice, many enzymes retain their catalytic activity in the crystalline state. By soaking in a substrate and flash-cooling the crystal in liquid nitrogen, intermediate species can be freeze-trapped for structure determination (Kovaleva & Lipscomb, 2007; Schlichting & Chu, 2000; Katona et al., 2007). Identification of the trapped species, however, is limited by the resolution of the diffraction data obtained and the occupancy of the trapped state. This is often insufficient to unambiguously assign the chemical intermediate observed in the electron density map, especially in cases in which there are no large conformational changes in the ligand and/or protein. This ambiguity highlights the need for spectroscopic methods complementary to X-ray crystallography in structural biology to identify the exact chemical state of the trapped species.

The most commonly accessed of these methods to date is single-crystal UV/visible spectroscopy (UV-SCS) (Pearson et al., 2004), although fluorescence (Royant et al., 2007; Bourgeois et al., 2002 Klink et al., 2006), Raman (Carpentier et al., 2007) and Fourier-transform infrared (Moukhametzianov et al., 2006) spectroscopy are also in use. Ideally, spectra are recorded at 100 K from the same crystal used for X-ray diffraction data collection, though this is not always possible. As a result of the increasing interest in this field, several single-crystal microspectrophotometers are now available worldwide at both synchrotron sources and home laboratories; for a recent review see De la Mora-Rey & Wilmot (2007). The synchrotron systems in particular are designed to operate both on- and off-line, allowing direct monitoring of crystal spectra during X-ray data collection.

As well as the ability to identify trapped reaction intermediates, on-line microspectrophotometers have also allowed the monitoring of the effects of X-ray exposure upon protein crystals. It is well established that excessive X-ray exposure results in eventual loss of diffraction signal (Henderson, 1990 ; Owen et al., 2006). However, both on-line single-crystal spectroscopy and X-ray absorption near-edge spectroscopy have revealed that chemical and conformational changes can occur in the crystal at much lower doses than are usually considered to cause radiation damage (Beitlich et al., 2007 ; Yano et al., 2005).

These changes are illustrated in Figure 1 and range from a near instantaneous generation of aqueous, or solvated, electrons (An external file that holds a picture, illustration, etc. Object name is s-16-00173-efi1.jpg) upon X-ray exposure, through fast changes at electron-sensitive centres such as redox sites or disulphide bonds at low X-ray dose, to the eventual decay of diffraction and crystal death. These radiation-induced changes have been observed in a variety of biological systems (for example, see Berglund et al., 2002 ; Adam et al., 2004 ; Hough et al., 2008 ; Pearson et al., 2007).

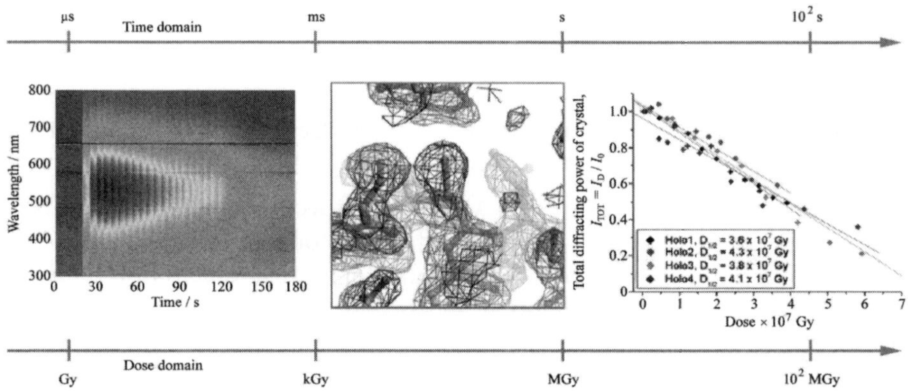

Figure 1. Manifestations of radiation damage over a wide range of time scales: UV–Vis absorption spectra of irradiated ethylene glycol [data collected at the ESRF using the arrangement described by Southworth-Davies & Garman (2007)] showing a rapid increase in absorption owing to e aq upon X-ray irradiation; disulphide bond breakage in low-dose composite data sets of DsbA (R. L. Owen, unpublished data); decay of diffracting power of holoferritin crystals as a function of X-ray dose (Owen et al., 2006).

Unfortunately only a small subset of proteins have functional groups amenable to UV–Vis absorption spectroscopy; additional complementary methods are therefore required for tracking X-ray-induced changes or obtaining electronic and vibrational information from crystals. Raman spectroscopy has the advantage of not requiring coloured functional groups, and can provide detailed information (Carey, 1999 ; Hildebrandt & Lecomte, 2000 ; Tuma, 2005). Raman spectroscopy is analogous to infrared absorption spectroscopy in that it probes the vibrational energy levels of molecules rather than the electronic transitions probed by UV–Vis absorption.

Vibrational spectroscopy provides information with a sensitivity beyond that usually achieved in a macromolecular diffraction experiment. Information on protonation states, van der Waals and electrostatic interactions can be directly measured rather than inferred, greatly facilitating the transition from structure to function (for an overview, see Siebert & Hildebrandt, 2007). For a multimode spectrometer targeting biological samples, infrared absorption has several disadvantages when compared with Raman spectroscopy. CCD detectors and lenses optimized for UV–Vis absorption are non-ideal for IR and vice versa, and therefore for multimode spectroscopy UV–Vis and Raman spectroscopies are more complementary. Infrared absorption is also limited by the so-called 'water problem': water has strong absorption bands in the infrared limiting the possible sample size and environment (Susi, 1969). While macromolecular Raman spectroscopy poses several challenges, primarily in the form of the extremely weak nature of the effect, obscuration of bands by fluorescence and difficulties in interpreting spectra, Raman spectroscopy nonetheless provides a rich source of complementary information for crystallographers and is gaining in popularity (Carey, 2006 ; Katona et al., 2007). Processes involving changes within a crystal, for example kinematic and ligand binding experiments, are greatly simplified by use of difference Raman spectroscopy easing difficulties in interpretation of spectra. Protein fluorescence can also be a valuable source of information on changes in local environment and conformation during X-ray data collection.

Some of the challenges associated with Raman spectroscopy mentioned above can be bypassed by exploiting resonance effects. Resonance Raman occurs when the frequency of the laser probe is tuned to that of an electronic transition in the molecule of interest. This highlights an area in which the UV–Vis absorption and Raman modes of a multimode spectrometer are complementary to each other, as well as to X-ray diffraction. The UV–Vis absorption spectrum of a molecule reveals the wavelengths at which Raman modes will be selectively enhanced under resonant conditions by a factor of up to 106, allowing bands to be clearly discerned above fluorescence and specific regions of a molecule to be probed. This enhancement means that Raman data acquisition times can be reduced to match those in macromolecular crystallography, if a laser wavelength matching an electronic transition of the biological molecule is available. Good Raman data can also be collected under non-resonant conditions, although these require acquisition times of the order of several minutes.

The most commonly accessed biological resonant modes are exhibited for iron- or copper-containing proteins at an excitation wavelength of 413.1 nm (Kr+ gas laser probe), for example myoglobin, cytochrome C oxidase, rubredoxins and azurins (Coyle et al., 2003 ; Konishi et al., 2004 ; van Amsterdam et al., 2002 ; Sanders-Loehr, 1988 ; Engler et al., 2000 ; Averill et al., 1987), or 514 nm (Ar+ gas laser probe), for example purple acid phosphatase and rubredoxin (Xiao et al., 2005 ; Averill et al., 1987). Some of these modes can also be accessed using longer laser wavelengths, for example cytochrome C oxidase at 580–615 nm (Bocian et al., 1979 ; Czernuszewicz et al., 1994) and rubredoxin at 647.1 nm (Czernuszewicz et al., 1994). Increasing the range of available laser wavelengths correspondingly increases the range of accessible metal–ligand modes; red laser lines, for example, make possible resonant scattering from rhodopsin (600 nm) (Mathies et al., 1976), methane monooxygenase (647.1 nm) (Liu et al., 1995) and galactose oxidase (659 nm) (Whittaker et al., 1989). Care must be taken, however, in resonance Raman experiments to avoid laser-induced changes (Tonge et al., 1993 ; Meents et al., 2007) as photochemical processes are enhanced when the laser wavelength matches an electronic transition (Turro, 1991). In the case of structural biology the primary advantage of moving to longer wavelengths is the accompanying decrease in protein fluorescence, so that non-resonance Raman bands are not obscured by this unwelcome effect (Carey, 2006).

Owing to the highly concentrated and spatially ordered nature of biological molecules in the crystalline state, Raman crystallography can provide superior spectra to solution phase spectroscopy. In particular, the small spectral changes associated with ligand binding are more readily followed (Carey & Dong, 2004 ; Altose et al., 2001 ; Helfand et al., 2003 ; and references above). As Raman scattering is a tensorial quantity (Tsuboi & Thomas, 1997), care must therefore be taken with alignment of the crystal axes of samples, since changes in orientation can cause the relative intensities of different bands to change. Raman spectroscopy has also recently been used to measure degradation of selenomethionine derivatives (Vergara et al., 2008), providing further motivation for the implementation of an instrument for on-line Raman spectroscopy.

In summary, Raman and UV–Vis absorption spectroscopies can provide information complementary to that obtained using X-ray diffraction, and are well placed to greatly facilitate the development of a new temporal dimension in macromolecular crystallography. The inherent difference between Raman spectroscopy and UV–Vis absorption, in that inelastic scattering of a monochromatic source of photons is of interest (Raman) rather than absorption of a polychromatic beam (UV–Vis absorption), means that multiple optical arrangements must be accommodated in a multimode spectrometer. How these arrangements have been resolved in a manner compatible with macromolecular X-ray diffraction and then integrated into the beamline environment at X10SA of the Swiss Light Source (SLS) is described below.

Materials and Methods

Experimental Design

In situ UV–Vis absorption spectroscopy has previously been successfully implemented at the SLS protein crystallography beamlines using an off-axis geometry (Beitlich et al., 2007). This arrangement resulted in several experimental limitations, some of which are common to most off-axis spectrometers currently in use. An off-axis arrangement greatly crowds the sample environment as, in addition to standard beamline components such as the collimator, alignment camera and illumination, two objective optics and an arc mount must be accommodated. While crowding can be alleviated to an extent by the use of long-working-distance objectives (McGeehan et al., 2009), spatial restrictions mean that it is impractical to permanently install such a set-up at a synchrotron beamline. The objectives must therefore be reinstalled and aligned both with respect to each other and the beamline before each spectroscopy run, frequently a time-consuming procedure. An off-axis geometry also results in difficulties in ensuring the same sample volume is probed by X-rays and spectroscopy, and, as a consequence of the sample geometry, is more prone to spectral artefacts caused by reflections of the probing light.

The use of an on-axis geometry circumvents these drawbacks and allows the instrument to be permanently installed at the beamline. This avoids the time-consuming alignment procedures previously associated with spectroscopic experiments and raises the possibility of in situ spectroscopy becoming as commonplace as taking a fluorescence scan or changing the wavelength for MAD data collection. Co-axial simultaneous micro-Raman and synchrotron microdiffraction at ID13 of the ESRF has already been introduced as a powerful tool (Davies et al., 2005), and here we detail the development of a novel solution for on-axis spectroscopy at beamline X10SA of the Swiss Light Source (Pohl et al., 2006).

The on-axis geometry of the SLS multimode microspectrophotometer (SLS-MS) (shown in Figure 2) is achieved with a design analogous to that of existing on-axis microscopes, i.e. with a drilled objective mounted on the exposure box. The majority of on-axis microscopes utilize a drilled objective followed by a drilled mirror to deflect light away from the X-ray axis towards further alignment optics and a CCD detector mounted below the beamline.

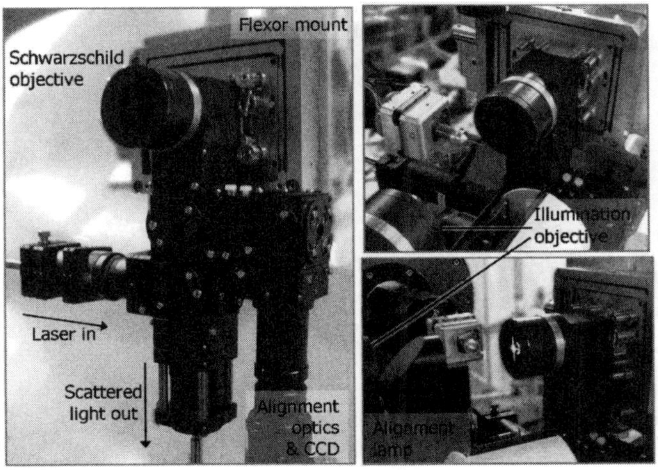

Figure 2. Left: the SLS multimode spectrometer mounted off-line, highlighting the extent of the system and arrangement employed. Top and bottom right: the spectrometer installed at X10SA. In both cases the illumination optic is shown in the far position for sample alignment. The box containing the scintillator, beam-defining apertures and collimator can be seen below the sample position. The cryostream has been removed for clarity.

For the SLS-MS, a high-magnification reflective Schwarzschild objective (Newport, 15× magnification, f = 13 mm, numerical aperture 0.4) is used. This has the advantage of freeing up a large amount of space around the sample environment as the same objective can be used for both sample alignment and spectroscopy. Reflective objectives combine several desirable characteristics for both sample alignment and spectroscopy, including zero chromatic aberration, a high laser power threshold and a long working distance. Light collected by the reflective objective is reflected in a direction 90° below the X-ray axis to the branched SLS-MS (Figure 3).

The branched design of the SLS-MS is shown in Figure 3, with the alignment and spectroscopy branches highlighted in aqua and blue/green, respectively, in Figure 3. In order to accommodate the multiple optical arrangements required by UV–Vis absorption and Raman spectroscopies, the spectroscopy branch is divided into two. For UV–Vis absorption spectroscopy, absorption of a polychromatic beam passing through the sample is of interest. Therefore an illuminating objective is required and only the collection path (blue) of the spectroscopy branch is utilized. For Raman spectroscopy, inelastic scattering of a monochromatic light source is of interest and a 180° scattering geometry is used. In this case an illuminating objective is not required and both the illumination (green) and collection (blue) paths of the spectroscopy branch are used (Figure 3).

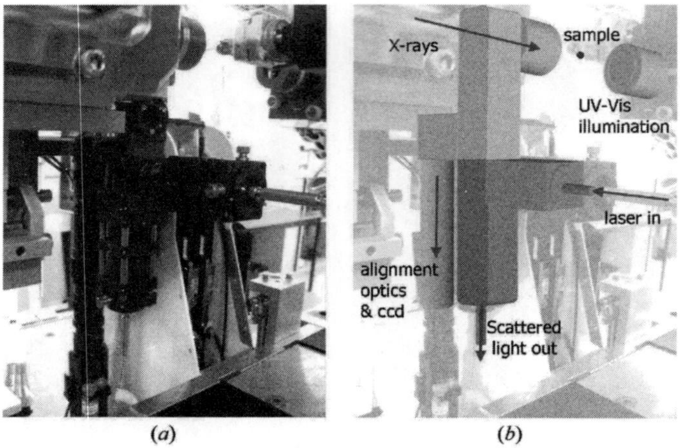

Figure 3. The branched nature of the SLS multimode spectrometer. The alignment branch (aqua) delivers light via infinity focused zoom optics to a CCD camera. The spectroscopy branch further divides into a branch for delivery of laser light for Raman spectroscopy (green) and a branch for collection of scattered light for all spectroscopies (blue).

The illumination objective required for UV–Vis absorption spectroscopy is provided by a second Schwarzschild objective mounted on a motorized stage [shown in Figures 2 and 3]. This can be driven between three positions: (i) spectroscopic data collection; (ii) sample mounting and alignment; (iii) X-ray data collection. The reproducibility of the movement of this stage will be discussed in the following section. The unequal splitting between both spectroscopic sub-branches and the alignment and spectroscopy branches is achieved via the use of pellicle beam-splitters. These divide light between the branches in the ratio 1:12 with the extremely thin (~2 μm) pellicle membranes eliminating refraction-based errors and ghost images.

The SLS Raman probe is a Kr+ gas laser (Coherent Innova 300C) allowing a range of wavelengths to be accessed;[1] currently laser clean-up and holographic notch filter sets are available for the 413.1, 647.1 and 752.5 nm lines of the laser. In order to change the lasing wavelength, the laser mirrors must be exchanged and re-aligned. This procedure takes ~30 min, with full laser powers achievable after a further 30 min as the laser warms up. Laser powers of the order of 10, 20 and 10 mW at the sample position are attainable at 413.1, 647.1 and 752.5 nm, respectively. As outlined above, this range of wavelengths allows a wide range of resonant modes to be accessed, while at the two longer wavelengths protein fluorescence is greatly reduced allowing these lines to be used for non-resonance Raman spectroscopy.

Both Raman and UV–Vis absorption spectra are recorded using an Andor 303i Czerny–Turner spectrograph and a Newton electron multiplying CCD (Andor technology). Two grating sets optimized for Raman data collection in the UV (413.1 nm Kr+ line) and red to near below-red (647.1 and 752.5 nm Kr+ lines) can be mounted within the spectrograph; for each grating set three line-spacings can be used allowing either 'global' spectra to be collected or a small region of interest to be investigated. Both the spectrograph and Kr+ gas laser are located in an optical hutch adjacent to the beamline allowing on-line experiments to make use of pre-aligned/optimized optical arrangements, and off-line experiments to operate independently of the beamline. Light is coupled between the spectroscopy hutch and beamline by means of 20 m optical fibres; for off-line experiments these can be replaced by the corresponding 2 m fibres. The demagnification ratio between a 50 μm fibre and the focal spot diameter is 0.69. The SLS-MS is not yet currently permanently installed at the beamline and so can be used for both on- and off-line measurements. Once the optical arrangements and design are finalized, and the system is permanently installed, all the components are available for constructing a duplicate off-line system.

Results

Optical and X-Ray Transmission Properties of a Drilled Schwarzschild Objective

The drilling of a 1 mm-diameter hole in the secondary mirror does not degrade the optical properties of the objective since the occluded region of this mirror is 4.6 mm in diameter. This free diameter, f d, is shown in Figure 4. Of potentially greater impact on the light throughput of the system is the hole in the 45° mirror. This is illustrated by ray-tracing the light path from two points in the focal plane through the system: the focal point and a point a distance δ off-axis (shown in blue and green, respectively, in Figure 4). Light from the focal point does not fall on this hole, although it is possible for light from the off-axis point to fall there, with the amount of light 'lost' increasing as a function of δ. If a value of 0.6 mm is taken for δ, corresponding to the full field of view of the objective, then the intensity of light lost is less than 2%. Holes of diameter 1 mm are large enough to allow X-rays to pass through the system even when the flexor mount is adjusted. This is illustrated in Figure 4; in this case a knife-edge was scanned across the sample position with either the Schwarzschild objective or normal on-axis viewing system lens mounted.

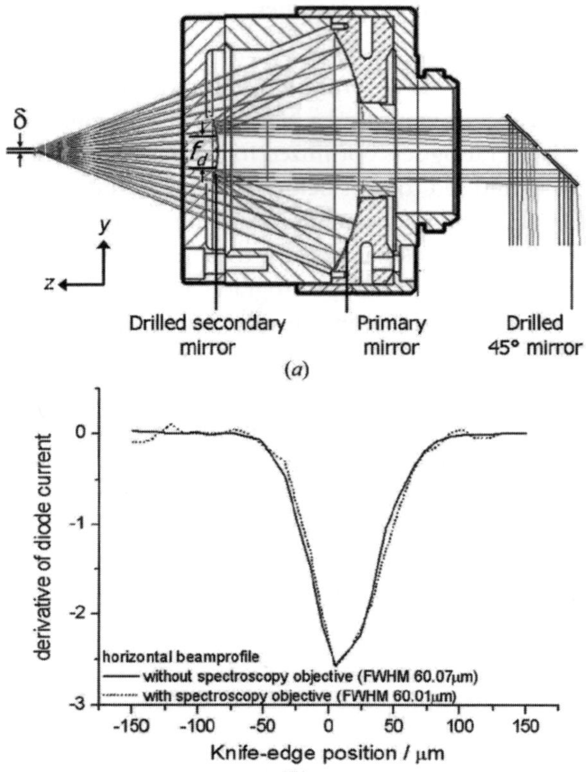

Figure 4. The effect of drilling on the optical and X-ray transmission properties of the Schwarzschild objective. (a) Ray traces calculated using the raytracing software Zemax (http://www.zemax.com/) from the focal point (blue) and a point a distance δ off-axis (green) through the system are shown. Translation of a knife-edge across the X-ray beam allows the profile to be determined with and without the spectroscopy objective in place; no change is observed in either (b) the horizontal or vertical (data not shown) directions.

The current recorded by a diode placed behind the knife-edge allowed the X-ray beam profile to be determined in the horizontal and vertical (data not shown) in both cases. No change in the beam properties was observed with the Schwarzschild objective mounted, even in the case of a defocused X-ray beam.

Alignment of Optical and X-ray Axes

The use of the same on-axis objective for sample alignment and spectroscopic data collection allows unambiguous alignment of the X-ray and visible optical axes. In order to achieve this, the flexor-mounted Schwarzschild objective is translated in z (Figure 4 for definition of axes) so that the centre of rotation of the sample is at the working distance of the objective. This can be achieved easily by use of a camera mounted above the beamline, and its view of the focusing of the xenon light by the Schwarzschild objective is shown in Figure 5; the presence of the cryostream allows the visible light to be observed. Placement of a scintillator at the sample position then allows the X-ray and optical axes to be made coincident at the sample position. This is achieved by adjustment of the 'pitch' and 'yaw' of the flexor mount of the collection optics (Figure 5). The UV–Vis illumination Schwarzschild objective can be translated in x, y and z using its motorized stage and multi-axis fibre mount (Newport) which allows pitch and yaw adjustment to the fibre input. The focal spot of this objective is aligned to the centre of rotation of the sample by use of a 12.5 μm pinhole placed at the sample position through which the intensity of transmitted light at the spectrometer is maximized.

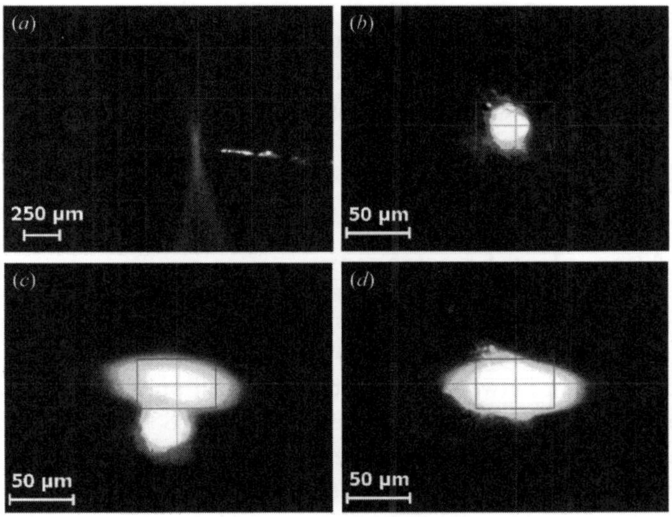

Figure 5. (a) Alignment of the focal point of the xenon lamp to the focal plane of the alignment camera using a camera mounted directly above the beamline and a pin at the sample position; the focus of the Schwarzschild objective can be clearly seen. The placement of a scintillator at the sample position allows the sample alignment camera to be used for alignment of the X-ray and optical axes: (b) X-ray shutter closed, optical shutter open; (c) both X-ray and optical shutters open with objective mis-aligned; (d) both X-ray and optical shutter open with the position of the yaw of the objective mount adjusted to maximize overlap of the X-ray and optical beams at the sample position.

The focal plane of the alignment system can be fine-tuned and made exactly coincident with the laser/visible-light focal spot by means of a Kepler-type arrangement of two lenses in the alignment branch of the spectrometer (F. Schwarz, personal communication). This arrangement also allows the field of view to be matched to the size of the alignment CCD. Owing to the high magnification

(15×) of the objectives, the field of view is limited. The full field of view is 1.2 mm, with uniform illumination possible over ~600 × 400 μm for sample alignment, and we have found this area to be sufficient in the case where standardized pins are used throughout an experimental run. In the case where non-standard pins are used, coarse sample alignment using the camera mounted above the beamline has been integrated into the sample alignment graphical user interface to complement on-axis alignment.

This use of the same on-axis optic for sample alignment and spectroscopy allows simple and unambiguous alignment of the X-ray and visible-light axes ensuring the same sample volume is always probed by both diffraction and spectroscopy. As the scintillator is mounted on a motorized stage (Figure 2) and can be easily moved to the sample position, this alignment can be checked at intervals throughout the experiment. The on-axis geometry also means that the optics can remain in place between experiments, eliminating the time-consuming alignment currently associated with spectroscopy at synchrotron sources. Tests of the reproducibility of the UV–Vis illumination objective position when driven between its three positions (spectroscopy, sample alignment and mounting, X-ray data collection) showed less than 3% change in transmitted intensity of the Xe lamp over ten iterations.

UV-Vis Absorption Spectroscopy

Test single-crystal UV–Vis absorption spectra have been recorded for a variety of systems (Figure 6). Owing to the transmission properties of the focusing objectives coupling light between the spectrometer and optical fibres, the available UV–Vis absorption range is approximately 325 nm to 850 nm. The transmission of the system is approximately linear between these wavelengths, and is ~18%. Longer wavelengths are accessible but intense peaks of the xenon lamp in that region make simultaneous collection of short- and long-wavelength data difficult, though neutral density and coloured filters can be automatically introduced to the light path to alleviate this.

The Andor Shamrock 303i spectrograph is equipped with motorized adjustable slits (10 μm–1.2 mm) and the exposure times can vary from a minimum of 20 μs to considerably longer exposure times if required. This flexible arrangement allows matching of the data acquisition parameters to the optical properties of the experimental system.

The SLS-MS has been used to investigate the sensitivity of two systems to radiation damage during X-ray exposure. The heme group of myoglobin is known to be sensitive to reduction by X-rays (Beitlich et al., 2007). A myoglobin single crystal was mounted at beamline X10SA and UV–Vis absorption spectra were recorded during X-ray exposure without rotation of the crystal (Figure 7). Four accumulations of 0.8 ms exposures were recorded every 0.1 s. During data collection the crystal was not rotated in order to avoid spectral changes owing to variation in the crystal orientation, apparent changes in the crystal thickness, and the effect of the loop entering the light path (Wilmot et al., 2002). As single-crystal spectra are anisotropic, crystals were pre-orientated to a 'sweet spot' where the anisotropic spectra best resembled the isotropic solution spectra. Rapid reduction occurs, as evidenced by a red shift in the Soret band and appearance of characteristic peaks at 550–560 nm (Figure 7), and is essentially complete within 8 s of initial X-ray exposure corresponding to an absorbed dose of ~3 MGy [absorbed dose calculated using RADDOSE (Murray et al., 2004).

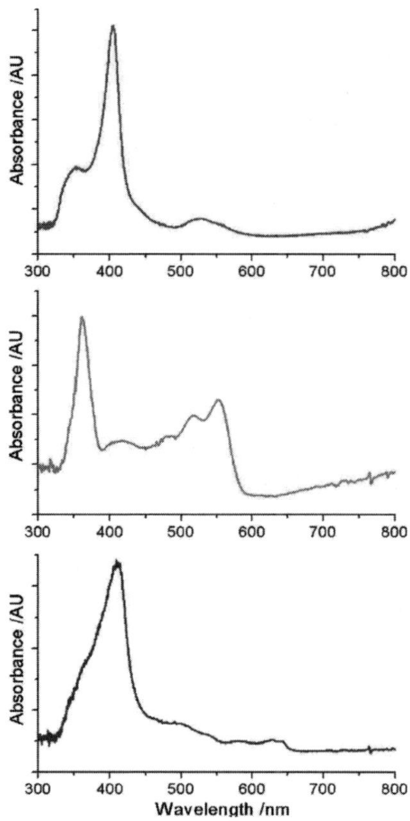

Figure 6. UV–Vis absorption spectra of cryocooled (100 K) cytochrome C thin film (top); vitamin B12 crystal (~10 × 60 × 150 mm) (middle); and met myoglobin (metMb) crystal (~10 × 80 × 150 mm) (bottom), crystallized as previously described (Beitlich et al., 2007), taken using the on-axis spectrometer. The discontinuities observed at 470 and 765 nm are due to intense peaks in the xenon lamp spectra at these wavelengths. Data were collected with accumulation times of 10 100 ms, 5 × 20 ms and 4 × 0.8 ms, respectively, using a 10 mm slit width and a grating with 150 lines mm 1 in all cases.

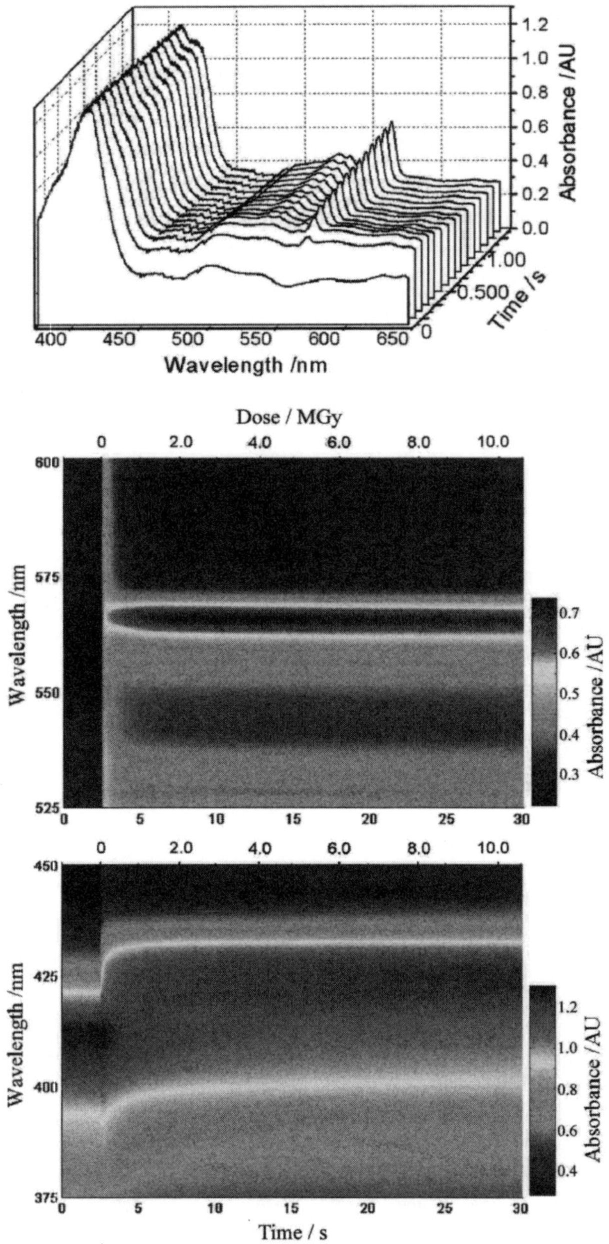

Figure 7. UV–Vis absorption spectra of a metMb crystal (~10 × 90 × 120 mm) highlighting X-ray-induced changes at 100 K. Top: interval line plot, with a spectrum plotted every 0.1 s. Middle: contour plot showing the evolution of bands at 555 and 565 nm. Bottom: contour plot showing red shift of the 410 nm Soret peak; the time (t ≅ 2 s) at which the X-ray shutter is opened can clearly be seen by a global intensity increase as observed in other systems (Dubnovitsky et al., 2005).

Radiation effects were investigated in crystals of a second system, a ubiquinone binding E. coli membrane protein, DsbB, involved in disulfide bond formation (Malojcic et al., 2008). DsbB is known to induce a red shift in ubiquinone upon binding (Inaba et al., 2004) and a characteristic red-shifted visible peak appears upon ubiquinone binding within the DsbA–DsbB–ubiquinone (DsbAB-Q8) complex. Unlike heme-containing proteins in which the Soret band has a large molar extinction coefficient, this peak has a small extinction coefficient (~4750 cm−1 M −1). However, despite the low intensity of this peak, it can be clearly observed in the SLS-MS with only 4 × 20 μs exposures (10 μm slits) (Figure 8).

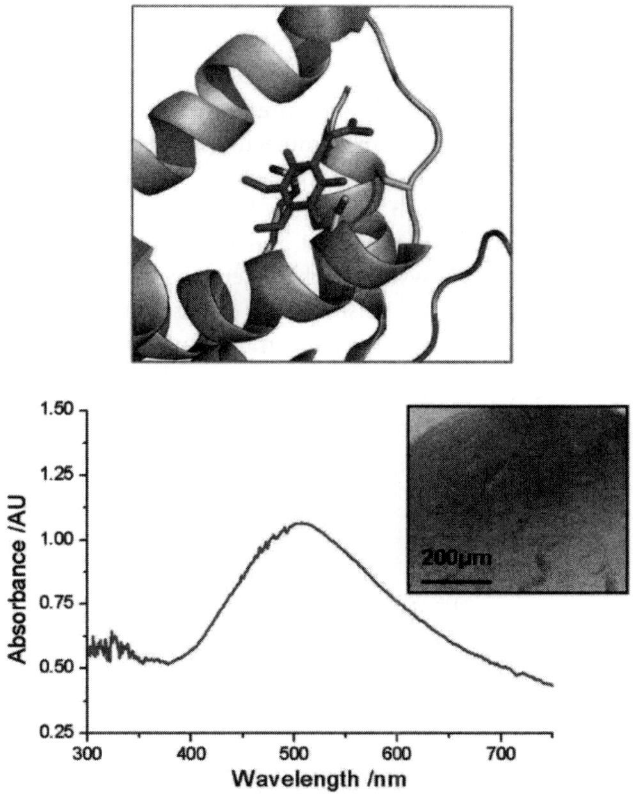

Figure 8. Top: a charge transfer interaction between wild-type DsbB and ubiquinone induces a pink colouration in the quinone group. Inset bottom: in situ UV–Vis absorption spectroscopy of DsbAB crystals confirm the presence of the charge transfer interaction in the form of (bottom) a broad absorption peak centred at 510 nm.

Figure 8 also highlights the value of the complementary information that UV–Vis absorption spectroscopy can provide. Owing to the limited diffracting power of DsbAB-Q8 crystals (3.7 Å for the best crystals in the most favourable orientation), it is not possible to identify unambiguously the presence of the charge transfer interaction in the crystalline state via X-ray diffraction alone (Malojcic et al., 2008). In situ UV–Vis absorption spectra allow spectroscopic confirmation of the presence of the interaction within the crystal, aiding interpretation of crystallographic data. In contrast to the example of myoglobin above, the UV–Vis absorption spectra of DsbAB-Q8 do not change as a function of X-ray dose, revealing the interaction to be radiation insensitive over the periods required to collect a complete crystallographic data set. The dynamic range of the instrument is highlighted by its capability to measure both the myoglobin Soret band reported above and the 510 nm DsbAB-Q8 quinone signal.

Raman Spectroscopy

The Raman capabilities of the SLS-MS are presently undergoing commissioning. We show here initial data from a small molecule standard (cylcohexane, Figure 9). Clear Raman peaks are visible, even for relatively short exposure times. Spectra were collected using the 647.1 nm line of the Kr+ laser with 4 × 5 s exposures, a spectrograph grating with 300 lines mm−1 and a slit width of 40 µm. The calculated spectral resolution of the system at this laser wavelength with this grating is ~10 cm−1. The data have been left unsmoothed or baseline corrected to highlight the quality of data obtained, and major spectral features are labelled. The results using these parameters are extremely encouraging when compared with other single-crystal Raman systems.

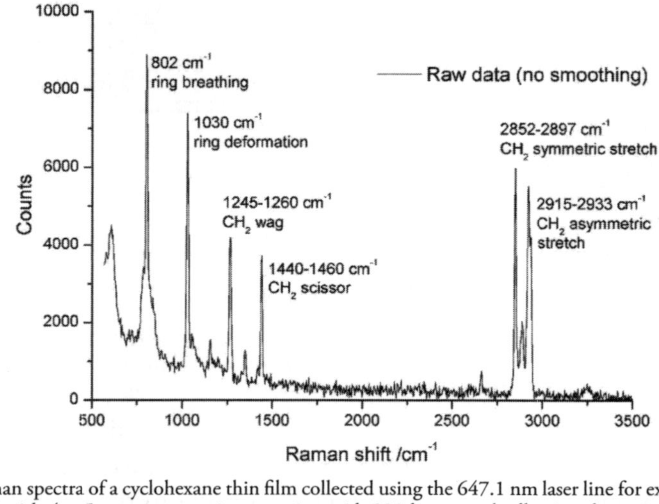

Figure 9. Raman spectra of a cyclohexane thin film collected using the 647.1 nm laser line for excitation. Spectra were collected with 4 × 5 s exposures using a grating with 300 lines mm^{-1}, allowing the entire spectrum to be collected simultaneously. Major spectral features are labelled.

X-ray Data Collection

Measurements at the sample position confirm that the reflecting objective does not reduce the available X-ray flux, with knife-edge scans and diode measurements showing identical beam profiles and fluxes with and without the spectrometer mounted (Fig. 4b). Diffraction tests with the microspectrophotometer mounted show that there is no effect on diffraction data quality (data not shown). As the illumination objective must be lowered in order to collect diffraction data, simultaneous UV–Vis absorption and X-ray diffraction data cannot be collected; this, however, is not a limitation as good quality UV–Vis absorption data can usually only be collected at a particular crystal orientation (Wilmot et al., 2002 ; Pearson et al., 2007). The 180° scattering geometry utilized for Raman spectroscopy means that simultaneous (resonance) Raman and X-ray diffraction experiments are possible using the SLS-MS since only a single spectroscopy objective is required, though it is anticipated that Raman and X-ray data collection will rather be interleaved for optimal crystal orientation during spectroscopic data collection.

Discussion

The SLS-MS is the first on-axis single-crystal spectrometer at a synchrotron macromolecular crystallography beamline. Its design allows it to be permanently installed at the beamline, removing the need for time-consuming installation and alignment associated with non-permanent off-axis systems. Designed as an integral part of the beamline, it has an uncrowded sample environment that allows other beamline components such as the robotic sample changer and fluorescence detector access to the sample, and also facilitating manual crystal mounting.

The facility to have both X-ray and spectroscopy probing the same sample volume in the SLS-MS circumvents a persistent problem with off-axis UV–Vis absorption single-crystal spectroscopy instruments. Unless the X-ray beam size is greater than or equal to the crystal size, an off-axis spectrometer will probe both X-ray irradiated and unirradiated volumes of the crystal. This is a major problem when attempting to follow changes resulting from radiation damage by UV–Vis absorption spectroscopy, although less of an issue for Raman spectroscopy if the spectrometer is correctly focused. The degree of change in the X-ray-illuminated volume may therefore be mis-estimated by off-axis UV–Vis absorption spectroscopy, resulting in incorrect decisions concerning the permissible dose a crystal can receive before the final electron density map is dominated by a radiation-damaged state. A second advantage related to the on-axis geometry is the ability to perform photoactivation experiments using the standard beamline set-up. The

SLS-MS on-axis arrangement and availability of a range of laser excitation wavelengths allows laser excitation of the sample volume to be probed by the X-rays, as well as permitting simultaneous spectroscopic characterization of the efficiency of laser excitation. In addition, linked gating of the laser and X-ray shutters is possible. This, combined with an eight-fold increase in the laser power at the sample position (by switching of the laser fibre input to the spectroscopy branch of the spectrometer) if photoactivation is the sole aim of the experiment, makes the SLS-MS particularly well suited to this type of experiment. For thick samples and/or samples with high extinction coefficients, classical orthogonal absorption experiments may provide more uniform sample excitation/illumination owing to the limited penetration of the laser beam (Anderson et al., 2004 ; Schotte et al., 2003 , 2004).

The SLS-MS is now in the final stages of commissioning before permanent installation at beamline X10SA. The current design, based on modular components, provides great flexibility for optimization and variation of the optical arrangement. It is anticipated that, once the optical arrangement is finalized, the body of the spectrometer will be machined as a single component which will give additional stability during regular user operation. Although the UV–Vis spectroscopy set-up is now stable and 'user ready,' the Raman spectroscopy requires further commissioning to optimize it for use with protein crystals. Preliminary studies have shown that, unlike commercially available Raman probes used in other systems, considerable inelastic fibre scattering effects as well as protein fluorescence obscure much of the low wavenumber range (<1000 cm−1). Additional optical components alleviating these effects are currently being commissioned for testing, as is improved optical fibre coupling to further increase the light throughput, sensitivity and spectral range of the spectrometer. The combination of the SLS-MS on-axis and the 180° Raman scattering geometries permits permanent provision of simultaneous collection of Raman spectroscopy and X-ray diffraction data at X10SA. Beamline software for both ease of use and avoidance of potential collisions of the UV–Vis absorption illumination optic with beamline components (for example beamstop, sample illumination lamp and detector) is currently being integrated into the experimental arrangement. Also under investigation is the use of a kappa geometry to allow the measurement of Raman bands in different crystal positions in order to derive information about the orientation of specific vibrations with respect to the crystal axes.

The 180° scattering geometry used by the SLS-MS for Raman spectroscopy is also well suited to other spectroscopies. Collection of both fluorescence (using either the available xenon lamp or Kr+ laser for excitation) and X-ray excited optical luminescence (XEOL) spectra are possible without further modification to the instrument design. Initial XEOL experiments have been carried out (data not shown) indicating that the spectrometer design is compatible with fluorescence-type experiments.

The development of the instrument described here allows a spectrometer capable of UV–Vis absorption and (resonance) Raman spectroscopy to be permanently integrated into the beamline environment, making the on-axis SLS multimode spectrometer an extremely attractive tool for obtaining complementary spectroscopic information and pursuing kinetic crystallography at macromolecular crystallography beamlines.

Acknowledgements

We would like to thank Friedrich Schwarz, Bob Shoeman and Ilme Schlichting for extensive discussions, testing of the spectrometer and kindly providing myoglobin crystals; and Goran Malojčić and Rudi Glockshuber for providing crystals of DsbAB.

Footnotes

1. For a full list of Kr+ lines, see http://facs.scripps.edu/Lasers.html.

References

1. Adam, V., Royant, A., Nivière, V., Molina-Heredia, F. P. & Bourgeois, D. (2004). Structure, 12, 1729–1740.
2. Altose, M. D., Zheng, Y., Dong, J., Palfey, B. A. & Carey, P. R. (2001). Proc. Natl. Acad. Sci. USA, 98, 3006–3011.
3. Amsterdam, I. M. C. van, Ubbink, M., van den Bosch, M., Rotsaert, F., Sanders-Loehr, J. & Canters, G. W. (2002). J. Biol. Chem.277, 44121–44130.
4. Anderson, S., Srajer, V., Pahl, R., Rajagopal, S., Schotte, F., Anfinrud, P., Wulff, M. & Moffat, K. (2004). Structure, 12, 1039–1045.
5. Averill, B. A., Davis, J. C., Burman, S., Zirino, T., Sanders-Loehr, J., Loehr, T. M., Sage, J. T. & Debrunner, P. G. (1987). J. Am. Chem. Soc.109, 3760–3767.
6. Beitlich, T., Kühnel, K., Schulze-Briese, C., Shoeman, R. L. & Schlichting, I. (2007). J. Synchrotron Rad.14, 11–23.
7. Berglund, G. I., Carlsson, G. H., Smith, A. T., Szoke, H., Henriksen, A. & Hajdu, J. (2002). Nature (London), 417, 463–468.

8. Bocian, D., Lemley, A., Peteren, N., Brudwig, G. & Chan, S. (1979). Biochemistry, 20, 4396–4402.
9. Bourgeois, D., Vernede, X., Adam, V., Fioravanti, E. & Ursby, T. (2002). J. Appl. Cryst.35, 319–326.
10. Carey, P. R. (1999). J. Biol. Chem.274, 26625–26628.
11. Carey, P. R. (2006). Annu. Rev. Phys. Chem.57, 527–554.
12. Carey, P. R. & Dong, J. (2004). Biochemistry, 43, 8885–8893.
13. Carpentier, P., Royant, A., Ohana, J. & Bourgeois, D. (2007). J. Appl. Cryst.40, 1113–1122.
14. Coyle, C. M., Vogel, K. M., Rush, T. S., Kozlowski, P. M., Williams, R., Spiro, T. G., Dou, Y., Ikeda-Saito, M., Olson, J. S. & Zgierski, M. Z. (2003). Biochemistry, 42, 4896–4903.
15. Czernuszewicz, R. S., Kilpatrick, L. K., Koch, S. A. & Spiro, T. G. (1994). J. Am. Chem. Soc.116, 7134–7141.
16. Davies, R., Burghammer, M. & Riekel, C. (2005). Appl. Phys. Lett.87, 264105.
17. De la Mora-Rey, T. & Wilmot, C. M. (2007). Curr. Opin. Struct. Biol.17, 580–586.
18. Dubnovitsky, A. P., Ravelli, R. B. G., Popov, A. N. & Papageorgiou, A. C. (2005). Protein Sci.14, 1498–1507.
19. Engler, N., Ostermann, A., Gassmann, A., Lamb, D. C., Prusakov, V. E., Schott, J., Schweitzer-Stenner, R. & Parak, F. G. (2000). Biophys. J.78, 2081–2092.
20. Helfand, M. S., Totir, M. A., Carey, M. P., Hujer, A. M., Bonomo, R. A. & Carey, P. R. (2003). Biochemistry, 42, 13386–13392.
21. Henderson, R. (1990). Proc. R. Soc. Lond. B, 241, 6–8.
22. Hildebrandt, P. & Lecomte, S. (2000). Encyclopedia of Spectroscopy and Spectrometry, Vol. I, edited by J. Lindon, G. Tranter and J. Holmes, pp. 88–97. London: Academic Press.
23. Hough, M. A., Antonyuk, S. V., Strange, R. W., Eady, R. R. & Hasnain, S. S. (2008). J. Mol. Biol.378, 353–361.
24. Inaba, K., Takahashi, Y.-H., Fujieda, N., Kano, K., Miyoshi, H. & Ito, K. (2004). J. Biol. Chem.279, 24906.
25. Katona, G., Carpentier, P., Niviere, V., Amara, P., Adam, V., Ohana, J., Tsanov, N. & Bourgeois, D. (2007). Science, 316, 449–453.

26. Klink, B. U., Goody, R. S. & Scheidig, A. J. (2006). Biophys. J.91, 981–992.

27. Konishi, K., Ishida, K., Oinuma, K.-I., Ohta, T., Hashimoto, Y., Higashibata, H., Kitagawa, T. & Kobayashi, M. (2004). J. Biol. Chem.279, 47619–47625.

28. Kovaleva, E. G. & Lipscomb, J. D. (2007). Science, 316, 453–457.

29. Liu, K., Valentine, A., Qiu, D., Edmondson, D., Appelman, E., Spiro, T. G. & Lippard, S. (1995). J. Am. Chem. Soc.117, 4997–4998.

30. Malojcic, G., Owen, R. L., Grimshaw, J. P. A. & Glockshuber, R. (2008). FEBS Lett.582, 3301–3307.

31. Mathies, R., Oseroff, A. R. & Stryer, L. (1976). Proc. Natl. Acad. Sci. USA, 73, 1–5.

32. McGeehan, J. E., Ravelli, R. B. G., Murray, J. W., Owen, R. L., Cipriani, F., McSweeney, S., Weik, M. & Garman, E. F. (2009). J. Synchrotron Rad.16, 163–172.

33. Meents, A., Owen, R. L., Murgida, D., Hildebrandt, P., Schneider, R., Pradervand, C., Bohler, P. & Schulze-Briese, C. (2007). AIP Conf. Proc.879, 1984–1987.

34. Moukhametzianov, R., Klare, J. P., Efremov, R., Baeken, C., Goeppner, A., Labahn, J., Engelhard, M., Bueldt, G. & Gordeliy, V. I. (2006). Nature (London), 440, 115–119.

35. Murray, J. W., Garman, E. F. & Ravelli, R. B. G. (2004). J. Appl. Cryst.37, 513–522.

36. Owen, R. L., Rudino-Pinera, E. & Garman, E. F. (2006). Proc. Natl. Acad. Sci. USA, 103, 4912–4917.

37. Pearson, A., Mozzarelli, A. & Rossi, G. (2004). Curr. Opin. Struct. Biol.14, 1–7.

38. Pearson, A. R., Pahl, R., Kovaleva, E. G., Davidson, V. L. & Wilmot, C. M. (2007). J. Synchrotron Rad.14, 92–98.

39. Pohl, E., Pradervand, C., Schneider, R., Tomizaki, T., Pauluhn, A., Chen, Q., Ingold, G., Zimoch, E. & Schulze-Briese, C. (2006). Synchrotron Radiat. News19, 24–26.

40. Royant, A., Carpentier, P., Ohana, J., McGeehan, J., Paetzold, B., Noirclerc-Savoye, M., Vernède, X., Adam, V. & Bourgeois, D. (2007). J. Appl. Cryst.40, 1105–1112.

41. Sanders-Loehr, J. (1988). Metal Clusters in Proteins, edited by L. Que. Washington: American Chemical Society.

42. Schlichting, I. & Chu, K. (2000). Curr. Opin. Struct. Biol.10, 744–752.

43. Schotte, F., Lim, M., Jackson, T. A., Smirnov, A. V., Soman, J., Olson, J. S., Phillips, G. N. Jr, Wulff, M. & Anfinrud, P. A. (2003). Science, 300, 1944–1947.
44. Schotte, F., Soman, J., Olson, J. S., Wulff, M. & Anfinrud, P. A. (2004). J. Struct. Biol.147, 235–246.
45. Siebert, F. & Hildebrandt, P. (2007). Vibrational Spectroscopy in Life Science. New York: Wiley.
46. Southworth-Davies, R. J. & Garman, E. F. (2007). J. Synchrotron Rad.14, 73–83.
47. Susi, H. (1969). Structure and Stability of Biological Macromolecules, edited by S. N. Timasheff and L. Stevens, pp. 575–663. New York: Dekker.
48. Tonge, P. J., Carey, P. R., Callender, R., Deng, H., Ekiel, I. & Muhandiram, D. R. (1993). J. Am. Chem. Soc.115, 8757–8762.
49. Tsuboi, M. & Thomas, G. J. (1997). Appl. Spectrosc. Rev.32, 263–299.
50. Tuma, R. (2005). J. Raman Spectrosc.36, 307–319.
51. Turro, N. J. (1991). Modern Molecular Photochemistry. New York: University Science Books.
52. Vergara, A., Merlino, A., Pizzo, E., D'Alessio, G. & Mazzarella, L. (2008). Acta Cryst. D64, 167–171.
53. Whittaker, M., DeVito, V., Asher, S. & Whittaker, J. (1989). J. Biol. Chem.264, 7104–7106.
54. Wilmot, C. M., Sjögren, T., Carlsson, G. H., Berglund, G. I. & Hajdu, J. (2002). Methods Enzymol.353, 301–318.
55. Xiao, Y., Wang, H., George, S. J., Smith, M. C., Adams, M. W. W., Jenney, F. E., Sturhahn, W., Alp, E. E., Zhao, J., Yoda, Y., Dey, A., Solomon, E. I. & Cramer, S. P. (2005). J. Am. Chem. Soc.127, 14596–14606.
56. Yano, J., Kern, J., Irrgang, K.-D., Latimer, M. J., Bergmann, U., Glatzel, P., Pushkar, Y., Biesiadka, J., Loll, B., Sauer, K., Messinger, J., Zouni, A. & Yachandra, V. K. (2005). Proc. Natl. Acad. Sci. USA, 102, 12047–12052.

High-Resolution 3D Structure Determination of Kaliotoxin by Solid-State NMR Spectroscopy

Jegannath Korukottu, Robert Schneider, Vinesh Vijayan, Adam Lange, Olaf Pongs, Stefan Becker, Marc Baldus and Markus Zweckstetter

ABSTRACT

High-resolution solid-state NMR spectroscopy can provide structural information of proteins that cannot be studied by X-ray crystallography or solution NMR spectroscopy. Here we demonstrate that it is possible to determine a protein structure by solid-state NMR to a resolution comparable to that by solution NMR. Using an iterative assignment and structure calculation protocol, a large number of distance restraints was extracted from 1H/1H mixing experiments recorded on a single uniformly labeled sample under magic angle spinning conditions. The calculated structure has a coordinate precision of 0.6 Å and 1.3 Å for the backbone and side chain heavy atoms, respectively, and deviates from the structure observed in solution. The approach is expected to be applicable to larger systems enabling the determination of high-resolution structures of amyloid or membrane proteins.

Introduction

Structural characterization of membrane proteins and many other biological systems by X-ray crystallography or solution NMR spectroscopy is difficult because of problems with crystallization, solubility or molecular size. Significant advances, however, have been made to construct three-dimensional (3D) molecular structures from solid-state NMR data obtained under Magic Angle Spinning (MAS) [1] conditions[2], [3], [4]. These efforts resulted in high-resolution 3D conformations for small peptides[5], [6], [7], [8] and the determination of medium-resolution backbone structures for a few solid-phase proteins.[9], [10], [11], [12].

Structure determination from solid-state NMR data typically follows the approach employed by solution-state NMR, namely assignment of backbone and side chain resonances using pulse sequences for sequential correlation of resonances, characterization of torsion angles and detection of tertiary contacts. Unless sample orientation provides a direct route to monitor molecular structure under MAS conditions[13], [14], the collection of medium and long-range distance constraints is most crucial. Ideally, these correlations are closely related to molecular structure, can be measured in high spectral resolution and lead to unequivocal assignments of structure-relevant correlations. Two strategies have been developed in this direction: (i) measurement of ^{13}C-^{13}C distances on ^{13}C block-labeled protein microcrystals[9] and (ii) extraction of ^{1}H-^{1}H-distance restraints from ^{13}C,^{13}C- and ^{15}N,^{13}C-encoded $^{1}H/^{1}H$ mixing experiments on a uniformly $^{13}C/^{15}N$-labeled sample[11].

Here we combine ^{13}C,^{13}C- and ^{15}N,^{13}C-encoded $^{1}H/^{1}H$ mixing experiments recorded on a uniformly $^{13}C/^{15}N$-labeled sample with a probabilistic assignment algorithm originally developed for the automatic assignment of 1H-1H correlations in Nuclear Overhauser Effect spectra recorded on proteins in solution[15]. We determine the high-resolution structure of the 38-residue scorpion toxin kaliotoxin (KTX) and show that the structure of KTX in the solid phase deviates from the one observed in solution.

Results and Discussion

Earlier, the backbone fold of the 38-residue potassium channel blocker toxin KTX in the solid phase was deduced from 28 manually assigned interresidue

CHHC correlations (ProteinDataBank (PDB) code: 1XSW)[11]. To define the structure of KTX at higher accuracy, a significantly higher number of medium and long-range correlations was required. For this aim, we set out to combine analysis tools originally developed for the assignment of internuclear correlations in liquid-state NMR spectra with solid-state NMR data. To avoid intermolecular contacts, the $^{13}C/^{15}N$-labeled protein was diluted six-fold (when compared to previous measurements [11]) by the addition of unlabeled KTX. In addition, mixing times in 2D CHHC experiments were reduced to 100, 175 and 250 μs to reduce the potential impact of spin diffusion. The three 2D CHHC spectra and one 2D NHHC spectrum were analyzed using PASD, a probabilistic assignment algorithm for automated structure determination [15]. Three successive PASD passes of cross peak assignment and simulated annealing were performed and each pass was started from a set of randomly generated coordinates. Calculations were carried out in torsion angle space using assigned distance restraints along with torsion angle restraints predicted from backbone chemical shifts using the program TALOS[16], [17],[18]. After completion of the PASD calculations, cross peak assignments were selected that had a good fit to the 1XSW backbone fold (PASD assignment likelihood of 1.0). Subsequently, a high-resolution structure was calculated on the basis of selected cross peak assignments using an optimized simulated annealing protocol.[19] These calculations were started from random initial coordinates, all verified distance restraints were active during the course of calculation and torsion angle restraints predicted by TALOS were included.

Previously, 15 long-range, 7 medium-range and 6 short-range correlations could be assigned [11]. Using the above described semi-automated approach a total of 260 ^1H-^1H distance correlations could be assigned unambiguously (Figure 1CD and Table 1). 62 of these were long-range, 33 medium-range and 165 sequential. The 3D solid-state structure of KTX that was calculated from the 260 distance restraints and 58 dihedral angle restraints is shown in Figure 2. The resulting ensemble of KTX structures tightly converged with a coordinate precision of 0.6 Å and 1.3 Å for backbone and side chain heavy atoms, respectively. Backbone and most side chains had a well-defined orientation except the N- and C- terminal residues and Asn30 located in the loop connecting the second and third β-strand of KTX (Figure 3). The high-resolution solid-state structure of KTX deviates by 2.4 Å from the backbone conformation (PDB code: 1XSW) obtained on the basis of 28 manually assigned distance restraints[11], which deviates by 2.7 Å from the solution structure. The most pronounced deviation between 1XSW and the high-resolution structure was observed at the N-terminus, where four residues were rotated by about 50°, such that the first β-strand was straight and not bent as seen in the high-resolution structures.

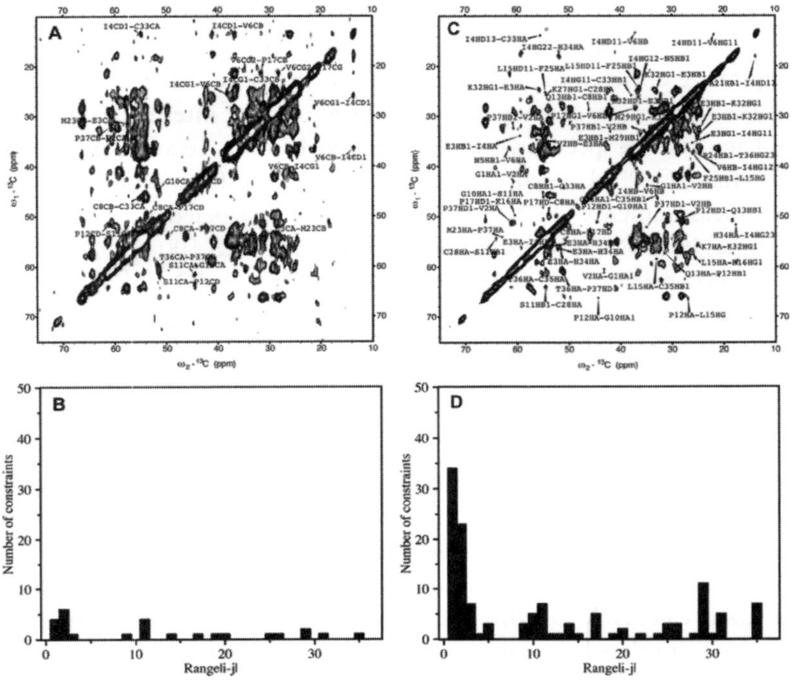

Figure 1. Comparison of interresidue correlations assigned earlier [11] (A and B) and assigned in this study (C and D) for KTX in the solid phase.
Signals assigned in the 2D CHHC spectrum of diluted U-[13C, 15N]-KTX recorded with a mixing time of 250 µs are labeled. (B) and (D) show the number of unambiguously assigned distance constraints as a function of residue difference i and j.

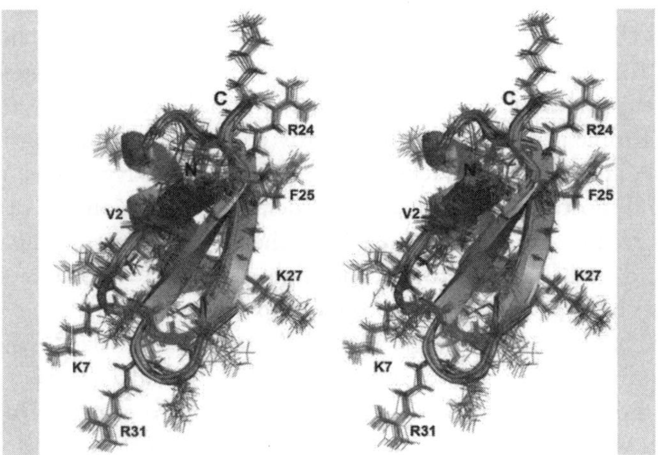

Figure 2. High-resolution 3D structure of KTX determined in the solid phase.
Stereo view of the 20 lowest-energy structures are shown.

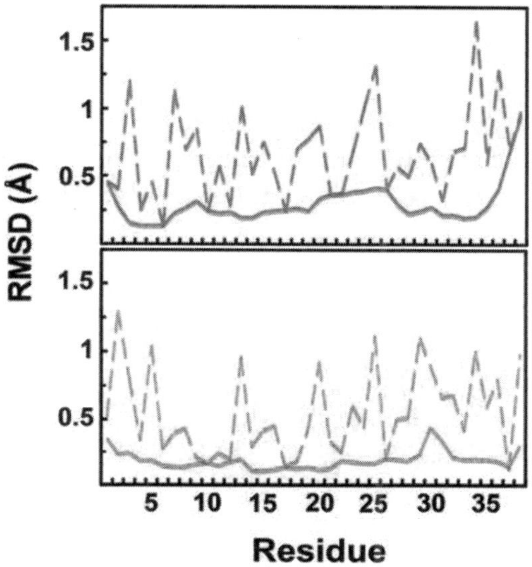

Figure 3. Coordinate precision of KTX in solution (A) and solid phase (B).
Shown are residue-based rms deviations of the coordinates of backbone atoms (solid line) and non-hydrogen side chain atoms (dashed line) within the ensemble of 20 lowest energy structures.

Table 1. Structural statistics for the 20 lowest-energy structures of KTX in solution and in the solid phase.

	Solution	Free
Proton-Proton distance correlations		
Total	314	260
Short range	199	165
Medium range	45	33
Long range	70	62
Distance violations (>0.5 Å)	0	2[b]
Dihedral angles	64	58
Dihedral violations (>5°)	0	0
Energy(kcal/mol)		
Total	−1203.3±60.1	−1365.1±62.3
Dihedral	−133.4±5.6	−119.3±5.1
NOE/CHHC	−166.8±34.3	−167.7±29.3
RMSD		
Dihedral	3.2±1.3	4.9±1.0
NOE/CHHC	0.08±0.02	0.09±0.01
Coordinate precision [a]		
Backbone atoms (Å)	0.7	0.6
All heavy atoms (Å)	1.6	1.3
Ramachandran statistics		
Most favored region (%)	86.7	84.0
Disallowed region (%)	3.3	3.7

[a]Defined as the average rmsd difference between the 20 structures and the mean coordinates.
[b]The two distance restraints G1(Hα1)-I4(Hα) and I4(Hγ11)-C35(Hα) were violated by 0.69 and 0.6 Å, respectively. The two restraints came from weak cross peak in the spectra and were assigned a distance range of 1.8–6.0 Å in the calculations.

Various tests were performed to probe the convergence of the structure calculations and support the accuracy of the high-resolution solid-state structure (see Materials and Methods): (i) use of CHHC spectra with longer mixing times and at six-fold higher concentration; (ii) use of different conformations for calculating likelihood estimates in PASD; (iii) influence of chemical shift tolerances; (iv) sensitivity towards distance ranges used for interresidue correlations; (v) dependence on the number of CHHC spectra; (vi) influence of disulphide bond restraints. In all cases, the backbone of the calculated structures deviated by less than 0.7 Å from the backbone of the structure shown in Figure 2.

Recently, a method for automatic assignment of cross peaks in ^{13}C-^{13}C correlation spectra was developed[20]. The approach called SOLARIA was used to analyze proton-driven spin diffusion spectra recorded on ^{13}C– block-labeled, microcrystalline preparations of the α-spectrin SH3 domain. In this study, only a modest improvement in the 3D backbone structure was observed. In contrast, our strategy based on C/NHHC correlations leads to an atomic resolution definition of both the backbone and the side-chain structure of KTX. We attribute these improvements to the higher fraction of long-range contacts in initial-rate N/CHHC spectra that allows for the same small distances boundaries[21], [22] during structure calculation as used in liquid-state NMR.

To enable a direct comparison, we determined the solution structure of KTX employing the identical strategy as used for KTX in the solid phase. 70 long-range, 45 medium-range and 199 sequential NOEs could be assigned unambiguously, closely resembling the amount and distribution of distance restraints obtained from 2D N/CHHC spectra for KTX in the solid phase (Table 1). The newly determined solution structure deviates by 0.6 Å from a previously determined solution structure of KTX (PDB code: 2KTX)[23].

The backbone of the high-resolution solid-state structure of KTX deviates by 1.3 Å from that observed in solution (Figure 4A). Structural differences were observed for the two N-terminal residues, the loop between the first β-strand and the α-helix, and the C-terminal β-sheet in particular next to G26 (Figure 4B). The structural differences are due to a combination of changes in interresidue cross peaks and in backbone dihedral angles. For residues 8–11, 23–25, K27 and M29, backbone dihedral angles predicted by TALOS from the solid-state chemical shifts clearly deviated from those predicted by TALOS from the solution-state chemical shifts (Figure 4B).

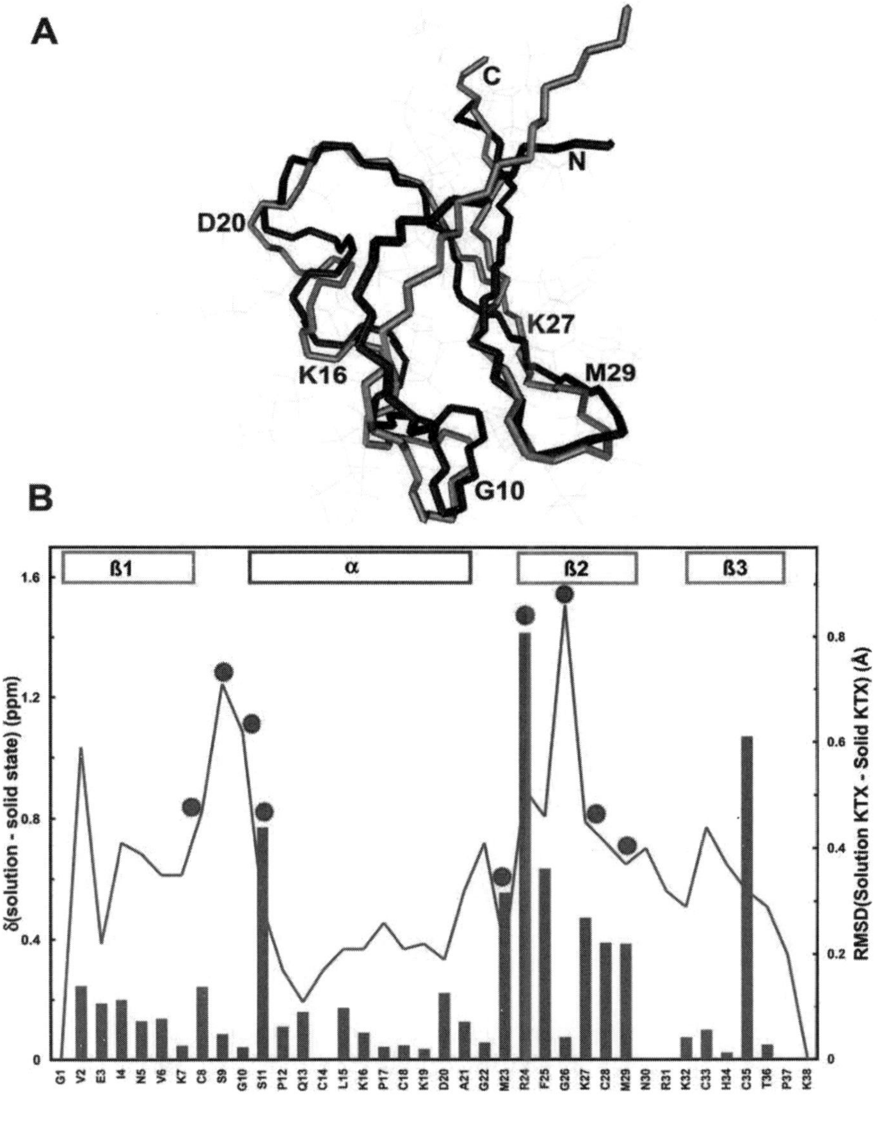

Figure 4. (A) Superposition of high-resolution solid-state structure (green) and solution structure (blue; PDB code: 2KTX) of KTX. (B) Comparison of averaged Cα/Cβ chemical shift differences (blue bars; calculated according to $0.256^*[\Delta\delta C\alpha^2+\Delta\delta C\beta^2]^{1/2}$) with rms deviation between the mean structures (blue line) of KTX in solution and in the solid-phase. Red dots mark residues, for which the backbone dihedral angles predicted by TALOS differ in solution and the solid state. Secondary structure is indicated.

Conclusion

Our study demonstrates that high-resolution 3D structures of globular proteins can be obtained from solid-state NMR data. The combination of ^{15}N,^{13}C-encoded $^{1}H/^{1}H$ mixing experiments with a probabilistic cross peak assignment algorithm is particularly powerful, as short distances between protons provide the principal source of long-range structural information. Depending on the molecule under investigation, the presented approach can be combined with other solid-state NMR spectroscopic methods. Applications to larger proteins may benefit from the use of block[9], modular[24] or stereo-array[25] isotope labeling, and allow the determination of high-resolution structures of amyloid or membrane proteins.

Materials and Methods

NMR Spectroscopy

Unlabelled and uniformly [13C,15N]-labeled KTX was prepared recombinantly as previously described[11], [34].

Solution-state NMR experiments were carried out at 298 K on a Bruker AVANCE 600 spectrometer. NMR samples contained 2 mM unlabelled KTX in 95% H_2O/5% D_2O, pH 7.5. 2D TOCSY (mixing time of 60 ms using MLEV17) and NOESY experiments (mixing time of 200 ms) were used to facilitate backbone assignment[26], [27]. The spectra were recorded using 362×724 complex data points in F1 and F2 dimensions with 32 scans per increment and a relaxation delay of 1.2 s. The spectral widths were 9615 and 9603 Hz in the F1 and F2 dimensions, respectively. $^{13}C\alpha$, $^{13}C\beta$ and ^{15}N chemical shifts were obtained from natural abundance, two-dimensional 1H-15N and 1H-13C HSQCs[28]. All data were processed using NMRPipe[29].

An anisotropic medium for measurement of residual dipolar couplings was prepared by addition of Pf1 filamentous phages (Asla, Riga, Latvia) to a concentration of 12 mg/ml[30]. To lower the electrostatic attraction between KTX and the highly negatively charged Pf1 phage, the ionic strength was raised to 500 mM. ^{1}HN-^{15}N splittings were measured under isotropic and anisotropic conditions using 2D IPAP-^{1}H-^{15}N HSQC experiments[31]. RDCs were extracted by subtraction of the 1JNH scalar coupling measured for the isotropic sample. Comparison of experimental RDCs (1DNH) with values back-calculated from the redetermined solution-state structure of KTX using singular value decomposition as implemented in PALES[32] resulted in a Pearson's correlation coefficient of 0.93. The magnitude Da and rhombicity R of the alignment tensor were back-calculated as Da = 2.5 Hz and R = 0.23. When best-fitting the experimental

RDCs to the published solution-state structure (PDB code: 2KTX), a Pearson's correlation coefficient of 0.85 was obtained.

Solid-state NMR data comprised three CHHC spectra (100, 175 and 250 µs (^{1}H,^{1}H) mixing) and one NHHC spectrum (100 µs (^{1}H,^{1}H) mixing) [11]. CHHC spectra were obtained on a uniformly [^{13}C,^{15}N]-labeled KTX sample diluted approximately 1:6 in unlabeled KTX, while the NHHC spectrum was recorded on an undiluted uniformly [^{13}C,^{15}N]-labeled sample. Data were recorded on a wide-bore Bruker 600 MHz instrument at 11 kHz MAS speed (CHHC spectra with 100 and 175 µs mixing time) and on a standard-bore Bruker 800 MHz instrument at 12.5 kHz MAS (CHHC spectrum with 250 µs mixing time and NHHC spectrum) using 4 mm triple-resonance (^{1}H,^{13}C,^{15}N) probes. Sample temperature was about 280 K in all cases. 1H field strengths used for 90° pulses and SPINAL64 [33] decoupling during evolution and detection periods were between 70 and 83 kHz. Short CP contact times of tHC = 100 µs or tHN = 250 µs enclosing the (^{1}H,^{1}H) transfer step were employed to ensure polarization transfer between directly bonded nuclei only. Spectra were recorded with 105×1024 (CHHC 100 and 175 µs), 140×1280 (CHHC 250 µs) or 40×1536 (NHHC) complex data points in F1 and F2 dimensions, respectively, with around 1024 (CHHC) or 2048 (NHHC) scans per increment. Spectral widths were 83 (CHHC) or 44 (NHHC) ppm in the indirect dimension and 310 to 355 ppm in the direct dimension, respectively. The recycle delay was set to 2s.

Automated Cross-Peak Assignment and Structure Calculation

Two-dimensional CHHC, NHHC and NOESY spectra were automatically peak picked using Sparky 3 (T. D. Goddard and D. G. Kneller, University of California, San Francisco). Diagonal peaks were manually removed. Peak intensities were classified into four ranges and converted into distance ranges of 1.8–2.7, 1.8–3.3, 1.8–5.0, and 1.8–6.0 Å. Lists of cross peaks were subjected to the automated cross-peak assignment and structure calculation algorithm PASD implemented in Xplor-NIH[15], [17]. For analysis of the solid-state spectra by PASD, ^{13}C and ^{15}N chemical shifts were labeled as if they were proton chemical shifts. Tolerances for matching chemical shifts to cross-peaks were 0.015 ppm in F2 and F1 for the NOESY spectrum, and 0.38 ppm and 0.60 ppm in the acquisition and indirect dimension of the N/CHHC spectra, respectively.

PASD was applied largely following published procedures[15]. In short, three successive passes of simulated annealing calculations in torsion angle space were carried out. Each pass was started from a set of randomly generated coordinates. The target function comprised a potential function for experimental distance restraints (e.g. obtained from NOEs or CHHC correlations), a quadratic van der

Waals repulsion term, a square-well potential for torsion angles and a torsion angle database potential of mean force. Pass 1 and 2 protocol comprised two high-temperature phases (4000 K) and a slow cooling phase (from 4000 to 100 K) with a linear NOE potential. Pass 3 comprised a single high-temperature phase (4000 K) followed by a cooling phase with a quadratic NOE potential. Final assignment likelihoods were determined at the end of pass 3 calculations. Calculations were carried out on a Linux cluster of 32 processors and took about two days for each structure.

PASD structures do not represent fully-refined NMR structures[15]. Therefore, we selected cross-peaks that were in agreement with the backbone fold of KTX determined previously (see below): for KTX(solution) and KTX(solid), 31% and 28%, respectively, of all long-range restraints, 83% and 80%, respectively, of the medium-range restraints, and 99% and 100%, respectively, of the sequential restraints had final restraint likelihoods of 1.0. Assignments obtained for these cross-peaks by PASD were verified by manual inspection of the 2D N/CHHC spectra for KTX(solid) or the NOESY spectrum for KTX(solution).

Convergence of Automated Cross-Peak Assignment and Structure Calculation

We performed several tests to probe the reliability of the solid-state 3D structure of KTX: (i) use of CHHC spectra recorded with longer mixing times and using undiluted 13C/15N-labeld KTX; (ii) use of different conformations for calculating likelihood estimates in the PASD analysis; (iii) influence of chemical shift tolerances; (iv) influence of distance ranges; (v) influence of disulphide bond restraints; (vi) combination of solid-state distance restraints with solution-state dihedral angles (and vice versa).

To i): In contrast to the measurements performed in this study, CHHC spectra were previously recorded on 13C/15N-labeld KTX that was not diluted by unlabeled protein [11]. In these spectra, intermolecular cross peaks may appear. In addition, the spectra had been recorded with mixing times of 250, 325 and 400 μs, increasing the risk of spin diffusion. Nevertheless, when using these three CHHC spectra together with the 2D NHHC spectrum, the resulting structure deviated by less than 0.7 Å (rms value for all N, Cα, CO backbone atoms) from the structure shown in Figure 2.

To ii): At the end of pass 1 and 2 the PASD algorithm calculates likelihood estimates that each particular assignment associated with a cross-peak is correct. The likelihoods are calculated using the ensemble of structures present at the end of the corresponding pass. Thus, they are a metric of how consistent a given assignment is with the ensemble of structures at the end of each calculation pass[15]. Here we have not used the ensemble of structures present at the end of pass 1 and 2 for calculation of likelihood estimates, but either the high-resolution structure of KTX obtained under different conditions or a medium-resolution backbone fold. This improved convergence in the structure calculations and was justified as we previously established that the fold of KTX in solution and in the solid phase is the same[11]. The PASD calculations of KTX(solid) were done once by using KTX(solution) (PDB CODE: 2KTX) for calculation of the likelihood estimates at the end of pass 1 and 2. Thus, we biased on purpose the calculation towards the solution-state structure. Then a second PASD calculation was done, in which the likelihood estimates were determined using the medium resolution backbone fold obtained previously for KTX in the solid phase (PDB CODE: 1XSW).[11] In all cases, the structures obtained from the two different PASD calculations were indistinguishable. This supports the relevance of the differences between the solution and solid-state structure. Note, that identical structure calculation protocols were used in all cases.

To iii): For the calculations reported in the main part of the manuscript, tolerances for matching chemical shifts to cross-peaks were set to 0.38 ppm and 0.60 ppm in the acquisition and indirect dimension, respectively. We repeated the structure calculations with chemical shift tolerances of 0.38 ppm and 0.4 ppm in the acquisition and indirect dimension, respectively. The resulting structure deviated by less than 0.7 Å (rms value for all N, Cα, CO backbone atoms) from the structure shown in Figure 2.

To iv): Peak intensities obtained from the 2D CHHC and NHHC spectra were classified into four ranges and converted into distance ranges of 1.8–2.7, 1.8–3.3, 1.8–5.0 and 1.8–6.0 Å, respectively. The classification was done independently for the four proton-proton correlation spectra (see main manuscript). To test the sensitivity of the solid-state structure to the used distance ranges, we repeated the structure calculations assigning to all N/CHHC correlations a distance range of 2.4–6.0 Å. The resulting structure deviated by less than 0.3 Å (rms value for all N, Cα, CO backbone atoms) from the structure shown in Figure 2.

To v): For both KTX(solution) and KTX(solid), structure calculations were performed without and with restraints for the three disulphide bonds. The resulting structures did not differ (backbone rms deviation below 0.5 Å) and only the results of calculations, in which the disulphide bonds were not enforced, were reported.

To vi): Are the structural differences due to an uncertainty in the analysis of N/CHHC spectra? To address this question, we recalculated the structure (using XPLOR-NIH and starting from an extended strand) using the same solid-state N/CHHC distance restraints, but supplementing them with the dihedral angles obtained by TALOS from the solution-state chemical shifts (instead of those obtained from the solid-state chemical shifts). The backbone of the resulting structure deviated by 0.5 Å from the high-resolution solid-state structure. The coordinate precision for backbone and all heavy atoms was 0.7 Å and 1.7 Å, respectively. However, two dihedral angle violations were introduced (for residues 2 and 24) and residue 24 moved into the disallowed region of the Ramachandran plot. In addition, the total energy increased from −1307±54 kcal/mol to −1032±48 kcal/mol, the dihedral angle energy from −110±6 kcal/mol to −16±36 kcal/mol and the distance restraint energy from −157±28 kcal/mol to −18±67 kcal/mol (when compared to the pure solid-state structure calculation). Similarly, when the solution-state distance restraints were combined with the solid-state dihedral angles, one dihedral angle violation (for S9) was introduced, the total energy was increased from 1203±60 kcal/mol to −1154±65 kcal/mol, the dihedral angle energy from −133±6 kcal/mol to −55±32 kcal/mol and the distance restraint energy from −167±34 kcal/mol to −31±53 kcal/mol (when compared to the pure solution-state structure calculation). The backbone of the resulting structure deviated by 0.6 Å from the high-resolution solution-state structure. The coordinate precision for backbone and all heavy atoms was 0.8 Å and 1.9 Å, respectively. These data demonstrate that the solid-state distance restraints are only in agreement with the solid-state backbone chemical shifts, and the solution-state distance restraints are only in agreement with the solution-state backbone chemical shifts.

Acknowledgements

We thank John Kuszewski and Christian Griesinger for discussions.

Authors' Contributions

Conceived and designed the experiments: MZ. Performed the experiments: JK AL VV. Analyzed the data: JK. Contributed reagents/materials/analysis tools: RS SB MB. Wrote the paper: OP MZ JK MB.

References

1. Andrew ER, Bradbury A, Eades RG (1958) Nuclear Magnetic Resonance Spectra from a Crystal rotated at High Speed. Nature 182: 1659–1659.
2. Griffin RG (1998) Dipolar recoupling in MAS spectra of biological solids. Nat Struct Biol 5: Suppl508–512.

3. Luca S, Heise H, Baldus M (2003) High-resolution solid-state NMR applied to polypeptides and membrane proteins. Accounts of Chemical Research 36: 858–865.

4. Tycko R (2001) Biomolecular solid state NMR: Advances in structural methodology and applications to peptide and protein fibrils. Annual Review of Physical Chemistry 52: 575–606.

5. Jaroniec CP, MacPhee CE, Bajaj VS, McMahon MT, Dobson CM, et al. (2004) High-resolution molecular structure of a peptide in an amyloid fibril determined by magic angle spinning NMR spectroscopy. Proc Natl Acad Sci USA 101: 711–716.

6. Nomura K, Takegoshi K, Terao T, Uchida K, Kainosho M (1999) Determination of the complete structure of a uniformly labeled molecule by rotational resonance solid-state NMR in the tilted rotating frame. Journal of the American Chemical Society 121: 4064–4065.

7. Petkova AT, Ishii Y, Balbach JJ, Antzutkin ON, Leapman RD, et al. (2002) A structural model for Alzheimer's beta -amyloid fibrils based on experimental constraints from solid state NMR. Proc Natl Acad Sci USA 99: 16742–16747.

8. Rienstra CM, Tucker-Kellogg L, Jaroniec CP, Hohwy M, Reif B, et al. (2002) De novo determination of peptide structure with solid-state magic-angle spinning NMR spectroscopy. Proc Natl Acad Sci USA 99: 10260–10265.

9. Castellani F, van Rossum B, Diehl A, Schubert M, Rehbein K, et al. (2002) Structure of a protein determined by solid-state magic-angle-spinning NMR spectroscopy. Nature 420: 98–102.

10. Zhou DH, Shea JJ, Nieuwkoop AJ, Franks WT, Wylie BJ, et al. (2007) Solid-State Protein-Structure Determination with Proton-Detected Triple-Resonance 3D Magic-Angle-Spinning NMR Spectroscopy. Angewandte Chemie International Edition 46: 8380–8383.

11. Lange A, Becker S, Seidel K, Giller K, Pongs O, et al. (2005) A concept for rapid protein-structure determination by solid-state NMR spectroscopy. Angew Chem Int Ed Engl 44: 2089–2092.

12. Zech SG, Wand AJ, McDermott AE (2005) Protein structure determination by high-resolution solid-state NMR spectroscopy: application to microcrystalline ubiquitin. J Am Chem Soc 127: 8618–8626.

13. Andronesi OC, Pfeifer JR, Al-Momani L, Ozdirekcan S, Rijkers DT, et al. (2004) Probing membrane protein orientation and structure using fast magic-angle-spinning solid-state NMR. J Biomol NMR 30: 253–265.

14. Glaubitz C, Watts A (1998) Magic angle-oriented sample spinning (MAOSS): A new approach toward biomembrane studies. J Magn Reson 130: 305–316.
15. Kuszewski J, Schwieters CD, Garrett DS, Byrd RA, Tjandra N, et al. (2004) Completely automated, highly error-tolerant macromolecular structure determination from multidimensional nuclear overhauser enhancement spectra and chemical shift assignments. J Am Chem Soc 126: 6258–6273.
16. Grishaev A, Bax A (2004) An empirical backbone-backbone hydrogen-bonding potential in proteins and its applications to NMR structure refinement and validation. J Am Chem Soc 126: 7281–7292.
17. Schwieters CD, Kuszewski JJ, Tjandra N, Clore GM (2003) The Xplor-NIH NMR molecular structure determination package. J Magn Reson 160: 65–73.
18. Cornilescu G, Delaglio F, Bax A (1999) Protein backbone angle restraints from searching a database for chemical shift and sequence homology. J Biomol NMR 13: 289–302.
19. Linge JP, Williams MA, Spronk CA, Bonvin AM, Nilges M (2003) Refinement of protein structures in explicit solvent. Proteins 50: 496–506.
20. Fossi M, Castellani F, Nilges M, Oschkinat H, van Rossum BJ (2005) SOLARIA: a protocol for automated cross-peak assignment and structure calculation for solid-state magic-angle spinning NMR spectroscopy. Angew Chem Int Ed Engl 44: 6151–6154.
21. Baldus M (2007) ICMRBS founder's medal 2006: Biological solid-state NMR, methods and applications. Journal of Biomolecular Nmr 39: 73–86.
22. Lange A, Seidel K, Verdier L, Luca S, Baldus M (2003) Analysis of proton-proton transfer dynamics in rotating solids and their use for 3D structure determination. J Am Chem Soc 125: 12640–12648.
23. Gairi M, Romi R, Fernandez I, Rochat H, Martin-Eauclaire MF, et al. (1997) 3D structure of kaliotoxin: is residue 34 a key for channel selectivity? J Pept Sci 3: 314–319.
24. Pickford AR, Campbell ID (2004) NMR studies of modular protein structures and their interactions. Chemical Reviews 104: 3557–3565.
25. Kainosho M, Torizawa T, Iwashita Y, Terauchi T, Ono AM, et al. (2006) Optimal isotope labelling for NMR protein structure determinations. Nature 440: 52–57.
26. Braunschweiler L, Ernst RR (1983) Coherence Transfer by Isotropic Mixing - Application to Proton Correlation Spectroscopy. Journal of Magnetic Resonance 53: 521–528.

27. Macura S, Ernst RR (1980) Elucidation of Cross Relaxation in Liquids by Two-Dimensional Nmr-Spectroscopy. Molecular Physics 41: 95–117.

28. Bax A, Ikura M, Kay LE, Torchia DA, Tschudin R (1990) Comparison of Different Modes of 2-Dimensional Reverse-Correlation Nmr for the Study of Proteins. Journal of Magnetic Resonance 86: 304–318.

29. Delaglio F, Grzesiek S, Vuister GW, Zhu G, Pfeifer J, et al. (1995) Nmrpipe - a Multidimensional Spectral Processing System Based on Unix Pipes. Journal of Biomolecular Nmr 6: 277–293.

30. Hansen MR, Mueller L, Pardi A (1998) Tunable alignment of macromolecules by filamentous phage yields dipolar coupling interactions. Nat Struct Biol 5: 1065–1074.

31. Ottiger M, Delaglio F, Bax A (1998) Measurement of J and dipolar couplings from simplified two-dimensional NMR spectra. Journal of Magnetic Resonance 131: 373–378.

32. Zweckstetter M (2008) NMR: prediction of molecular alignment from structure using the PALES software. Nat. Protoc. 3: 679–690.

33. Fung BM, Khitrin AK, Ermolaev K (2000) An Improved broadband decoupling sequence for liquid crystals and solids. Journal of Magnetic Resonance 142: 97–101.

34. Lange A, Giller K, Hornig S, Martin-Eauclaire MF, Pongs O, et al. (2006) Toxin-induced conformational changes in a potassium channel revealed by solid-state NMR. Nature 440: 959–962.

Kissing G Domains of MnmE Monitored by X-Ray Crystallography and Pulse Electron Paramagnetic Resonance Spectroscopy

Simon Meyer, Sabine Böhme, André Krüger, Heinz-Jürgen Steinhoff, Johann P. Klare and Alfred Wittinghofer

ABSTRACT

MnmE, which is involved in the modification of the wobble position of certain tRNAs, belongs to the expanding class of G proteins activated by nucleotide-dependent dimerization (GADs). Previous models suggested the protein to be a multidomain protein whose G domains contact each other in a nucleotide dependent manner. Here we employ a combined approach of X-ray crystallography and pulse electron paramagnetic resonance (EPR) spectroscopy to show that large domain movements are coupled to the G protein cycle of MnmE. The X-ray structures show MnmE to be a constitutive homodimer where the highly mobile G domains face each other in various orientations but are not in close contact as suggested by the GDP-AlFx structure of the isolated domains. Distance measurements by pulse double electron-electron resonance (DEER) spectroscopy show that the G domains adopt an open conformation in the nucleotide free/GDP-bound and an open/closed two-state equilibrium in the GTP-bound state, with maximal distance variations of 18 Å. With GDP and AlFx, which mimic the transition state of the phosphoryl transfer reaction, only the closed conformation is observed. Dimerization of the active sites with GDP-AlFx requires the presence of specific monovalent cations, thus reflecting the requirements for the GTPase reaction of MnmE. Our results directly demonstrate the nature of the conformational changes MnmE was previously suggested to undergo during its GTPase cycle. They show the nucleotide-dependent dynamic movements of the G domains around two swivel positions relative to the rest of the protein, and they are of crucial importance for understanding the mechanistic principles of this GAD.

Author Summary

MnmE is an evolutionary conserved G protein that is involved in modification of the wobble U position of certain tRNAs to suppress translational wobbling. Despite high homology between its G domain and the small G protein Ras, MnmE displays entirely different regulatory properties to that of many molecular switch-type G proteins of the Ras superfamily, as its GTPase is activated by nucleotide-dependent homodimerization across the nucleotide-binding site. Here we explore the unusual G domain cycle of the MnmE protein by combining X-ray crystallography with pulse electron paramagnetic resonance (EPR) spectroscopy, which enables distance determinations between spin markers introduced at specific sites within the G domain. We determined the structures of the full-length MnmE dimer in the diphosphate and triphosphate states, which represent distinct steps of the G domain cycle, and demonstrate that the G domain cycle of MnmE comprises large conformational changes and domain movements of up to 18 Å, in which the G domains of the dimeric protein traverse from a GDP-bound open state through an open/closed equilibrium in the triphosphate state to a closed conformation in the transition state, so as to assemble the catalytic machinery.

Abbreviations

5-F-THF, 5-formyl-tetrahydrofolate; AlFx, aluminium tri- or tetrafluoride; cmnm, carboxymethylaminomethyl; DEER, double electron-electron resonance; DTE, dithioerythritol; EPR, electron paramagnetic resonance; GAD, G protein activated by nucleotide-dependent dimerization; GAP, GTPase activating protein; GppCp, guanosine-5′-(β,γ-methylene)triphosphate; GppNHp, guanosine 5′-imidotriphosphate; mGDP, 2′-/3′-O-(N′-Methylanthraniloyl)-GDP; MME, monomethyl ether; MTSSL, (1-oxyl-2,2,5,5-tetramethyl-3-pyrroline-3-methyl) methanethiosulfonate spin label; PEG, polyethylene glycol

Introduction

Cells devote substantial biosynthetic effort and resources to posttranscriptional modification of tRNAs [1]. A frequent feature of tRNAs in all domains of life are modified nucleosides in the anticodon region and especially at the wobble position (position 34) [2], which prestructure the anticodon domain to insure correct codon binding during translation [3]. MnmE is an evolutionary conserved G protein found in bacteria, fungi, and humans, which together with the protein GidA catalyzes the formation of a carboxymethylaminomethyl-group (cmnm) at the 5 position of the wobble uridine (U34) of tRNAs reading 2-fold degenerated codons ending with A or G, i.e., tRNAArg(UCU), tRNAGln(UUG), tRNAGlu(UUC), tRNALeu(UAA), and tRNALys(UUU) [4]–[6]. This modification (cmnm5U34) together with a thiolation at the 2 position favours the interaction with A and G, but suppresses base-pairing with C and U [3],[7]–[10]. By controlling rare codon recognition and reading frame maintenance, hypermodified U34 moreover plays a regulatory role in gene expression [11]. Eucaryotic homologues of MnmE and GidA (termed MSS1 and Mto1, respectively, in yeast) are targeted to mitochondria [12],[13], and the human homologues (termed hGTPBP3 and Mto1, respectively) have been implicated in the development of severe mitochondrial myopathies such as MERRF (myoclenic epilepsy ragged red fibres), MELAS (mitochondrial encephalomyopathy lactic acidosis stroke), and nonsyndromic deafness [14]–[18].

The crystal structure of MnmE from Thermotoga maritima reveals a three-domain protein consisting of an N-terminal tetrahydrofolate-binding domain, a central helical domain, and a canonical Ras-like G domain inserted into the helical domain [19]. The asymmetric unit of these crystals contained one MnmE molecule and the N-terminal domain of a second proteolysed MnmE chain interacting with the N-terminal domain of the first molecule, suggesting that MnmE is a dimer in solution (Figure 1A) [19]. By superposition of the first MnmE chain on the second N-terminal domain a model for the full-length homodimer was generated in which the two G domains face each other with a distance of almost 50 Å between the two P-loops (Figure 1A) [19].

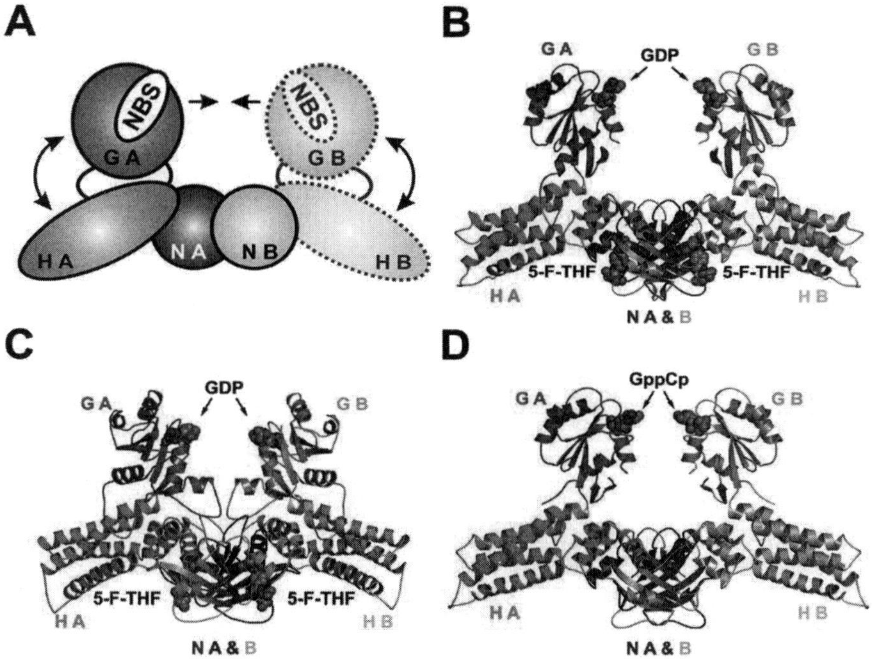

Figure 1. X-ray structures of full-length MnmE dimers.
(A) Model of dimeric MnmE obtained from the partial structure of nucleotide-free MnmE from T. maritima, where only the N-terminal domain (NB), but not the helical (HB), or G domain (GB) of molecule B were present in the crystal. The model was obtained by superimposition of molecule A on the N domain of B and the expected positions of the nucleotide binding sites (denoted as NBS) in this model are indicated. (B–D) Ribbon models of X-ray structures of CtMnmE·GDP (B), No MnmE·GDP (C), and Ct MnmE·GppCp (dimer a) (D), with colors of the N, H, and G domains as indicated, and the protomers A and B.

In contrast to Ras-like small G proteins that require a guanine nucleotide exchange factor (GEF) protein to drive the nucleotide exchange and a GTPase activating protein (GAP) to stimulate hydrolysis [20],[21], MnmE displays lower affinities towards nucleotides and a higher intrinsic K+-stimulated GTP hydrolysis [19],[22]–[24]. A G domain dimerization across the nucleotide binding site has been proposed on the basis of biochemical data and the crystal structure of the isolated MnmE G domains in complex with GDP-aluminium tri- or tetra-fluoride (AlFx) (a mimic of the transition state of GTP hydrolysis [25]) [22]. The G domains dimerize via their switch regions to position an invariant Glu-residue

(E282) for optimal orientation of a water molecule for the nucleophilic attack of the γ-phosphate group [22]. Dimerization stabilizes a highly conserved loop in switch I, the so-called K-loop, to coordinate K+ in a position analogous to the positive charge of the arginine finger in the Ras-RasGAP system. This explains why K+ is required both for the GTPase stimulation and for G domain dimerization [22]. On the basis of the common feature that the G domain cycle is regulated by homodimerization, MnmE has been categorized as G protein activated by nucleotide-dependent dimerization (GAD) [26], together with the signal recognition particle (SRP) and its receptor (SR) [27],[28], the regulator of Ni insertion into hydrogenases HypB [29], the dynamins [30], the human guanylate binding protein hGBP1 [31], the chloroplast import receptors Toc33/34 [32],[33], the septins [34], and the Roc-COR tandem found to be mutated in Parkinson disease [35]. It has been postulated that nucleotide-dependent G domain dimerization activates the GTPase and the distinct biological functions of these proteins, although the mechanisms of coupling G domain dimerization to biological function within this class are diverse and incompletely understood [26].

So far, neither the structural model of the full-length MnmE dimer nor dimerization of the G domains in the context of the full-length dimer have been proven directly. With the architecture of the proposed dimer model, dimerization of the G domains would require large domain movements suggesting that large conformational rearrangements of the protein are coupled to its GTPase cycle [22]. Here we study these GTPase-coupled rearrangements by trapping the protein in various steps of its GTPase cycle by X-ray crystallography and pulse double electron-electron resonance (DEER) spectroscopy in combination with site-directed spin labeling [36]–[38]. The distance distributions obtained for spin labeled sites in the G domains of MnmE allow us to characterize the G domain movements during the GTPase cycle of MnmE.

Results

Crystal Structures of Full Length MnmE Bound to GDP and GppCp

Various MnmE homologous have been screened for crystallization coditions in the presence of GDP, GDP-AlFx and guanosine-5′-(β,γ-methylene)triphosphate (GppCp), and K+ and were found to crystallize readily in diverse conditions, but only in three cases—Chlorobium tepidum MnmE (CtMnmE) in the presence of K+, GDP, or GDP-AlFx; Nostoc MnmE (NoMnmE) in the presence of K+, GDP, or GDP-AlFx; and CtMnmE in the presence of K+ and GppCp-crystals with sufficient diffraction quality were obtained. In the case of CtMnmE, a polyethylene glycol (PEG) 6000/NaCl-condition produced diffraction quality crystals in the presence of GPD and GDP-AlFx. Crystals had the same unit cell parameters and the same space group and are thus isomorphous. NoMnmE crystals with sufficient diffraction were obtained in a PEG 550 monomethyl ether (MME) condition. As with CtMnmE, crystals obtained in the presence of GDP-AlFx or GDP were isomorphous. Structure determination showed in both cases that the crystals contained the GDP-bound form of MnmE, despite the presence of AlFx. Quality of crystals grown in the presence of GDP-AlFx were somewhat better, hence their datasets were used for structure determination.

CtMnmE·GDP and NoMnmE·GDP (grown in presence of AlFx) crystallized in the space groups I4(1)22 and P4(3)2(1)2, respectively, each with one full length protomer in the asymmetric unit. In both cases homodimers are formed via crystallographic symmetry by means of the N-terminal domains (Figure 1B and 1C). Apart from the location of G domains, the structure is very similar to the dimer model proposed for nucleotide-free MnmE (Figure 1A) [19]. Strikingly, two molecules of 5-formyl-tetrahydrofolate (5-F-THF) were identified in the structure of NoMnmE·GDP, which were apparently copurified from the bacterial expression system. This suggests a high affinity for 5-F-THF and supports the recently proposed enzymatic mechanism whereby the C1 group of the cmnm modification is donated by THF [19],[39]. The cofactor is bound as previously described for the complex prepared in situ [19], with two folate binding sites within the dimer interface of the N-terminal domains. CtMnmE·GDP crystals were incubated with a 5-F-THF-containing cryoprotectant prior to data collection and in the crystal structure 5-F-THF is found in identical positions as in the NoMnmE·GDP-dimer and in the TmMnmE-dimer.

In the case of CtMnmE·GppCp, the crystallographic asymmetric unit contained three protomers (chains A, B, C). Molecules B and C form a dimer within the asymmetric unit, while protomer A forms a dimer with its crystallographic symmetry mate (shown in Figure 1D). No density is found for the G domain of molecule C, but crystals applied on an SDS-page confirmed an intact protein (unpublished data). Thus two dimeric structures of CtMnmE·GppCp were analyzed, i.e., the dimer generated by protomer A and a symmetry related chain A (termed "dimer A") and the dimer generated by protomer B and a second protomer B docked onto chain C (termed "dimer B").

The overall homodimer architecture found in the three structures resembles the proposed model obtained from a partial dimer (Figure 1A), with the G domains facing each other with their nucleotide binding sites (Figure 1B–1D). However, even though triphosphate analogues such as GppCp or AlFx and GDP were used in the crystallization trials, the G domains were separated from each other by large distances. They do not display any structural contacts between each other nor to the N-terminal or helical domains. In all the structures, nucleotides are far apart from each other, with distances of 38 to 56 Å between the first P-loop glycines' Cα atom (GxxxxGKS motif).

The Mobile G Domains

In each structure, the G domain adapts the canonical Ras-fold with either both switch regions (CtMnmE·GDP, CtMnmE·GppCp) or switch II (NoMnmE·GDP) disordered and thus not resolved. Nucleotides are bound in a way typical for Ras-like G domains. In CtMnmE·GDP however, no Mg2+ is coordinated to the phosphates, and switch I-contacts to GDP are absent. In NoMnmE·GDP, two Zn2+ atoms from the crystallisation condition, localized by their anomalous signal, are coordinated to the G domain. One of these is coordinated to helix Gα4 and is involved in crystal contacts (see below), the other occupies the usual Mg2+-binding site at the β-phosphate of GDP. As Switch I is resolved, but does not contact the bound GDP and since there is no indication for a physiological role of Zn2+, we consider this to be a crystallographic artefact also observed in the nucleotide binding pockets of other small G proteins [40].

For conventional G proteins regulated by GAPs [20] as well as for G proteins activated by dimerization [26], AlFx-in the γ-phosphate binding site mimics the transition state of the phosphor transfer reaction and is considered the litmus test for correct assembly of the active site. In the case of MnmE, this is thought to be achieved by dimerization and close juxtaposition of the two G domains across the nucleotide binding site, as observed for the isolated G domains [22]. Although both GDP-bound structures have been obtained using GDP and AlFx in the crystallization trials, no electron density for AlFx could be observed. One would thus conclude that close contact between the G domains is not possible in the full-length protein or that the G domains are too mobile for fixation in the crystal and/or that the crystal lattice forces do not allow the close state to occur.

Another possibility would be that crystallisation conditions with high concentrations of precipitants inhibit formation of the closed state of the G domains. Indeed we can show by a previously established fluorometric assay, by which an increase of the fluorescence of 2′-/3′-O-(N′-methylanthraniloyl)-GDP (mGDP) bound to MnmE upon addition of AlFx in the presence of K+ is attributed to G domain dimerization [22], that in the presence of any of the precipitants used for crystallisation, dimerization of the G domains is severely inhibited in the full length protein. This explains why despite the presence of AlFx in the crystallisation trials only the GDP-bound conformations are found. In the crystals, the G domains are thus trapped in an open state that does not allow tight binding of AlFx, into the γ-phosphate binding site.

Superposition of the five available homodimer structures (CtMnmE·GDP, NoMnmE·GDP, CtMnmE·GppCp dimers of molecules A and B, T. maritima MnmE dimer model, generated with pdb 1XZP) reveals that the N-terminal domains align quite well and only minor displacements are present for the helical domains (Figure 2A; Table 1). Strikingly, the superposition shows large rotational and translational displacements of the G domains (Figures 2A–2C), which are reflected in their higher root mean square deviation (RMSD) values (Table 1) leading to separation of nucleotide binding sites between, for instance, CtMnmE·GDP and NoMnmE·GDP by 18 Å (Cα-Cα distance of the first P-loop glycines) (Figure 2A). This becomes clearly visible in the displacement of the G domain β-sheets and of helix Gα6 (Figure 2B and 2C). A video generated from the five homodimer structures makes the drastic displacements of the G domains evident and highlights the dynamic character of the G domains.

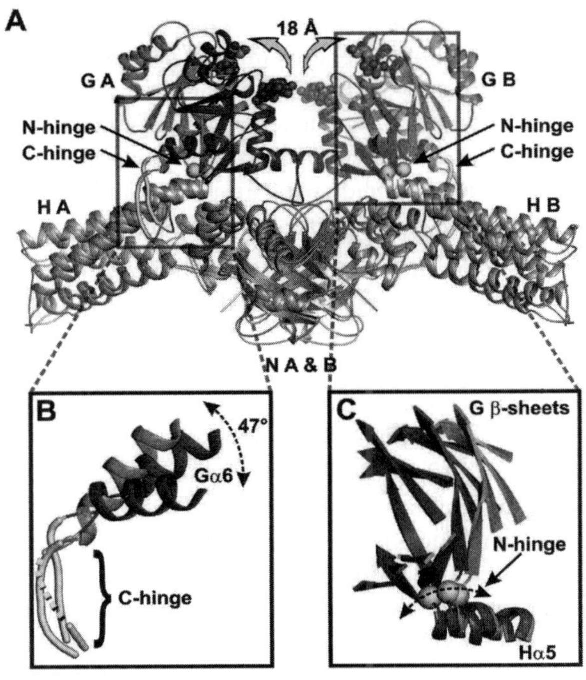

Figure 2. Orientations of the G domain.
(A) Superimposition of CtMnmE·GDP (green) and NoMnmE•GDP (blue) dimers (displayed as ribbon models) via the N-terminal and helical domains with domains labeled as in Figure 1, highlighting the G domains, the relative movements of the nucleotides (displayed as spherical models), and the N- (yellow spheres) and C-hinge (yellow tubes), shown in detail in (B, C). (B, C) Superimposition of the C-hinge (B) and the N-hinge adjacent to helix Hα5 of the H domain (C) of the Ct and NoMnmE structures (coloring as in [A]) together with the corresponding parts of CtMnmE·GppCp, (chain A, red), highlighting the relative movements of the last helix of the G domain, Gα5 (B), and of the G domain β-sheet (C). The part of the C-hinge in CtMnmE·GppCp not resolved in the X-ray structure is depicted as dashed yellow line (B).

Table 1. Average RMSD of each N-terminal domain, helical domain, and G domain from a superposition of the five MnmE structures (CtMnmE·GDP, CtMnmE·GppCp A and B, T. maritima MnmE [pdb 1XZP], NoMnmE·GDP) with NoMnmE·GDP as reference structure.

Average RMSD/Å to NoMnmE·GPD			
Domain	**N-Terminal Domain**	**Helical Domain**	**G Domain**
CtMnmE·GDP	1.37	1.79	9.47
CtMnmE·GppCp, chain A	1.04	1.64	7.62
CtMnmE·GppCp, chain B	1.28	2.31	9.84
T. maritima MnmE	1.59	2.61	6.53

The different orientations indicate that the G domains are highly flexible with regard to the rest of the protein probably due to the rather loose connections between G and helical domains. A conserved glycine residue is situated between helix Hα5 and the first strand of the G domain β-sheet (Figure 2C), which because of its higher conformational freedom could function as a hinge ("N-hinge"). A second hinge point ("C-hinge") is where a not-well-ordered loop attaches the C-terminal end of the G domain after Gα6 to the helical domain (Figure 2B). The angle by which Gα6 is shifted spans up to 47°. In the crystal structure of CtMnmE·GppCp this loop region is not resolved underlining its high flexibility.

Although crystals grew under many more conditions, crystals diffracting to reasonable resolution were only obtained in the cases reported here. This result is most likely due to the fact that in these cases, crystal contacts trap the G domains in defined orientations, whereas in the weakly diffracting crystals the G domains are only loosely packed causing lattice disorder. The G domains in the CtMnmE·GDP and NoMnmE·GDP structures pack against symmetry mates with contact areas of 376 Å2 and 488 Å2. In NoMnmE·GDP a Zn2+-ion tightly links the G domain to symmetry mates, while in CtMnmE·GDP the G domains fix each other by a toothing upside-down arrangement. Crystal contacts of G domains A and B in the CtMnmE·GppCp structure comprise areas of 845 Å2 and 987 Å2, respectively. Docking the G domain of molecule B (or A) into the asymmetric unit of the CtMnmE·GppCp structure to the position expected for the G domain of molecule C would create a much smaller hypothetical crystal contact area of only 18 Å2 (or 131 Å2). Thus we would expect that the G domain of molecule C is present in the crystal but, due to its high mobility and absence of sufficient crystal contacts, is not visible in the electron density map. This is similar to the recent structure of the Roco protein, which is also a GAD protein. There, the second G domain of the constitutive dimer is present in the crystal but can not be identified in the electron density map [35].

G Domain Mobility Measured by DEER

To test whether the "open" G domain arrangement found in our GDP- and GppCp-bound structures is representative for the conformation in solution and to identify and characterize the putative transition state with closed G domains, which could not be obtained by crystallization, we applied four-pulse DEER spectroscopy [36]–[38], to measure distances between nitroxide spin labels in the G domains of full-length EcMnmE in different steps of the GTPase cycle. Positions mutated to cysteine for spin labeling with (1-oxyl-2,2,5,5-tetramethyl-3-pyrroline-3-methyl) methanethiosulfonate spin label (MTSSL) are Glu287, close to the top of the G domain in Gα2, Ser278 in switch II, and Asp366, located in Gα6, and, as shown in Figure 3, result in the introduction of two symmetry-related spin labels in the functional MnmE dimer. As a possible "negative control" we also spin labeled position Ile105 in the N-terminal domain, for which no distance changes are expected. The Cβ-Cβ distances between these sites derived from the structures of the open and the model of closed state are listed in Table 2. To avoid unwanted side effects of cysteine substitutions, only nonconserved, surface-exposed residues have been selected. Furthermore, mutant proteins were assayed for K+-stimulated GTPase activity with and without attached MTSSL-label. No impairment of GTPase activity in comparison to wild type could be observed by the mutation itself or the introduction of the spin label. Since efficient GTPase activity in the presence of K+ is strictly dependent on correct K+-binding and G domain dimerization [22], we can conclude that the structural and functional aspects of G domain dimerization and GTPase activity of the mutants are preserved in the proteins used for DEER.

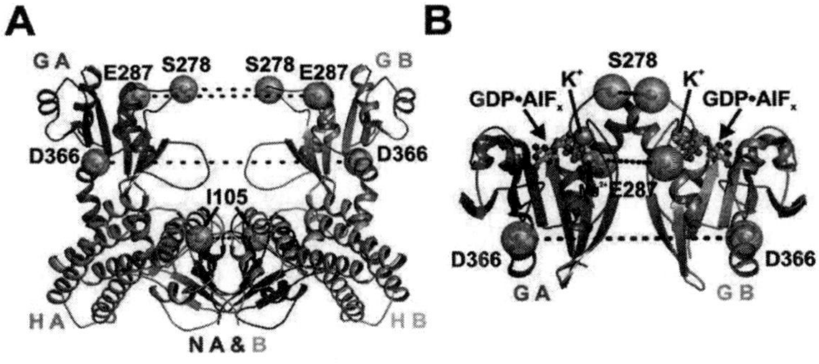

Figure 3. Spin label sites in the MnmE dimer.
Position of residues that were mutated to Cys and spin labeled (yellow spheres), with dashed lines indicating distances between residues in the open (A) and closed (B) conformation. Domains are labeled as in Figure 1. In (B) GDP-AlFx is displayed as stick model, Mg2+ as grey sphere and K+ as blue spheres. Cβ-Cβ distances were calculated from the respective residues for the open conformation represented from the model in Figure 1a (generated with pdb 1XZP) and the closed conformation obtained from the structures of the G domain in the GDP-AlFx state (pdb 2GJ8), as summarized in Table 2.

Table 2. Cβ-Cβ distances between pair of residues mutated to Cys for MTSSL labeling measured in various MnmE dimer crystal structures and maxima in distance distributions for the pair of spin labels from experimentally determined DEER distance distributions.

Residue[a] Mutated to Cys	**Nucleotide State**	**Cβ-Cβ Distance from X-ray Structures/Å**	**Maximum in DEER Distance Distribution/Å[b]**
E287R1, Gα2	apo	53[c]	55
	GDP	—	53
	GppNHp	—	37, 55
	GDP-AlF$_x$	28[d]	36
E366R1, Gα6	apo	62[c]	67
	GDP	**57**[e], 63[f]	65
	GppNHp	49[g], 53[h]	47, 63
	GDP-AlF$_x$	47[d]	**48**, 58
S278R1, switch II	apo	22[c]	25–50 (**46**)
	GDP	—	25–50 (**47**)
	GppNHp	—	27, **43**
	GDP-AlF$_x$	18[d]	28
I105R1, N-terminal domain	apo	37[c]	29
	GDP	36[e], 37[f]	—
	GppNHp	36[g], 36[h]	—
	GDP-AlF$_x$	—	29

Note that not all residues selected for spin labeling are resolved in all X-ray structures.
[a]Numbering according to *E. coli* MnmE sequence.
[b]Major maxima are highlighted in bold.
[c]*T. maritima* homodimer model (generated with pdb 1XZP).
[d]From *E. coli* G domain dimer (pdb 2GJ8).
[e]From CtMnmE·GDP.
[f]From NoMnmE·GDP.
[g]From CtMnmE·GppCp, dimer A.
[h]From CtMnmE·GppCp, dimer B.

Nucleotide Free and GDP-Bound State

Figure 4A illustrates the results of the DEER measurements in the presence of 100 mM KCl, where the left panel shows the background-corrected dipolar evolution data, the centre panel the respective dipolar spectra, and the right panel the corresponding distance distributions (obtained by Tikhonov regularization; see Methods), which are summarized in Table 2. The DEER analysis of mutant E287R1 (R1 denotes the MTSSL side chain), close to the top of the G domain in Gα2, indicates one major peak centered at a distance of 55 Å for the apo- and 53 Å for the GDP-bound state. This distances correspond well to the Cβ-Cβ distances in the TmMnmE crystal structure model of 53 Å (the corresponding residues in the CtMnmE and NoMnmE structures are not resolved) and is therefore in agreement with an open conformation of the G domains. For D366R1 (situated at Gα6), a well-defined interspin distance distribution centered at 67 Å in the apo- state and 65 Å in the GDP-bound state could be observed in good agreement with the distances obtained from the TmMnmE dimer model (62 Å) and NoMnmE·GDP (63 Å), suggesting again an open conformation of the G domains. The corresponding Cβ-Cβ distance in CtMnmE·GDP dimer is somewhat shorter (57 Å), which is due to the different orientation of G domains in this structure (Figure 2A) and to the different tilting of Gα6 (Figure 2B). From the E287R1 and D366R1 data in the apo- and GDP-bound states, we conclude that instead of a continuum of freely moving orientations, the MnmE G domains seem to have defined major orientation reflected by the distance distributions.

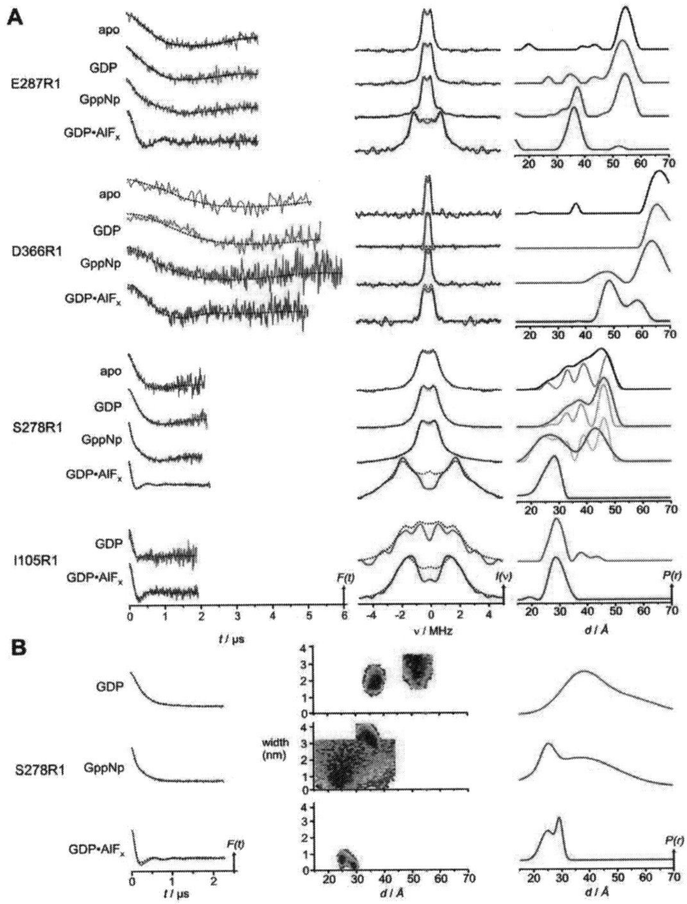

Figure 4. DEER characterization of nucleotide-dependent domain movements of MTSSL labeled MnmE (E287R1, D366R1, S278R1, and I105R1).
(A) Left panel, background corrected dipolar evolution data for the apo, GDP, GppNHp, and GDP-AlFx state of the respective MnmE mutants as indicated. Centre panel: dipolar spectra (Fourier transformation of the dipolar evolution data in the left panel). Right column: distance distributions obtained by Tikhonov regularization. All plots are normalized by amplitude. Broken lines in the left and center panel are fits to the data obtained by Tikhonov regularization. For S278R1 in apo, GDP, and GDP-AlFx state, alternative fits and resulting distance distributions obtained with smaller regularization parameters α, are shown in corresponding pale colours. (B) Data for S278R1 in the GDP-, GppNHp-, and GDP-AlFx state analyzed assuming a sum of Gaussian distributed conformers. Left panel: background corrected dipolar evolution data. Centre panel: goodness-of-fit (χ2) surfaces, created by random sampling of distance and width for each Gaussian population in the distance distributions shown in the right panel. Plots in the left and right column are normalized by amplitude. Broken lines in the left panel are fits to the data.

In contrast, the analysis for S278R1 (switch II region) by Tikhonov regularization did not allow discrimination between a continuum of distances ranging from 25 Å to 50 Å with increasing probabilities for larger distances (shown in dark colours) or three to four distinct distances corresponding to different protein and/or spin label conformers (shown in pale colours). To clarify this issue, we additionally fitted the GDP data with a Monte Carlo/SIMPLEX algorithm assuming a sum of Gaussian-distributed conformers contributing to the dipolar evolution data (Figure 4B) [41]. The experimental data were satisfactorily reproduced by a distance distribution with two Gaussian populations, which are well defined as judged by the $\chi 2$ surfaces, summing up to a broad distribution in the range 30–50 Å. Possible explanations for such a continuum in the distance distribution could be (i) that the labeled position is located in the switch II region, which is flexible in the free and GDP-bound states, in line with the X-ray results, or (ii) that the spin label side chains are not restricted in their conformational space and populate multiple rotamers, or (iii) a combination of (i) and (ii). A general continuum of G domain orientations can be excluded from the results for positions E287R1 and D366R1. Control measurements of K+-stimulated GTPase activity make severe structural perturbations appear unlikely. Instead the deviation from the Cβ-Cβ distance of 22 Å in the TmMnmE dimer model is probably due to a switch II conformation induced by crystal packing forces. It has been observed before, that even in structures of the same G protein-nucleotide complex different switch II conformations were induced by crystal packing forces [42]. Nevertheless, the most pronounced distances between 40–50 Å as well as the minor fractions situated between 30 and 40 Å observed by DEER are in strong agreement with an open state of the G domains as observed in the apo- and GDP-bound crystal structures.

GppNHp-Bound State

In the presence of the nonhydrolizable GTP analogue guanosine 5′-imidotriphosphate (GppNHp) the distance distributions comprise two fractions with different interspin distances for all three labeled positions. One larger distance (E287R1, 55 Å; D366R1, 63 Å; and S278R1, 43 Å) corresponds to the open state of the G domains as observed for the nucleotide-free and GDP-bound forms, whereas the other distance, contributing about 30% to the distance distribution (average value calculated from the area under the distance distribution curve) is characterized by significantly shorter distances (E287R1, 37 Å; D366R1, 47 Å; and S278R1, 27 Å), clearly indicating the presence of a second conformation, where the two G domains are in close proximity. As for the GDP-bound state, the GppNHp data for S278R1 were additionally fitted assuming a sum of

Gaussian distributions. Despite differences especially in the distribution width for the two populations, this approach also reveals the presence of the two conformations of the G domains. In the X-ray structure of the AlFx-complexed G domain dimer, the Cβ-Cβ distance of the S278- and E287-pair are 18 Å and 28 Å, respectively and thus somewhat shorter as compared to the GppNHp DEER data (S278R1, 27 Å; E287R1, 37 Å). However the MTSSL-side chain itself has an average length of 7 Å between the nitroxyl-radical and the Cβ-atom [43]. This can increase the measured distance up to 14 Å for a pair of MTSSL side chains, depending on their rotamer orientation. The longer distances of the short distance maxima in the GppNHp-distance distributions of S278R1 and E287R1 measured in solution are thus most likely the result of a closed conformation of G domains, where the MTSSL side chains protrude away from the symmetry axis of the G domain dimer. Overall, the GppNHp measurements lead us to conclude that in the presence of GppNHp two conformations are in thermal equilibrium. In the crystal structure of GppCp-bound MnmE the G domains are found in the open state, indicating that this equilibrium is shifted towards the open state under the crystallization conditions.

GDP-AlFx–Bound State

In the presence of the transition state mimic GDP-AlFx, S278R1 and E287R1 show a single population maximum, with defined distances of 28 Å and 36 Å, respectively, in line with a closed conformation (Figure 4). The observed distances are close to the observed Cβ-Cβ distances in the crystal structure of the GDP-AlFx–bound G domain dimer structure (S278, 18 Å; E287 28 Å), with deviations due to spin label conformations as discussed above. Compared to the distances characterizing the closed conformation in the presence of GppNHp, the distance distributions for the transition state mimic are sharper and the maxima are slightly shifted. For E287R1 it decreases by about 1–2 Å and for position S278R1 the broad distribution between 20 and 30 Å converts to a more defined but asymmetric distribution with a major distance of 28 Å, which is well reproduced also by the Monte Carlo approach (Figure 4B). For position D366R1 two major fractions with inter spin distances of 58 and 48 Å are visible, presumably due to two different rotamer populations of the spin label side chain. The maximum at 48 Å corresponds nicely to the Cβ-Cβ distance in the GDP-AlFx–bound G domain structure, whereas the 58Å distance likely represents an MTSSL-rotamer population pointing away from each other. As is obvious from the distance distributions for the GppNHp and the GDP-AlFx state, the closed state in the presence of GDP-AlFx slightly differs from that in the presence of GppNHp suggesting that on the reaction pathway from the triphosphate state to the GTPase competent conformation further rearrangements in the active site of the G domains take place. Overall the distance maxima are shifted to shorter distances in the GDP-AlFx state as compared to the apo-, GDP- and GppNHp distances. This shows that the G domains adapt a closed conformation as observed in the GDP-AlFx-complexed G domain structure.

Position Ile105 in the N-Terminal Domain

To explore whether G domain dimerisation leads to domain rearrangements in the N-terminal dimerization domain, a spin label was introduced at position Ile105 (Figure 3A). A comparison of the distance distributions obtained for the GDP state (open conformation) and GDP-AlFx state (closed conformation) does not show any significant differences concerning the major population in the distance distribution with an average distance of 29 Å for both nucleotide states (Figure 4A; Table 2), indicating, that closing of the G domains does not significantly disturb the overall integrity of the N-terminal domains. The deviation to the corresponding Cβ-Cβ distances in the various dimer models (36 Å, 37 Å) are likely due to spin label rotamer conformations.

Cation Dependence of G Domain Dimerization

Previous studies have shown K+ ions to activate the MnmE GTPase. This follows from the finding that dimerization of the MnmE G domains and GDP-AlFx complex formation strictly require K+, which is bound in the dimer interface (Figure 3B), such that its position overlaps with that of an Arg finger required for the GAP-mediated GTP hydrolysis on Ras-like G proteins [20],[22]. Moreover, GTPase activity and AlFx-induced dimerization are at least partially stimulated by cations with an ionic radius comparable to K+ (1.38 Å) such as Rb+ (1.52 Å) and, to a lesser extent, NH4+ (1.44 Å), whereas Na+ (0.99 Å) and Cs+ (1.67 Å) do not show this effect [22]. Consistent with this, Rb+ and NH4+ were also found to be coordinated to the K+ binding site in two MnmE G domain dimer structures GDP complexed with AlFx (pdb 2GJ9 and 2GJA) [22]. To analyze the cation dependency of G domain dimerization in full-length protein in solution, we determined distance distributions for the sites S278R1 and E287R1 in the apo, GDP, GDP-AlFx, and GppNHp bound state in the presence of various cations, i.e., Na+, K+, Rb+, Cs+, and for S278R1 additionally in the presence of NH4+ for the GDP and GDP-AlFx state (Figure 5).

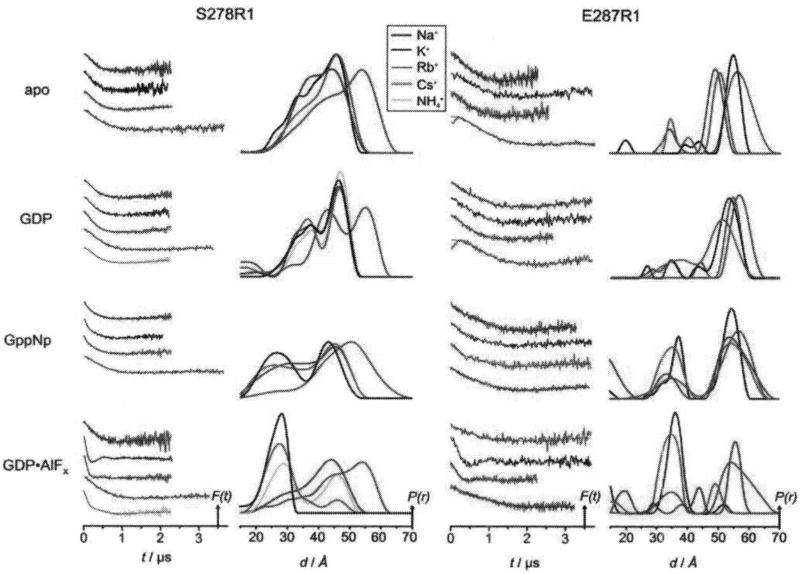

Figure 5. Cation dependency of DEER distance distributions, for MnmE mutants S278R1 (left) and E287R1 (right).
For each mutant, the left column shows the background corrected dipolar evolution data and the fit obtained by Tikhonov regularization (broken line) and the right column the corresponding distance distribution. The evolution data and the respective distance distributions are colored according to the cation present in the experiment (red, Na+; black, K+; blue, Rb+; green, Cs+; and pale green, NH4+ [only for S278R1, GDP, and GDP-AlFx]). The area under the distance distribution corresponds to the number of interacting spins, derived from the modulation amplitude of the background corrected dipolar evolution data.

The distance distributions and dipolar time traces show that in the presence of GDP-AlFx only K+ is capable for shifting the equilibrium completely towards the closed G domain dimer. The ability of the respective cations to stabilize G domain dimerization follows the order K+>Rb+>NH4+>Cs+≈Na+, clearly correlated with their ionic radii and their ability to stimulate GTP hydrolysis [22]. In the presence of GppNHp, we observe the same order of cations with regard to their capability for shifting the equilibrium towards the closed state. Notably, Cs+, which is completely unable to stabilize G domain dimerization, seems to have an influence on switch II conformational dynamics and on the overall orientation of the G domains, as seen from the significantly broadened and shifted distance distributions compared to those for the other cations.

Discussion

Understanding how GADs use the GTPase cycle as the driving force to perform a variety of functions like insertion of signal sequences into the ER translocon by the SRP/SR system [44], tRNA modification by MnmE [19],[22], kinase activation by the Parkinson kinase LRRK2 [45], or metal ion delivery to hydrogenases [46] is a crucial step for elucidating the diverse mechanism by which these proteins operate. Although within this class of proteins MnmE is one of the structurally and biochemically best characterized and a model for the GTPase cycle dependent G domain dimerization has been proposed [22], neither the structural model of the full-length MnmE dimer nor dimerization of the G domains in the full-length dimeric protein have been proven directly.

Here we have applied a combined approach of X-ray crystallography and pulse electron paramagnetic resonance (EPR) spectroscopy to study the behavior of the G domains in full-length MnmE in different steps of the GTPase cycle. We were able to solve the first X-ray structures of full-length MnmE in complex with nucleotides. The structures confirm the previously postulated homodimer model [19] according to which MnmE constitutively dimerizes via its N-terminal domain whereas the G domains, separated by a large distance of approximately 48 Å (measured from Cα of the first glycine of the P-loop), face each other with their nucleotide binding sites. The distance distributions obtained by DEER of MnmE in the apo, GDP, and GppNHp state reveal that the G domains are far apart also in solution excluding that the open conformations in the crystal structures are crystallographic artefacts. Comparison of the different full-length structures reveals that the G domains are present in drastically different orientations suggesting them to be highly mobile elements capable of moving independently with regard to the other domains. As judged from the X-ray structure, they need to overcome a 20–30-Å distance gap on formation of the GDP-AlFx complex [22]. In contrast to the X-ray data, the DEER distance distributions suggest the presence of one defined orientation for the open state in solution, arguing that the different G domain orientations in the X-ray structures result from crystal packing forces. That reasonable diffraction data can only be obtained when the G domains are stabilized by packing interaction is a further indication for their high mobility. Moreover, for CtMnmE we find different orientations between the GDP- and GppCp-bound structures and even between different molecules in one asymmetric unit of CtMnmE·GppCp.

Although the crystals for all structures presented here were grown in the presence of K+ and triphosphate or a transition state mimic to induce G domain dimerization, the structures show the G domains in an open state, suggesting that the closed state is not stable under crystallization conditions. We can demonstrate indeed using a fluorometric assay with mant-GDP, that close juxtaposition of G domains with AlFx is inhibited in the presence of crystallization precipitants.

The interspin distances between the spin labeled G domains obtained by DEER directly prove for the first time that the G domains contact each other in the presence of triphosphate or transition state analogs. A notable feature of the GppNHp-bound state is the coexistence of an open and closed state, pointing out that a triphosphate analog is not sufficient to fully stabilize the closed state. However, recently a stabilizing effect of GidA on the closed state of the G domains was shown, indicating that regulation of the MnmE G domain cycle is coupled to other components of the tRNA-modification system [6]. Unlike the results from X-ray structures, the EPR data, under low salt and in the absence of PEG, do not show a continuum of conformations but rather particular conformations in the open and closed state not observable in the X-ray experiment. We further show that only the presence of GDP, AlFx and K+ is capable of stabilizing the closed state, and that this effect is specific, since the effect is absent with Na+ and Cs+ and is smaller with similar size cations such as Rb+ and NH4+.

In summary, we were able to directly demonstrate the conformational changes MnmE was suggested to undergo during its GTPase cycle [6],[19],[22],[39],[47]. Dimerization of the MnmE G domains is accompanied by large domain movements of up to 20 Å from the open to the closed state, which is an apparently unique feature of MnmE with regard to other GADs, suggestive for a functional or regulatory coupling of these domain movements to the tRNA-modification reaction. For the architecturally similar Roc-COR tandem (see above), the G domains in the nucleotide free state are already in close proximity [35], rendering similar extensive domain rearrangements unlikely. Yet such drastic rearrangements are not untypical for NTPases, as for example Hsp90, which constitutively dimerizes via its C-terminal domain, undergoes dramatic domain movements during its ATPase cycle involving juxtaposition of its N-domains in the triphosphate state [48],[49].

MnmE forms a heterotetrameric complex with GidA [50], which is stabilized in the triphosphate state [6],[39], and tRNA modification was suggested to be exerted by this complex rather than by the individual proteins [5],[50], which was recently proven by an in vitro modification assay [6]. Furthermore active GTP-turnover rather than simple GTP-binding was shown to be essential for the modification reaction [6],[47] and in particular, nucleotide dependent G domain dimerization is tightly coupled to the tRNA-modification process both in vitro and in vivo [6]. According to a proposed reaction mechanism, the reaction itself does not require energy, but rather comprises several steps at presumably different, spatially separated active sites, requiring tight regulation [19],[39]. We thus speculate that G domain dimerization during GTP hydrolysis is required for orchestration of the multistep tRNA-modification reaction [6]. The exact link between G domain dimerization, GTP hydrolysis, conformational changes, and tRNA modification is focus of current investigations.

Materials and Methods

Proteins

C. tepidum and Nostoc sp. 7120 MnmE (CtMnmE, NoMnmE) were cloned into pET14b (Novagene) and expressed as N-terminal His-tagged proteins in Escherichia coli BL21-DE3. Cells were lysed in 50 mM Tris (pH 7.5), 100 mM NaCl, 5 mM MgCl2 (= buffer A) with 20 mM imidazole, 5 mM β-mercaptoethanol, 150 μM PMSF, and the proteins were purified by Ni-NTA, thrombin-cleavage of the His-Tag, and gel filtration on Superdex 200 in buffer A with 5 mM dithioerythritol (DTE). Cloning, expression, and purification of E. coli MnmE and mutants and preparation of nucleotide-free MnmE was carried out as described elsewhere [39].

Crystallography

Crystals were obtained by hanging-drop vapour diffusion. For CtMnmE·GDP crystals, 1 μl each of 50 mg/ml protein in 50 mM Tris (pH 7.5), 100 mM KCl, 5 mM MgCl2, 5 mM DTE (buffer B) plus 5 mM GDP, 5 mM AlCl3, 50 mM NaF, and precipitant (100 mM Tris-HCl [pH 8.5], 2.250 M NaCl, 15% [w/v] PEG 6000) were mixed. After 3 d the reservoir was changed to 100 mM Tris-HCl (pH 8.5), 2.250 M NaCl, 30% PEG 6000, and equilibrated for 2 more days. Crystals were soaked with precipitant supplemented with 12% glycerol and 5 mM 5-F-THF for 30 min and flash-frozen in liquid nitrogen. For NoMnmE·GDP, 1 μl of 20 mg/ml protein in buffer B with 5 mM GDP, 5 mM AlCl3, 50 mM NaF, and precipitant (100 mM Tris [pH 7.5], 22% [w/v] PEG 550 MME, 10 mM ZnSO4) were mixed and grown at 20°C. After 2 d crystals were cryo-dipped into reservoir solution with 28% (w/v) PEG 550 MME and flash-frozen into liquid nitrogen. For CtMnmE·GppCp, 40 mg/ml nucleotide free protein in buffer B with 5 mM GppCp was mixed (1:1) with 100 mM MES, 46 mM NaOH, 12% PEG 4000, 40 mM NaCl, and crystals were grown at 20°C. After 2–3 d, crystals were flash-frozen in reservoir containing 20% glycerol. All datasets were collected at 100 K on beamline PX2 (SLS, Villingen) at wavelengths of 0.98003, 0.9796, and 1.28186 Å (Zn2+-edge) for CtMnmE·GDP, CtMnmE·GppCp, and NoMnmE·GDP, respectively. All datasets were processed, indexed, and scaled with XDS [51].

Initial phases were obtained by molecular replacement with the N-terminal and the helical domain of T. maritima MnmE (pdb 1XZP) with MOLREP [52]. Coot [53] and REFMAC [54],[55] were used for model building and translation, libration, screw rotation (TLS)-refinement including NCS restraints and NCS-averaged maps in the case of CtMnmE·GppCp. Crystallographic simulated annealing of models was carried out with CNS [56]. Structural representations were prepared with pymol (www.pymol.org). For NoMnmE·GDP, Zn2+ atom positions were located by their anomalous signal. For CtMnmE·GppCp, a positive peak in the FO-FC-map close to the β- and γ-phosphate in the nucleotide binding site of G domain A was assigned to Mg2+, on the basis of its position at the usual Mg2+-site in G protein structures. Structures were analyzed by PROCHECK [57] revealing for all three structures 100% of torsion angles within the allowed Ramachandran regions. Data collection and refinement statistics are listed in Table 3. Structures were aligned with coot [53] and Superpose of the CCP4-package [54]. Crystal contact areas were calculated using the PROTORP server [58].

Table 3. Data collection and refinement statistics.

Name	CtMnmE·GDP[a]	CtMnmE·GppCp[a]	NoMnmE·GDP[a]
Pdb-code	3GEE	3GEI	3GEH
Data collection			
Dataset type	Native	Native	Native
Space group	I4(1)22	C222(1)	P4(3)2(1)2
Cell dimensions			
A, b, c (Å)	130.804, 130.804, 200.611	139.882, 224.572, 156.788	124.279, 124.279, 174.701
α, β, γ (°)	90.0, 90.0, 90.0	90.0, 90.0, 90.0	90.0, 90.0, 90.0
Resolution (Å)	20.00–2.95 (3.00–2.95)	20.00–3.40 (3.42–3.40)	20.00–3.20 (3.30–3.20)
R_{merge}	7.7 (71.7)	12.9 (65.4)	12.7 (48.5)
I/σ	24.49 (2.39)	12.13 (2.02)	8.61 (2.18)
Completeness (%)	98.9 (99.7)	99.2 (100.0)	99.4 (99.9)
Redundancy	7.16 (7.28)	7.45 (7.61)	7.78 (8.03)
Refinement			
Resolution (Å)	19.90–2.95 (3.03–2.95)	19.94–3.40 (3.49–3.40)	20.00–3.20 (3.28–3.20)
n Reflections	17,528	32,320	21,913
R_{work}/R_{free}	0.23/0.27	0.24/0.27	0.24/0.27
n Atoms	3,321	8,780	3,428
Protein	3,259	8,715	3,364
Ligand/Ion	62	65	64
B-factors (Å²)	106.21	123.56	52.69
Protein	105.96	124.35	51.85
Ligand/Ion	115.84	164.78	97.10
Root mean square deviations			
Bond lengths (Å)	0.006	0.008	0.007
Bond angles (°)	1.147	1.198	1.185

Values in parentheses are for the highest-resolution shell.
[a]Data from one crystal.

Fluorometric Detection of AlFx-Complex Formation

10 μM of nucleotide-free E. coli MnmE loaded with 0.5 μM of mGDP were incubated in 50 mM TriS-HCl (pH 7.5), 100 mM KCl (or NaCl), 5 mM MgCl2, 10 mM NaF with or without the precipitants 15% PEG 6000, 2,250 mM NaCl, or both or 22% PEG 550 MME at 20°C. The fluorescence of mGDP bound to

MnmE, excited at 366 nm and detected at 450 nm, was monitored over time in a Fluoromax 2 spectralfluorimeter (Spex Industries). To initiate AlFx-complex formation, 1 mM AlCl3 was added and the fluorescence was continuously monitored. For analysis, fluorescence amplitudes were normalized to the amplitude before addition of AlCl3.

Spin Labeling

Purified, nucleotide-free Cys-mutants of E. coli MnmE-C451S (Table 2) were pretreated with DTE (4°C). After removal of DTE protein solutions were incubated with 1–5 mM MTSSL (Toronto Research, Alexis) for 16 h (4°C). Excess MTSSL was removed by gel filtration. Labeling efficiencies have been determined to be >80% in all cases.

Steady State GTPase Measurements

GTPase reactions were started by adding 0.5 µM of wild type or mutant MTSSL-labeled or nonlabelled MnmE protein to 186 µM of GTP in 50 mM Tris-HCl (pH 7.5), 100 mM KCl, 5 mM MgCl2, and performed at 20°C. At time points 0, 1, 2, 3, 5, 7, and 10 min aliquots were taken and analyzed for their nucleotide content by HPLC as described elsewhere [22]. For comparison, vapp was determined as the absolute value of the slope of a linear fit of GTP consumption over time, normalized to the total amount of enzyme for a range in which 10% of initial GTP was consumed.

Pulse EPR Measurements

Pulse EPR experiments (DEER) were accomplished at X-band frequencies (9.3–9.4 GHz) with a Bruker Elexsys 580 spectrometer equipped with a Bruker Flexline split-ring resonator ER 4118X-MS3 and a continuous flow helium cryostat (ESR900, Oxford Instruments) controlled by an Oxford Intelligent temperature controller ITC 503S. Buffer conditions for the EPR experiments were 200–500 µM protein in 100 mM KCl (or NaCl, RbCl, CsCl, NH4Cl), 50 mM Tris-HCl, 5 mM MgCl2 (pH 7.4) with 5% (v/v) ethylene glycol (for H2O buffer) or 12.5% (v/v) glycerol-d8 (for D2O buffer), and 1 mM GDP, 1 mM GppNHp or 1 mM GDP, 1 mM AlCl3, 10 mM NaF, respectively.

All measurements were performed using the four-pulse DEER sequence: $\pi/2(\nu_{obs})-\tau_1-\pi(\nu_{obs})-t'-\pi(\nu_{pump})-(\tau_1+\tau_2-t')-\pi(\nu_{obs})-\tau_2-echo$ [59]. A two-step phase cycling (+ ⟨x⟩, - ⟨x⟩) was performed on . Time t′ is varied, whereas τ1 and τ2 are kept constant, and the dipolar evolution time is given by . Data were analyzed only for t>0. The resonator was overcoupled to Q~100; the pump frequency ʋpump was set to the center of the resonator dip and coincided with the maximum of the nitroxide EPR spectrum, whereas the observer frequency ʋobs was 65 MHz higher, coinciding with the low field local maximum of the spectrum. All measurements were performed at a temperature of 50 K with observer pulse lengths of 16 ns for $\pi/2$ and 32 ns for π pulses and a pump pulse length of 12 ns. Proton modulation was averaged by adding traces at eight different $\tau 1$ values, starting at and incrementing by . For proteins in D2O buffer with deuterated glycerol used for their effect on the phase relaxation, corresponding values were and . Data points were collected in 8-ns time steps or, if the absence of fractions in the distance distribution below an appropriate threshold was checked experimentally, in 16- or 32-ns time steps. The total measurement time for each sample was 4–24 h. Analysis of the data was performed with DeerAnalysis2006.1/2008 [60]. Additionally, the data was fitted assuming a sum of Gaussian distributed conformers utilizing the program DEFit 3.9 [41], which employs a Monte Carlo/SIMPLEX algorithm to find a distance distribution to which the corresponding dipolar evolution function represents the best fit to the experimental data.

Accession Codes

Protein Data Bank (PDB) (http://www.rcsb.org/pdb): Coordinates und structure factors have been deposited with accession codes 3GEE (CtMnmE·GDP), 3GEI (CtMnmE·GppCp), and 3GEH (NoMnmE·GDP).

Acknowledgements

SM thanks I. Vetter and A. Scrima for valuable advice on the structural refinement, W. Versées for meaningful discussions, and the X-ray community of the Max-Planck-Institute of Molecular Physiology for data collection. We gratefully acknowledge P. Stege and C. Koerner for excellent technical assistance.

Authors' Contributions

The author(s) have made the following declarations about their contributions: Conceived and designed the experiments: JPK AW. Performed the experiments: SM SB AK JPK. Analyzed the data: SM SB JPK AW. Contributed reagents/materials/analysis tools: HJS. Wrote the paper: SM SB HJS JPK AW. Performed the protein preparations and biochemical experiments, performed the crystallization experiments, X-ray data analysis, and solving of the X-ray structures: SM.

Performed and analyzed the EPR experiments: SB JPK. Designed the mutants: SM AK JPK. Provided the EPR equipment: H-JS. Performed the protein preparations and biochemical experiments: AK.

References

1. Iwata-Reuyl D (2008) An embarrassment of riches: the enzymology of RNA modification. Curr Opin Chem Biol 12: 126–133.
2. Sprinzl M, Vassilenko K. S (2005) Compilation of tRNA sequences and sequences of tRNA genes. Nucleic Acids Res 33: D139–D140.
3. Agris P. F, Vendeix F. A, Graham W. D (2007) tRNA's wobble decoding of the genome: 40 years of modification. J Mol Biol 366: 1–13.
4. Elseviers D, Petrullo L. A, Gallagher P. J (1984) Novel E. coli mutants deficient in biosynthesis of 5-methylaminomethyl-2-thiouridine. Nucleic Acids Res 12: 3521–3534.
5. Bregeon D, Colot V, Radman M, Taddei F (2001) Translational misreading: a tRNA modification counteracts a +2 ribosomal frameshift. Genes Dev 15: 2295–2306.
6. Meyer S, Wittinghofer A, Versees W (2009) G-domain dimerization orchestrates the tRNA wobble modification reaction in the MnmE/GidA complex. J Mol Biol 392: 910–922.
7. Yarian C, Townsend H, Czestkowski W, Sochacka E, Malkiewicz A. J, et al. (2002) Accurate translation of the genetic code depends on tRNA modified nucleosides. J Biol Chem 277: 16391–16395.
8. Sakamoto K, Kawai G, Watanabe S, Niimi T, Hayashi N, et al. (1996) NMR studies of the effects of the 5′-phosphate group on conformational properties of 5-methylaminomethyluridine found in the first position of the anticodon of Escherichia coli tRNA(Arg)4. Biochemistry 35: 6533–6538.
9. Agris P. F, Sierzputowskagracz H, Smith W, Malkiewicz A, Sochacka E, et al. (1992) Thiolation of uridine carbon-2 restricts the motional dynamics of the transfer-rna wobble position nucleoside. J Am Chem Soc 114: 2652–2656.
10. Yokoyama S, Nishimura S (1995) Modified nucleosides and codon recognition. In: Söll D, RajBhandary U. L, editors. tRNA: structure, biosynthesis and function. Washington (D.C.): American Society for Microbiology. pp. 207–223.
11. Gustilo E. M, Vendeix F. A, Agris P. F (2008) tRNA's modifications bring order to gene expression. Curr Opin Microbiol 11: 134–140.

12. Decoster E, Vassal A, Faye G (1993) Mss1, a nuclear-encoded mitochondrial gtpase involved in the expression of cox1 subunit of cytochrome-c-oxidase. J Mol Biol 232: 79–88.

13. Colby G, Wu M, Tzagoloff A (1998) MTO1 codes for a mitochondrial protein required for respiration in paromomycin-resistant mutants of Saccharomyces cerevisiae. J Biol Chem 273: 27945–27952.

14. Li X, Guan M. X (2002) A human mitochondrial GTP binding protein related to tRNA modification may modulate phenotypic expression of the deafness-associated mitochondrial 12S rRNA mutation. Mol Cell Biol 22: 7701–7711.

15. Li X, Li R, Lin X, Guan M. X (2002) Isolation and characterization of the putative nuclear modifier gene MTO1 involved in the pathogenesis of deafness-associated mitochondrial 12 S rRNA A1555G mutation. J Biol Chem 277: 27256–27264.

16. Suzuki T, Suzuki T, Wada T, Saigo K, Watanabe K (2002) Taurine as a constituent of mitochondrial tRNAs: new insights into the functions of taurine and human mitochondrial diseases. EMBO J 21: 6581–6589.

17. Villarroya M, Prado S, Esteve J. M, Soriano M. A, Aguado C, et al. (2008) Characterization of human GTPBP3, a GTP-binding protein involved in mitochondrial tRNA modification. Mol Cell Biol 28: 7514–7531.

18. Bykhovskaya Y, Mengesha E, Wang D, Yang H, Estivill X, et al. (2004) Phenotype of non-syndromic deafness associated with the mitochondrial A1555G mutation is modulated by mitochondrial RNA modifying enzymes MTO1 and GTPBP3. Mol Genet Metab 83: 199–206.

19. Scrima A, Vetter I. R, Armengod M. E, Wittinghofer A (2005) The structure of the TrmE GTP-binding protein and its implications for tRNA modification. EMBO J 24: 23–33.

20. Vetter I. R, Wittinghofer A (2001) The guanine nucleotide-binding switch in three dimensions. Science 294: 1299–1304.

21. Bos J. L, Rehmann H, Wittinghofer A (2007) GEFs and GAPs: critical elements in the control of small G proteins. Cell 129: 865–877.

22. Scrima A, Wittinghofer A (2006) Dimerisation-dependent GTPase reaction of MnmE: how potassium acts as GTPase-activating element. EMBO J 25: 2940–2951.

23. Cabedo H, Macian F, Villarroya M, Escudero J. C, Martinez-Vicente M, et al. (1999) The Escherichia coli trmE (mnmE) gene, involved in tRNA modification, codes for an evolutionarily conserved GTPase with unusual biochemical properties. EMBO J 18: 7063–7076.

24. Yamanaka K, Hwang J, Inouye M (2000) Characterization of GTPase activity of TrmE, a member of a novel GTPase superfamily, from Thermotoga maritima. J Bacteriol 182: 7078–7082.
25. Wittinghofer A (1997) Signaling mechanistics: aluminum fluoride for molecule of the year. Curr Biol 7: R682–R685.
26. Gasper R, Meyer S, Gotthardt K, Sirajuddin M, Wittinghofer A (2009) It takes two to tango: regulation of G proteins by dimerization. Nat Rev Mol Cell Biol 10: 423–429.
27. Focia P. J, Shepotinovskaya I. V, Seidler J. A, Freymann D. M (2004) Heterodimeric GTPase core of the SRP targeting complex. Science 303: 373–377.
28. Egea P. F, Shan S. O, Napetschnig J, Savage D. F, Walter P, et al. (2004) Substrate twinning activates the signal recognition particle and its receptor. Nature 427: 215–221.
29. Gasper R, Scrima A, Wittinghofer A (2006) Structural insights into HypB, a GTP-binding protein that regulates metal binding. J Biol Chem 281: 27492–27502.
30. Praefcke G. J, McMahon H. T (2004) The dynamin superfamily: universal membrane tubulation and fission molecules? Nat Rev Mol Cell Biol 5: 133–147.
31. Ghosh A, Praefcke G. J, Renault L, Wittinghofer A, Herrmann C (2006) How guanylate-binding proteins achieve assembly-stimulated processive cleavage of GTP to GMP. Nature 440: 101–104.
32. Koenig P, Oreb M, Hofle A, Kaltofen S, Rippe K, et al. (2008) The GTPase cycle of the chloroplast import receptors Toc33/Toc34: implications from monomeric and dimeric structures. Structure 16: 585–596.
33. Koenig P, Oreb M, Rippe K, Muhle-Goll C, Sinning I, et al. (2008) On the significance of Toc-GTPase homodimers. J Biol Chem 283: 23104–23112.
34. Sirajuddin M, Farkasovsky M, Hauer F, Kuhlmann D, Macara I. G, et al. (2007) Structural insight into filament formation by mammalian septins. Nature 449: 311–315.
35. Gotthardt K, Weyand M, Kortholt A, Van Haastert P. J, Wittinghofer A (2008) Structure of the Roc-COR domain tandem of C. tepidum, a prokaryotic homologue of the human LRRK2 Parkinson kinase. EMBO J 27: 2239–2249.
36. Steinhoff H. J (2004) Inter- and intra-molecular distances determined by EPR spectroscopy and site-directed spin labeling reveal protein-protein and protein-oligonucleotide interaction. Biol Chem 385: 913–920.

37. Jeschke G, Polyhach Y (2007) Distance measurements on spin-labelled biomacromolecules by pulsed electron paramagnetic resonance. Phys Chem Chem Phys 9: 1895–1910.

38. Schiemann O, Prisner T. F (2007) Long-range distance determinations in biomacromolecules by EPR spectroscopy. Q Rev Biophys 40: 1–53.

39. Meyer S, Scrima A, Versees W, Wittinghofer A (2008) Crystal structures of the conserved tRNA-modifying enzyme GidA: implications for its interaction with MnmE and substrate. J Mol Biol 380: 532–547.

40. Prehna G, Stebbins C. E (2007) A Rac1-GDP trimer complex binds zinc with tetrahedral and octahedral coordination, displacing magnesium. Acta Crystallogr D Biol Crystallogr 63: 628–635.

41. Sen K. I, Logan T. M, Fajer P. G (2007) Protein dynamics and monomer-monomer interactions in AntR activation by electron paramagnetic resonance and double electron-electron resonance. Biochemistry 46: 11639–11649.

42. Sprang S. R (1997) G protein mechanisms: insights from structural analysis. Annu Rev Biochem 66: 639–678.

43. Rabenstein M. D, Shin Y. K (1995) Determination of the distance between two spin labels attached to a macromolecule. Proc Natl Acad Sci USA 92: 8239–8243.

44. Egea P. F, Stroud R. M, Walter P (2005) Targeting proteins to membranes: structure of the signal recognition particle. Curr Opin Struct Biol 15: 213–220.

45. Lewis P. A (2009) The function of ROCO proteins in health and disease. Biol Cell 101: 183–191.

46. Maier T, Lottspeich F, Bock A (1995) GTP hydrolysis by HypB is essential for nickel insertion into hydrogenases of Escherichia coli. Eur J Biochem 230: 133–138.

47. Yim L, Martinez-Vicente M, Villarroya M, Aguado C, Knecht E, et al. (2003) The GTPase activity and C-terminal cysteine of the Escherichia coli MnmE protein are essential for its tRNA modifying function. J Biol Chem 278: 28378–28387.

48. Pearl L. H, Prodromou C (2006) Structure and mechanism of the Hsp90 molecular chaperone machinery. Annu Rev Biochem 75: 271–294.

49. Wandinger S. K, Richter K, Buchner J (2008) The Hsp90 chaperone machinery. J Biol Chem 283: 18473–18477.

50. Yim L, Moukadiri I, Bjork G. R, Armengod M. E (2006) Further insights into the tRNA modification process controlled by proteins MnmE and GidA of Escherichia coli. Nucleic Acids Res 34: 5892–5905.

51. Kabsch W (1993) Automatic processing of rotation diffraction data from crystals of initially unknown symmetry and cell constants. J Appl Crystallogr 26: 795–800.

52. Vagin A, Teplyakov A (1997) MOLREP: An automated program for molecular replacement. J Appl Crystallogr 30: 1022–1025.

53. Emsley P, Cowtan K (2004) Coot: model-building tools for molecular graphics. Acta Crystallogr D Biol Crystallogr 60: 2126–2132.

54. Bailey S (1994) The Ccp4 Suite - programs for protein crystallography. Acta Crystallogr D Biol Crystallogr 50: 760–763.

55. Murshudov G. N, Vagin A. A, Dodson E. J (1997) Refinement of macromolecular structures by the maximum-likelihood method. Acta Crystallogr D Biol Crystallogr 53: 240–255.

56. Brunger A. T, Adams P. D, Clore G. M, DeLano W. L, Gros P, et al. (1998) Crystallography & NMR system: a new software suite for macromolecular structure determination. Acta Crystallogr D Biol Crystallogr 54: 905–921.

57. Laskowski R. A, Macarthur M. W, Moss D. S, Thornton J. M (1993) Procheck - a program to check the stereochemical quality of protein structures. J Appl Crystallogr 26: 283–291.

58. Reynolds C, Damerell D, Jones S (2009) ProtorP: a protein-protein interaction analysis server. Bioinformatics 25: 413–414.

59. Pannier M, Veit S, Godt A, Jeschke G, Spiess H. W (2000) Dead-time free measurement of dipole-dipole interactions between electron spins. J Magn Reson 142: 331–340.

60. Jeschke G, Chechik V, Ionita P, Godt A, Zimmermann H, et al. (2006) DeerAnalysis2006 - a comprehensive software package for analyzing pulsed ELDOR data. Appl Magn Reson 30: 473–498.

In-Cell Biochemistry Using NMR Spectroscopy

David S. Burz and Alexander Shekhtman

ABSTRACT

Biochemistry and structural biology are undergoing a dramatic revolution. Until now, mostly in vitro techniques have been used to study subtle and complex biological processes under conditions usually remote from those existing in the cell. We developed a novel in-cell methodology to post-translationally modify interactor proteins and identify the amino acids that comprise the interaction surface of a target protein when bound to the post-translationally modified interactors. Modifying the interactor proteins causes structural changes that manifest themselves on the interacting surface of the target protein and these changes are monitored using in-cell NMR. We show how Ubiquitin interacts with phosphorylated and non-phosphorylated components of the receptor tyrosine kinase (RTK) endocytic sorting machinery: STAM2 (Signal-transducing adaptor molecule), Hrs (Hepatocyte growth factor regulated substrate) and the STAM2-Hrs heterodimer. Ubiquitin binding mediates the processivity of a large network of interactions required for proper functioning of the RTK sorting machinery. The results are consistent with a weakening of the network of interactions when the interactor proteins are phosphorylated. The methodology can be applied to any stable target molecule and may be extended to include other post-translational modifications such as ubiquitination or sumoylation, thus providing a long-awaited leap to high resolution in cell biochemistry.

Introduction

The ultimate goal of all structural and biochemical research is to understand how macromolecular interactions give rise to and regulate biological activity within a natural environment, i.e. in living cells. The challenge is formidable due to the complexity of living matter and the relative scarcity of appropriate in vivo methods. The specific interactions between proteins that participate in signal transduction processes are further complicated by post-translational modifications (PTM) of protein structure. PTM is a mechanism for regulating cellular processes; such modifications can alter the strength or number of interactions in which these proteins engage and/or redirect them to subcellular compartments as required for proper functioning [1]. For example, endocytosis of receptor tyrosine kinases (RTKs) requires tyrosine phosphorylation and monoubiquitination of the receptor and downstream components to sort endocytosed cargo for subsequent degradation (down regulation) or recycling to the cell surface [2], [3], [4].

To examine the effect of PTM on protein-protein interactions that occur along this signaling pathway, we developed an in-cell biochemical methodology that, in combination with recently developed STINT-NMR (Structural Interactions using NMR spectroscopy) [5], allows us to directly observe the structural changes in the interaction surface of a target protein that result from phosphorylating interactor proteins. Since protein phosphorylation does not occur in E. coli, we can use these cells as a test tube for performing in-cell biochemistry, introducing the target protein, interactor proteins and kinase activity on separately inducible plasmids (Fig. 1ab). By employing a target protein that does not interact with endogenous bacterial proteins, the structural details of the specific interaction between these molecules can be observed in the absence and presence of post-translational modification within a cellular environment.

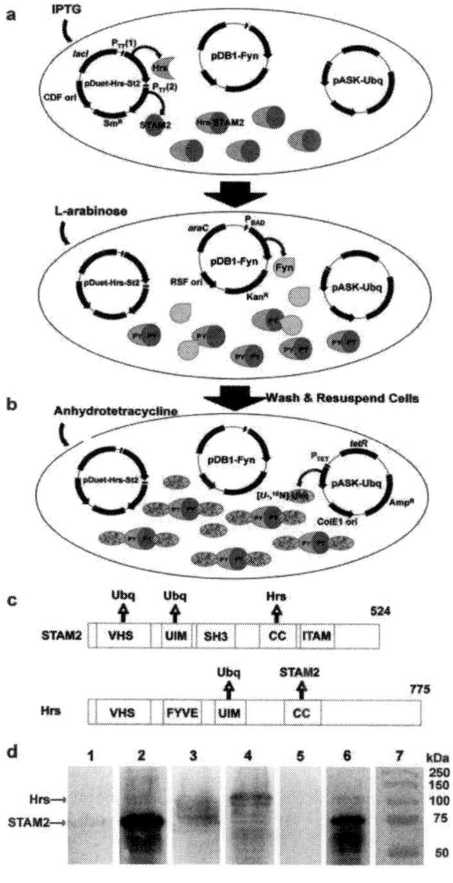

Figure 1. Sequential overexpression and in-cell post-translational modification of interacting proteins for STINT-NMR.

E. coli are transformed with up to three compatible plasmids and grown overnight in LB-glucose medium containing antibiotics. The cells are washed and resuspended in label-free medium. a) IPTG is used to induce overexpression of interactor proteins, STAM2 and Hrs, which form heterodimers. L-arabinose is then used to induce overexpression of Fyn kinase, which phosphorylates tyrosine residues (PY) on the interactors. b) The cells are washed and resuspended in labeling medium. Anhydrotetracycline is used to induce overexpression of uniformly labeled [U-15N] Ubiquitin target protein, which binds to the interactors. Samples are taken as the concentration of target increases. Changes in the target protein structure are monitored using in-cell NMR spectroscopy. A sample of labeled target containing no interactor is prepared separately as a reference. N.b. The experiment can be performed using a single interactor protein and without post-translational modification. The protocol can also be reversed, overexpressing labeled target first, followed by interactor(s) and post-translational modification (see Materials and Methods). c) Domain structure of STAM2 and Hrs. VHS (Vps27-Hrs-Stam domain); UIM (ubiquitin interacting motif); SH3 (src homology domain 3); CC (coiled coil domain); FYVE (FYVE-finger domain); up arrows indicate Ubiquitin, STAM2 or Hrs binding domain. d) Western blot of overexpressed interactor proteins from whole cells in the absence and presence of overexpressed Fyn kinase, probed with anti-tyrosine phosphate HRP-conjugate antibody. Lane 1: STAM2; lane 2: STAM2 & Fyn kinase; lane 3: Hrs; lane 4: Hrs & Fyn kinase; lane 5: STAM2 & Hrs; lane 6: STAM2, Hrs & Fyn kinase; Lane 7: MW.

STINT-NMR is an in-cell technique[6], [7] for examining the structural changes in a target protein resulting from protein-protein interactions [5], [8]. Protein overexpression is first induced in uniformly labeled medium [U, ^{15}N-] to produce a target protein containing NMR-active nuclei. The cells are washed and resuspended in non-labeling medium to induce overexpression of the interactor protein(s). We then use ^{15}N-edited heteronuclear single quantum coherence (^{1}H{^{15}N}-HSQC) NMR experiments to monitor changes in the chemical shifts of target backbone amide nuclei as the concentration of the interactor(s) increases. In cases where the target binds to a high molecular weight interactor protein, we monitor changes in the NMR spectrum of the free target since only this species gives rise to visible peaks. Depending on the chemical exchange rate between the free and bound states of the target, affected NMR peaks, corresponding to backbone amides, can either shift, broaden their line shape or disappear completely, thereby delineating the intermolecular interaction surface between the target and the interactor(s). By changing the isotopic composition on induction, we selectively label the target while the interactor proteins remain cryptic, thus reducing NMR spectral complexity. The resulting NMR data provide a complete titration of the interaction and identify the amino acids that comprise the interaction surface of a target protein. It is important to note that in-cell titrations lack the precision of binding isotherms that are obtained in vitro because of the variable levels of protein expression that are inherent when using living cell, and thus, are largely qualitative. However, in each case, the same structural endpoint is attained. Tight temporal control over protein expression allows us to perform in-cell biochemistry, such as phosphorylation, by expressing a kinase domain capable of post-translationally modifying the interactor protein, off an inducible plasmid (Fig. 1ab).

To demonstrate the efficacy of this methodology, we examined changes in the binding surface of a Ubiquitin (Ubq) target that result from phosphorylating the endocytic proteins, STAM2 (Signal-transducing adaptor molecule) and Hrs (Hepatocyte growth factor-regulated tyrosine kinase substrate), both of which form homodimers with molecular weights of ~115 kDa and ~150 kDa, respectively. The STAM2-Hrs heterodimer (MW ~132 kDa) directs trafficking of endocytosed, monoubiquitinated RTKs to the cell surface for recycling or to lysosomes for degradation. Several lines of evidence, mostly from the study of yeast membrane proteins, suggest that receptor sorting through endocytosis and subsequent degradation is controlled by ubiquitination of both the internalizing receptors and components of the endocytic machinery [9].

In addition to monoubiquitination, endocytic proteins undergo tyrosine phosphorylation, in some cases mediated by Src-family kinases [10], in response to cytokines or growth factor binding to RTKs. Indeed, phosphorylation of tyrosine residues on STAM2 and Hrs occurs during endocytosis. These two post-translational modifications seem to be linked since, in at least some cases, tyrosine phosphorylation is monoubiquitination-dependent [11]. The precise biological role of phosphorylation is unclear but, at least for Hrs, it was shown to be important in the cellular localization of the protein [10]. The molecular machinery that sorts endocytosed membrane proteins into the degradative pathway and away from the default recycling pathway is an area of intense research. This is because down regulation of receptor signaling provides a means to attenuate proliferation and minimize cell growth. The loss of control of down regulation can lead to rampant growth characteristic of most cancers.

STAM2 and Hrs have a modular domain architecture that allows them to participate in multiple protein-protein interactions and to bind the surrounding lipid bilayer[12] (Fig. 1c). Free Ubiquitin and monoubiquitinated proteins bind to the Ubiquitin interacting motif (UIM) present on STAM2 and Hrs [13], and to the N-terminal VHS domain of STAM2 [14]. STAM2 and Hrs are also monoubiquitinated, possibly at multiple sites, in a signal-dependent manner that requires the function of the UIM. Thus, UIM-containing monoubiquitinated protein may further assemble other monoubiquitinated and UIM-containing proteins in receptor-bound macromolecular complexes that contribute to the events leading to intracellular signal transduction and to receptor sorting in endocytic vesicles.

The main objectives of the experiments presented here are to demonstrate that we can regulate the post-translational modification of overexpressed proteins in bacterial cells, a process we dub in-cell biochemistry, and to identify structural changes in protein interaction surfaces due to presence of PTMs using in-cell NMR spectroscopy. These objectives were accomplished by successfully tyrosine phosphorylating STAM2 and Hrs (Fig. 1d), both of which undergo this post-translational modification to function in the endocytotic pathway, and by identifying changes in the interaction surfaces of Ubiquitin when bound to these post-translationally modified proteins by using STINT-NMR.

Results

In-Cell Spectrum of Ubiquitin Is Distinct from that Obtained In Vitro

There are noticeable differences between the solution and in-cell NMR spectra of free Ubiquitin, with only 86% of the Ubiquitin peaks observed in solution found within 0.1 ppm of those observed in cells [15]. To rule out the possibility that the NMR spectrum of Ubiquitin is due to extracellular protein [16], [17], after obtaining the in-cell NMR spectrum, the cells were centrifuged and the supernatant was examined. No NMR spectrum of Ubiquitin was observed above noise level, implying that there is no leakage or cell lysis occurring during the time it takes to acquire the NMR spectrum.

Binding Non-Phosphorylated Interactor Proteins Creates Unique Contact Surfaces on Ubiquitin

To dissect changes in the interaction surface of Ubiquitin we created complexes of [U-, 15N]-Ubiquitin-STAM2, [U-, 15N]-Ubiquitin-Hrs, and [U-, 15N]-Ubiquitin-STAM2-Hrs (Fig. 2). To study Ubiquitin binding to STAM2, we induced

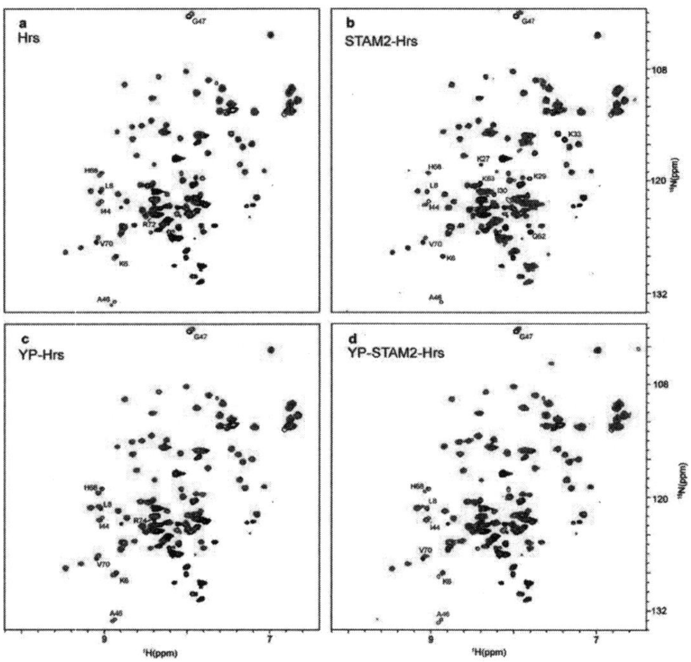

Figure 2. NMR-spectra of Ubiquitin-ligand complexes.
1H{15N}HSQC spectra of E. coli after 3-h of [15N]-Ubiquitin overexpression (black), overlaid with spectra (red) obtained from E. coli after 2-h of [15N]-Ubiquitin overexpression and: a) 4-h of Hrs overexpression; b) 4-h of STAM2 & Hrs co-overexpression; c) 4-h of Hrs and 2-h of Fyn kinase co-overexpression; d) 4-h of Hrs & STAM2 and 2-h of Fyn kinase co-overexpression. Individual peaks exhibiting either a chemical shift change >0.1 ppm or significant differential broadening (>30% change in intensity) are labeled with corresponding assignments. The strong peaks in the spectra between 8.5 and 7.8 ppm correspond to various metabolites of [U-15N] ammonium ion. NMR experiments were acquired at T = 298 K on Bruker Avance 700 MHz NMR spectrometer equipped with a cryoprobe. 1H{15N}-edited HSQC data were recorded with 16 transients as 512{128} complex points, apodized with a squared cosine-bell window function and zero-filled to 1k{256} points prior to Fourier transformation. The corresponding sweep widths were 12 and 35 ppm in the 1H and 15N dimensions, respectively. The Q49 peak is obscured by peaks from the [U-15N] ammonium ion metabolites and is not labeled. Ubiquitin ligands are indicated in each panel.

overexpression of [U-, 15N]-Ubiquitin prior to or following 3 or 4 hours of STAM2 overexpression. Over this time, the concentration of STAM2 was determined to range from 50–500 μM in these cells. As the concentration of STAM2 increased, we observed consistent broadening of selected Ubiquitin peaks in the NMR spectrum[5]. The NMR solution structure of Ubiquitin, which consists of 76 amino acids, is well-known [15]. Mapping the differentially broadened peaks onto the three-dimensional structure of Ubiquitin results in two distinct surfaces that define the interface between Ubiquitin and STAM2[5] (Fig. 3a). The chemical shift changes of Ubiquitin that we observed affect only surface residues, implying that there are no conformational changes accompanying complex formation. The first interface corresponds to the UIM binding surface and consists of residues K6, L8, I44, A46, G47, H68 and V70; the second interface corresponds to the binding surface for the VHS domain of STAM2 and consists of K11, I13, K27, K29, K33 and K63. High affinity binding by STAM2 to Ubiquitin requires both the UIM and the VHS [14]. Since the overall binding affinity of Ubiquitin for STAM2 is ~10 μM versus ~100 μM for the UIM alone [5], we conclude that the second interacting surface is responsible for increasing the overall affinity of the interaction.

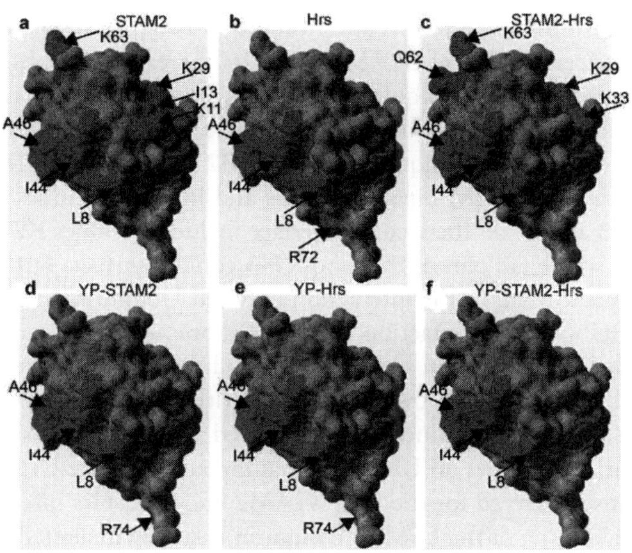

Figure 3. Interaction surface maps of Ubiquitin-ligand complexes.
Interaction surface of Ubiquitin mapped onto the three-dimensional structure of Ubiquitin (PDB code 1D3Z). Individual residues exhibiting either a chemical shift change >0.05 ppm or significant differential broadening are indicated in red. All perturbed residues lie on the Ubiquitin surface and, therefore, reflect changes in the interaction surface of the molecule rather than changes in tertiary or quaternary structure. a) STAM2-Ubq interaction; b) Hrs-Ubq interaction; c) STAM2-Hrs-Ubq interaction; d) phosphorylated STAM2-Ubq interaction (YP-STAM2); e) phosphorylated Hrs-Ubq interaction (YP-Hrs); f) phosphorylated STAM2-Hrs-Ubq interaction (YP-STAM2-Hrs). Ubiquitin ligands are indicated in each panel.

To study Ubiquitin binding to Hrs, we induced overexpression of [U-, 15N]-Ubiquitin prior to or following 3 or 4 hours of Hrs overexpression. Over this time, the concentration of Hrs was determined to range from 50–360 μM in these cells. As the concentration of Hrs increased, we observed changes in the chemical shifts of selected Ubiquitin peaks (Fig. 2a). Mapping the chemically-shifted peaks onto the three-dimensional structure of Ubiquitin results in a distinct interface between Hrs and Ubiquitin (Fig. 3b) corresponding to the UIM binding surface and consisting of residues K6, L8, I44, A46, G47, H68 and V70. In addition, R72, located near the C-terminus, was affected, resulting in an extension of the UIM binding patch. The reported binding affinity of Ubiquitin for Hrs is ~200 μM; this is slightly weaker than the binding of an isolated UIM to Ubiquitin [18]. The UIM of Hrs is a variant of the conventional UIM helix, dubbed a double UIM (DUIM), and is capable of binding two Ubiquitin molecules [19]. The small changes in the NMR spectra observed for the interaction between the UIM of STAM2 and the DUIM of Hrs may result from an altered structure for the DUIM.

To study Ubiquitin binding to the STAM2-Hrs heterodimer, we induced overexpression of [U-, 15N]-Ubiquitin prior to or following 3 or 4 hours of STAM2 and Hrs co-overexpression. The concentrations of STAM2 and Hrs were determined to range from 50 to 280 μM and from 20 to 100 μM, respectively, in these cells. As the concentrations of STAM2 and Hrs increased, we observed consistent broadening of selected Ubiquitin peaks (Fig. 2b). Mapping the differentially broadened peaks onto the three-dimensional structure of Ubiquitin results in two distinct interfaces between Ubiquitin and STAM2-Hrs (Fig. 3c). The first interface corresponds to the UIM binding surface and includes residues K6, L8, I44, A46, G47, H68 and V70. The second interface includes residues K27, K29, K33, Q62 and K63, which are part of the Ubq-VHS contact surface, but not K11 and I13, which are perturbed in the interaction between Ubiquitin and STAM2. The change in the Ubq-VHS interaction surface is subtle, truncated at one end and slightly extended at the other, in a manner that is characteristic of the ternary complex and may be a result of STAM2 reorienting within the heterodimer relative to STAM2 alone since these residues are not affected when Hrs binds to Ubiquitin. Overall, the surface residues on Ubiquitin that interact with the UIM are virtually the same as those observed for the Ubq-STAM2 and Ubq-Hrs interactions, indicating that the binding of the UIM to Ubiquitin is largely unaltered.

Binding Phosphorylated Interactor Proteins Results in the Loss of Contact Surfaces on Ubiquitin

STAM2 and Hrs were phosphorylated by inducing overexpression of the constitutively active Src-family tyrosine-kinase, Fyn, for the final 2 hours of STAM2,

Hrs, or STAM2-Hrs overexpression. The extent of STAM2 and Hrs phosphorylation were examined by using Western blots (Fig. 1d). Mass spectroscopic analysis of STAM2 revealed that Y291, Y371 and Y374 are phosphorylated by Fyn. A major site of phosphorylation on STAM2 is reported to be Y192 [20], however, the absence of detectable phosphorylation at Y192 likely reflects the inability of Fyn to phosphorylate at that position. This is not unexpected since distinct non-receptor tyrosine kinases couple EGF, HGF and PDGF stimulation with tyrosine phosphorylation of STAM2 and Hrs in vivo [10]. Y371 and Y374 are located in the ITAM (immunoreceptor tyrosine-based activation motif) domain, which contains sites for phosphorylation by Src kinase. The ITAM domain has been identified as necessary for tyrosine phosphorylation of STAM2 by Jak1 [21].

We did not confirm phosphorylation sites in Hrs since these sites have been well-established. The major sites of EGF-dependent phosphorylation on human Hrs are Y329 and Y334 [22]. Tyrosine to phenylalanine mutations at these residues do not affect Ubq-Hrs binding, Hrs ubiquitination, or the activity of Hrs in the sorting mechanism [22]. Furthermore, an intact UIM is required for Hrs phosphorylation [22]. Extensive tyrosine phosphorylation of Hrs was not confirmed (Fig. 1d) suggesting that Fyn kinase is likely not the optimal kinase to phosphorylate Hrs. The extent of tyrosine phosphorylation on Hrs appeared to be enhanced by the presence of Ubiquitin (not shown). Indeed, it was previously noted that Hrs phosphorylation may require direct interaction between Ubiquitinated receptors and Hrs [10].

The STINT-NMR spectrum of Ubiquitin interacting with phosphorylated STAM2 reveals primarily changes in chemical shifts that affect a smaller number of residues than were perturbed in the Ubq-STAM2 interaction, and the loss of the interaction surface attributed to the VHS domain of STAM2 (Fig. 3d). That the NMR spectrum represents an interaction between Ubiquitin and phosphorylated STAM2 is implicit since a population of >70–80% modified interactor is required to generate a unique set of chemical shifts. We estimate that the affinity of Ubiquitin for phosphorylated STAM2 is >100 μM (refer to Materials and Methods), which is comparable to that of the isolated UIM. One additional residue, R74, was perturbed as a result of phosphorylated STAM2 binding to Ubiquitin suggesting a slight rearrangement of the UIM and possibly a change in the overall binding affinity.

To verify that the Ubiquitin residues perturbed when STAM2 is phosphorylated are due to the post-translational modification, we mutated the ITAM tyrosine residues, 371 and 374, to phenylalanines (Y371/4F-STAM2). When this mutant is overexpressed and its interaction with Ubiquitin examined by using STINT-NMR, the resulting spectra and surface interaction maps for both the unphosphorylated and phosphorylated states are largely identical to those obtained using unphosphorylated wild-type STAM2. Small differences between the spectra are likely due to the fact that at least one additional Fyn-dependent tyrosine phosphorylation site (Y291) was still present in the mutant. We conclude that phosphorylating the ITAM tyrosines weakens the binding between Ubiquitin and STAM2, and infer that these residues may be critical for the processivity of the internalizing pathway, attenuating the binding affinity of Ubiquitin for STAM2 in response to post-translational modification.

In spite of the fact that we did not confirm extensive phosphorylation of Hrs by Fyn kinase (Fig. 1d), the NMR spectrum of Ubiquitin interacting with Hrs changes when Fyn kinase is expressed in the same cells providing indirect evidence for Hrs phosphorylation. The STINT-NMR spectrum of Ubiquitin interacting with phosphorylated Hrs (Fig.2c) shows primarily changes in chemical shifts that affect a larger number of residues than were perturbed in the Ubq-Hrs interaction (Fig. 2a). When these changes are mapped to the Ubiquitin surface, K6, L8, I44, A46, G47, Q49, H68, V70 and R74, are seen to participate in the interaction, whereas R72 is no longer part of the interaction surface (Fig. 3e). The change in the chemical shift of R74, a residue that was also implicated in the interaction between Ubiquitin and phosphorylated STAM2, suggests that a similar alteration of the Ubq-UIM interaction occurs when Ubiquitin binds to either phosphorylated molecule. We estimate that the overall binding affinity of Ubiquitin for phosphorylated Hrs is >100 μM, which is comparable to the affinity of Ubiquitin for unphosphorylated Hrs and the isolated UIM. Thus, phosphorylating tyrosines on Hrs does not appreciably alter the binding of Ubiquitin to Hrs.

The STINT-NMR spectrum of Ubiquitin interacting with phosphorylated STAM2-Hrs reveals changes in chemical shifts for a number of residues (Fig. 2d) and the loss of the interacting surface attributed to the VHS domain of STAM2 (Fig. 3f). The residues associated with UIM binding to Ubiquitin (K6, L8, I44, A46, G47, H68 and V70) are perturbed. However, residue R74, which is chemically shifted in the interaction between Ubiquitin and phosphorylated STAM2 or phosphorylated Hrs alone, is not perturbed. Ubiquitin appears to interact with the phosphorylated ternary complex in much the same way that it interacts with phosphorylated STAM2 and phosphorylated Hrs, involving contact with only the UIMs of both interactor proteins and a commensurate weakening of the binding due to the loss of the second interaction surface.

Discussion

There are advantages to using in-cell rather than in vitro NMR experiments to study interacting proteins: 1) the proteolytic machinery in cells is tightly regulated and this regulation is lost in lysates, which can result in proteolysis of the sample; 2) in-cell protein overexpression results in higher local concentrations of interacting partners than in lysates, thus increasing the likelihood of detecting weak interactions; 3) interactions occur within a cellular environment, which may confer biologically relevant structural conformations that cannot be duplicated in vitro.

Hrs and, to a lesser extent, STAM2, are somewhat labile. Therefore working with purified protein and performing in vitro binding assays are not feasible for either of these species. Furthermore, in vitro binding assays will not provide any structural details about the interaction, but will merely provide a more precise estimate of binding affinities. While the affinities that we estimated are consistent with those reported from in vitro studies(Shekhtman, 2002), we are primarily concerned with the interaction surfaces on Ubiquitin and how modulating these surfaces affects the processivity of the internalizing pathway.

Applying in-cell biochemistry using NMR spectroscopy to the ternary Ubiquitin-Hrs-STAM2 complex showed that the interaction surface is significantly modulated by the phosphorylation state of two STAM2 tyrosines, Y371 and Y374, located in the conserved ITAM domain. Since the ITAM domain is in immediate proximity to the STAM2 homo- and hetero-dimerization domain GAT [23], we expect that a change in the phosphorylation state of Y371 and Y374 will cause intermolecular rearrangements of the Ubiquitin-binding domains, VHS and UIM, leading to the observed loss of the VHS binding surface.

Our data cannot distinguish whether Ubiquitin binds separately to the VHS and to the UIM (two Ubiquitins bound per STAM2) or if the two binding motifs interact cooperatively to increase the overall affinity of Ubiquitin over that of the binding to either motif separately. During endocytosis, the bound Ubiquitin may come from the free cellular pool or from monoubiquitinated substrates. The latter form of the ligand may serve to "cross link" the higher order species formed, and may bind with a different affinity from free ubiquitin.

The study of biochemistry inside living cells entails utilizing complementary methods to resolve processes on different scales from micrometers to Angstroms. Light and fluorescence microscopy were successfully applied to study the localization and compartmentalization of macromolecules in vivo on the micrometer scale. Förster resonance energy transfer (FRET) extended our ability to identify protein-protein and protein-nucleic acid interactions on a scale of tens of nanometers and facilitated estimates of binding affinities and stoichiometries [24]. The advent of in-cell biochemistry using STINT-NMR, which allows us to modify and examine protein-protein interaction surfaces at the level of single amino acid residues, has pushed the limits of resolution to the subnanometer scale. Though in its infancy, this technique has the potential to open a window for investigating life processes inside a living cell at a level of detail never seen before.

Materials and Methods

Sequential Over-Expression and Labeling

E.coli strain BL21(DE3) codon+[Novagen] was co-transformed with pASK-Ubq and: pCDF-ST2 (Ubq-STAM2 interaction); or pRSF-Hrs (Ubq-Hrs interaction); or pCDFDuet-Hrs-ST2 (Ubq-STAM2-Hrs interaction); or pCDF-ST2 and pDB1-Fyn (Ubq-phosphorylated STAM2 interaction); or pCDFDuet-Hrs and pDB1-Fyn (Ubq-phosphorylated Hrs interaction); or pCDFDuet-Hrs-ST2 and pDB1-Fyn (Ubq-phosphorylated STAM2-Hrs interaction).

Cells were grown overnight at 37°C to an OD600 of ≥1.6 in Luria-Bertani medium (LB) supplemented with 150 mg/L of carbenicillin for cultures containing pASK-Ubq, 35 mg/L of kanamycin for cultures containing pRSF-Hrs or pDB1-Fyn, and 50 mg/L of streptomycin for cultures containing pCDF-ST2 or pCDFDuet-Hrs or pCDFDuet-Hrs-ST2. Cultures containing pDB1-Fyn were supplemented with 0.2% glucose to suppress fyn transcription from the PBAD promoter.

Two protocols were employed: In the first, the Ubiquitin target was overexpressed and labeled followed by overexpression of the interactor(s) and Fyn kinase; in the second, the interactor(s) and Fyn kinase were overexpressed followed by overexpression and labeling of the Ubiquitin target. The first protocol yielded a high concentration ratio of labeled target to interactor, while the second yielded a low ratio of target to interactor. Both protocols were required to assess the endpoint of the structural transition.

Protocol 1: Expression of [U-15N]-Ubiquitin

Cells from the overnight culture were washed once with minimal medium (M9) salts and re-suspended to an OD600 of ~0.5 in M9 medium containing the appropriate antibiotics, [U-15N] ammonium chloride (0.7 g/L) as the sole nitrogen source and either 0.2% glucose (for cultures containing pDB1-Fyn) or 0.4% glycerol as the sole carbon source. N.b. For all induced cultures we substituted ampicillin (100 mg/L) for carbenicillin. The cells were incubated at 37°C for 10–15 minutes and Ubiquitin overexpression was induced by adding 2 mg/mL anhydrotetracycline in dimethylformamide to a final concentration of 0.2 μg/mL. Ubiquitin overexpression was allowed to proceed for up to 4 hours.

Following the first induction, a 100 mL sample of culture was collected, the cells were centrifuged, washed twice with 50 mL of 10 mM potassium phosphate buffer [pH 7], resuspended with 1 mL 10 mM potassium phosphate buffer [pH 7.0] containing 10% glycerol and stored at –80°C for subsequent NMR analysis.

This control sample was used to assess the extent of overexpression and quality of labeling for a given experiment.

Expression of STAM2, Hrs and Fyn Kinase Domain

Following Ubiquitin overexpression and labeling, the culture was centrifuged and washed once with M9 salts before resuspending a sufficient number of cells to yield an OD600 of ~0.5 in LB medium supplemented with the appropriate antibiotics. The culture was incubated at 37°C for 10–15 minutes and 0.5 M IPTG was added to a final concentration of 0.5 mM to induce individual over-expression of STAM2 or Hrs, or 2 mM to induce simultaneous overexpression of STAM2 and Hrs; induction was allowed to proceed for 3 or 4 hours. In phosphorylation experiments, following 1 or 2 hours of IPTG induction, 20% L-arabinose was added to a final concentration of 0.2%, and overexpression of the Fyn kinase was allowed to proceed for 2 hours. 100 mL samples were taken, centrifuged, washed twice with 10 mM potassium phosphate buffer [pH 7], resuspended with 1 mL 10 mM potassium phosphate buffer [pH 7.0] containing 10% glycerol and stored at –80°C for subsequent NMR analysis. The use of a cryoprotectant is critical to eliminate cell lysis or breakage due to repeated freeze-thawing.

Protocol 2: Expression of STAM2, Hrs and Fyn Kinase Domain

Cells from the overnight culture were washed once with minimal medium (M9) salts and re-suspended to an OD600 of ~0.5 in LB medium supplemented with the appropriate antibiotics. The culture was incubated at 37°C for 10–15 minutes and 0.5 M IPTG was added to a final concentration of 0.5 mM to induce individual over-expression of STAM2 or Hrs, or 2 mM to induce simultaneous over-expression of STAM2 and Hrs; induction was allowed to proceed for 3 or 4 hours. In phosphorylation experiments, following 1 or 2 hours of IPTG induction, 20% L-arabinose was added to a final concentration of 0.2%, and overexpression of the Fyn kinase was allowed to proceed for 2 hours.

Separately, a sufficient volume of overnight culture was centrifuged, washed once with minimal medium (M9) salts and re-suspended to an OD600 of ~0.5 in M9 medium containing the appropriate antibiotics, [U-15N] ammonium chloride (0.7 g/L) as the sole nitrogen source and either 0.2% glucose (for cultures containing pDB1-Fyn) or 0.4% glycerol or as the sole carbon source. The cells were incubated at 37°C for 10–15 minutes and Ubiquitin over-expression was induced by adding 2 mg/mL anhydrotetracycline in dimethylformamide to a final concentration of 0.2 μg/mL. After 4 hours of induction, a 100 mL sample of culture was collected, centrifuged, washed twice with 50 mL of 10 mM potassium phosphate buffer [pH 7], resuspended with 1 mL 10 mM potassium phosphate buffer [pH 7.0] containing 10% glycerol and stored at -80°C for subsequent NMR analysis. This control sample was used to assess the extent of overexpression and quality of labeling for a given experiment.

Expression of [U-15N]-Ubiquitin

Following overexpression of the interactor (with or without post-translational modification), the culture was centrifuged and washed once with M9 salts before resuspending a sufficient number of cells to yield an OD600 of ~0.5 in M9 medium containing the appropriate antibiotics, [U-^{15}N] ammonium chloride (0.7 g/L) as the sole nitrogen source and either 0.2% glucose (for cultures containing pDB1-Fyn.) or 0.4% glycerol as the sole carbon source. The cells were incubated at 37°C for 10–15 minutes and Ubiquitin over-expression was induced by adding 2 mg/mL anhydrotetracycline in dimethylformamide to a final concentration of 0.2 μg/mL. Ubiquitin over-expression was allowed to proceed for 2, 3 or 4 hours. 100 mL samples were taken, centrifuged, washed twice with 10 mM potassium phosphate buffer [pH 7], resuspended with 1 mL 10 mM potassium phosphate buffer [pH 7.0] containing 10% glycerol and stored at -80°C for subsequent NMR analysis. The use of a cryoprotectant is critical to eliminate cell lysis or breakage due to repeated freeze-thawing.

Quantitation of Intracellular STAM2 and Hrs

To estimate the concentration of overexpressed STAM2 and Hrs present in cells, 10% SDS-PAGE was performed. Each gel contained a concentration range of purified STAM2 to generate a standard curve and samples from induced cells for each of the overexpression time points used in the STINT-NMR titration experiments. Gels were stained with SYPRO Ruby Gel Stain (Bio-Rad), which provides a linear fluorescent response over a wide range of protein concentration, and scanned using a Typhoon Trio Variable Mode Imager (Amersham). The scanned gels were analyzed using ImageQuant5.2 software (Molecular Dynamics). Blocked bands were corrected for background and the unknown samples were also corrected by subtracting the fluorescent optical density of a sample that was not induced for protein overexpression. We assumed 4.4×108 cells per mL per optical density unit at 600 nm and a cell volume of 4.2×10-15 L, modeled as a sphere of 2 μm in diameter. The experiments were performed in duplicate.

NMR Spectroscopy

[U-^{15}N] labeled cells were re-suspended in 0.5 mL of NMR buffer (10 mM potassium phosphate, pH 7.0, 90%/10% H_2O/D_2O) and transferred to an NMR tube. To rule out the possibility that the visible NMR spectrum was due to extracellular proteins due to leakage from the cells, we sedimented the cells from the NMR sample and acquired the $^1H\{^{15}N\}$-HSQC spectrum of the resultant supernatant. No protein NMR signal was visible above the noise level. All NMR experiments were performed using a Bruker Avance 700 MHz NMR spectrometer equipped with a cryoprobe. The cryoprobe affords a four-fold increase in sensitivity allowing data collection within ~1 hr for an individual experiment; this is critical to minimize cell leakage [17]. We used a watergate version of the $^1H\{^{15}N\}$-HSQC spectrum. $^1H\{^{15}N\}$-edited HSQC data were recorded with 32 transients as 512{64} complex points, apodized with a squared cosine-bell window function and zero-filled to 1k{128} points prior to Fourier transformation. The corresponding sweep widths were 12 and 35 ppm in the ^1H and ^{15}N dimensions, respectively. Chemical shifts of [U-, ^{15}N]-Ubiquitin inside the cell are slightly different from purified Ubiquitin. We reassigned the backbone chemical shifts of Ubiquitin using the clarified lysate of [U-, ^{13}C, ^{15}N]-Ubiquitin and a standard suite of triple resonance experiments [25]. During in-cell titration experiments, we measured the change in the chemical shifts of amide nitrogens and covalently attached amide protons according to the equation: $\Delta\Omega = \sqrt{\delta_H^2 + (0.25 * \delta_N)^2}$, where $\delta_{H(N)}$ represents a change in hydrogen and nitrogen chemical shifts. Even without quantifying protein concentrations present in the cell, in-cell NMR spectroscopy allows us to make a crude estimate of the protein binding affinities. Depending on the magnitude of the chemical shift change $\Delta\Omega$ and the rate constant, k_{off}, between bound and free states, chemical exchange can result in gradual changes of chemical shifts when $\Delta\Omega << k_{off}$ (fast exchange), line broadening when $\Delta\Omega \leq k_{off}$ (intermediate exchange) or the appearance of new peaks when $\Delta\Omega >> k_{off}$ (slow exchange). Assuming that the binding reaction is diffusion limited and the average change of the chemical shift is ~0.01 ppm, the fast exchange regime will occur when the dissociation constant, K_d, is larger than 100 μM and intermediate or slow exchange will occur when the dissociation constant is less than or equal to 10 μM.

Acknowledgements

Mass spectroscopy analysis was performed by University at Albany Proteomics Facility.

Authors' Contributions

Conceived and designed the experiments: AS. Performed the experiments: DB. Analyzed the data: AS DB. Wrote the paper: AS DB.

References

1. Seet BT, Dikic I, Zhou MM, Pawson T (2006) Reading protein modifications with interaction domains. Nat Rev Mol Cell Biol 7: 473–483.
2. Haglund K, Di Fiore PP, Dikic I (2003) Distinct monoubiquitin signals in receptor endocytosis. Trends Biochem Sci 28: 598–603.
3. Clague MJ, Urbé S (2001) The interface of receptor trafficking and signalling. J Cell Sci 114: 3075–3081.
4. Marmor MD, Yarden Y (2004) Role of protein ubiquitylation in regulating endocytosis of receptor tyrosine kinases. Oncogene 23: 2057–2070.
5. Burz DS, Dutta K, Cowburn D, Shekhtman A (2006) Mapping structural interactions using in-cell NMR spectroscopy (STINT-NMR). Nat Methods 3: 91–93.
6. Serber Z, Dötsch V (2001) In-cell NMR spectroscopy. Biochemistry 40: 14317–14323.
7. Selenko P, Serber Z, Gadea B, Ruderman J, Wagner G (2006) Quantitative NMR analysis of the protein G B1 domain in Xenopus laevis egg extracts and intact oocytes. Proc Natl Acad Sci USA 103: 11904–11909.
8. Burz DS, Dutta K, Cowburn D, Shekhtman A (2006) In-cell NMR for protein-protein interactions (STINT-NMR). Nat Protoc 1: 146–152.
9. Hicke L, Riezman H (1996) Ubiquitination of a yeast plasma membrane receptor signals its ligand-stimulated endocytosis. Cell 84: 277–287.
10. Row PE, Clague MJ, Urbé S (2005) Growth factors induce differential phosphorylation profiles of the Hrs-STAM complex: a common node in signalling networks with signal-specific properties. Biochem J 389: 629–636.
11. Abella JV, Peschard P, Naujokas MA, Lin T, Saucier C, et al. (2005) Met/Hepatocyte growth factor receptor ubiquitination suppresses transformation and is required for Hrs phosphorylation. Mol Cell Biol 25: 9632–9645.
12. Komada M, Kitamura N (2005) The Hrs/STAM complex in the downregulation of receptor tyrosine kinases. J Biochem (Tokyo) 137: 1–8.
13. Polo S, Sigismund S, Faretta M, Guidi M, Capua MR, et al. (2002) A single motif responsible for ubiquitin recognition and monoubiquitination in endocytic proteins. Nature 416: 451–455.

14. Mizuno E, Kawahata K, Kato M, Kitamura N, Komada M (2003) STAM proteins bind ubiquitinated proteins on the early endosome via the VHS domain and ubiquitin-interacting motif. Mol Biol Cell 14: 3675–3689.

15. Bax A, Tjandra N (1997) High-resolution heteronuclear NMR of human ubiquitin in an aqueous liquid crystalline medium. J Biomol NMR 10: 289–292.

16. Pielak GJ (2007) Retraction. Biochemistry 46: 8206.

17. Cruzeiro-Silva C, Albernaz FP, Valente AP, Almeida FC (2006) In-Cell NMR spectroscopy: inhibition of autologous protein expression reduces Escherichia coli lysis. Cell Biochem Biophys 44: 497–502.

18. Shekhtman A, Cowburn D (2002) A ubiquitin-interacting motif from Hrs binds to and occludes the ubiquitin surface necessary for polyubiquitination in monoubiquitinated proteins. Biochemical and Biophysical Research Communications 296: 1222–1227.

19. Hirano S, Kawasaki M, Ura H, Kato R, Raiborg C, et al. (2006) Double-sided ubiquitin binding of Hrs-UIM in endosomal protein sorting. Nat Struct Mol Biol 13: 272–277.

20. Steen H, Kuster B, Fernandez M, Pandey A, Mann M (2002) Tyrosine phosphorylation mapping of the epidermal growth factor receptor signaling pathway. J Biol Chem 277: 1031–1039.

21. Pandey A, Fernandez MM, Steen H, Blagoev B, Nielsen MM, et al. (2000) Identification of a novel immunoreceptor tyrosine-based activation motif-containing molecule, STAM2, by mass spectrometry and its involvement in growth factor and cytokine receptor signaling pathways. J Biol Chem 275: 38633–38639.

22. Urbé S, Sachse M, Row PE, Preisinger C, Barr FA, et al. (2003) The UIM domain of Hrs couples receptor sorting to vesicle formation. J Cell Sci 116: 4169–4179.

23. Prag G, Watson H, Kim YC, Beach BM, Ghirlando R, et al. (2007) The Vps27/Hse1 complex is a GAT domain-based scaffold for ubiquitin-dependent sorting. Dev Cell 12: 973–986.

24. Giepmans BN, Adams SR, Ellisman MH, Tsien RY (2006) The fluorescent toolbox for assessing protein location and function. Science 312: 217–224.

25. Etezady-Esfarjani T, Herrmann T, Horst R, Wuthrich K (2006) Automated protein NMR structure determination in crude cell-extract. J Biomol NMR 34: 3–11.

Structure of the Dimeric N-Glycosylated Form of Fungal Β-N-Acetylhexosaminidase Revealed by Computer Modeling, Vibrational Spectroscopy, and Biochemical Studies

Rüdiger Ettrich, Vladimír Kopecký Jr, Kateřina Hofbauerová, Vladimír Baumruk, Petr Novák, Petr Pompach, Petr Man, Ondřej Plíhal, Michal Kutý, Natallia Kulik, Jan Sklenář, Helena Ryšlavá, Vladimír Křen and Karel Bezouška

ABSTRACT

Background

Fungal β-N-acetylhexosaminidases catalyze the hydrolysis of chitobiose into its constituent monosaccharides. These enzymes are physiologically important

during the life cycle of the fungus for the formation of septa, germ tubes and fruit-bodies. Crystal structures are known for two monomeric bacterial enzymes and the dimeric human lysosomal β-N-acetylhexosaminidase. The fungal β-N-acetylhexosaminidases are robust enzymes commonly used in chemoenzymatic syntheses of oligosaccharides. The enzyme from Aspergillus oryzae was purified and its sequence was determined.

Results

The complete primary structure of the fungal β-N-acetylhexosaminidase from Aspergillus oryzae CCF1066 was used to construct molecular models of the catalytic subunit of the enzyme, the enzyme dimer, and the N-glycosylated dimer. Experimental data were obtained from infrared and Raman spectroscopy, and biochemical studies of the native and deglycosylated enzyme, and are in good agreement with the models. Enzyme deglycosylated under native conditions displays identical kinetic parameters but is significantly less stable in acidic conditions, consistent with model predictions. The molecular model of the deglycosylated enzyme was solvated and a molecular dynamics simulation was run over 20 ns. The molecular model is able to bind the natural substrate—chitobiose with a stable value of binding energy during the molecular dynamics simulation.

Conclusion

Whereas the intracellular bacterial β-N-acetylhexosaminidases are monomeric, the extracellular secreted enzymes of fungi and humans occur as dimers. Dimerization of the fungal β-N-acetylhexosaminidase appears to be a reversible process that is strictly pH dependent. Oligosaccharide moieties may also participate in the dimerization process that might represent a unique feature of the exclusively extracellular enzymes. Deglycosylation had only limited effect on enzyme activity, but it significantly affected enzyme stability in acidic conditions. Dimerization and N-glycosylation are the enzyme's strategy for catalytic subunit stabilization. The disulfide bridge that connects Cys448 with Cys483 stabilizes a hinge region in a flexible loop close to the active site, which is an exclusive feature of the fungal enzymes, neither present in bacterial nor mammalian structures. This loop may play the role of a substrate binding site lid, anchored by a disulphide bridge that prevents the substrate binding site from being influenced by the flexible motion of the loop.

Background

Fungal β-N-acetylhexosaminidases catalyze the hydrolysis of chitobiose into its constituent monosaccharides. These enzymes are physiologically important during the life cycle of the fungus for the formation of septa, germ tubes and fruit-bodies [1-3]. These processes are important in control of fungal and insect pests [4] and are relevant to human diseases [5], lending considerable interest in the catalytic mechanism of these enzymes. The enzymes are also used in chemoenzymatic synthesis of biologically interesting oligosaccharides based on their effective transglycosylation of β-GlcNAc and β-GalNAc [6-9].

Crystal structures are known for several β-N-acetylhexosaminidases from the glycohydrolase 20 family including the monomeric bacterial enzymes from Serratia marcescens [10,11] and Streptomyces plicatus [12,13]. The catalytic domain of β-N-acetylhexosaminidase is an α/β TIM-barrel. Crystallization with substrate analogs showed the conserved residues Asp^{539}-Glu^{540} to be close to the binding site and thus predict them to play a key role in chitobiose hydrolysis with Glu^{540} acting as a proton donor to the substrate, while Asp^{539} restrains its acetamido group in a specific orientation by hydrogen bonding with N2 of the nonreducing sugar [11]. β-N-acetylhexosaminidase from Streptomyces plicatus has been co-crystallized with the cyclic intermediate analogue N-acetylglucosamine-thiazoline. The pyranose ring of the analogue is bound in the active site in a conformation close to that of a $^4C_1$ chair. Within the substrate-binding pocket, Tyr^{393} and Asp^{313} appear important for positioning the 2-acetamido group of the substrate for nucleophilic attack at the anomeric center and for dispersing the positive charge distributed into the oxazolinium ring upon cyclization [12]. Experiments with two mutated forms of the enzyme (Asp^{313}Ala and Asp^{313}Asn) provided evidence that Asp^{313} stabilizes the transition states, and assists to correctly orient the 2-acetamido group for catalysis [13]. Recently, the structure of dimeric human lysosomal β-N-acetylhexosaminidase has been solved providing new insight into the mechanism of Sandhoff disease [14]. Most mutations associated with late-onset Sandhoff disease reside near the subunit interface, and are thus proposed to interfere with the correct formation of the enzyme dimer [14].

The fungal β-N-acetylhexosaminidases are robust enzymes commonly used in our laboratories in chemoenzymatic syntheses of oligosaccharides [6,7,9]. We have previously reported the remarkable inducibility of a fungal β-N-acetylhexosaminidase from Aspergillus oryzae by GlcNAc [15]. This enzyme was purified to homogeneity from the culture medium, and its sequence was determined using both direct protein sequencing and DNA sequencing of a genomic clone containing the hexA gene [16]. In the present work, to initiate structural studies of this enzyme we performed sequence alignment and homology modeling, and constructed a molecular model of the enzyme and of its complex with the natural substrate chitobiose and with its non-cleavable analog GlcNAcβ1→4ManNAc [17]. Several experimental approaches provide experimental verification of the overall features suggested by the model. Disulfide bridging, secondary structure, and the mode of subunit assembly were determined experimentally and correlated with the model. We also constructed a model of the N-glycosylated enzyme, and compared it with kinetic properties of the native and deglycosylated enzymes.

Results

Molecular Models of β-N-Acetylhexosaminidase

We have used the primary structure of the fungal β-N-acetylhexosaminidase determined in our laboratory [16] to perform homology modeling using the solved structures of these enzymes from Serratia marcescens [11], Streptomyces plicatus [18] and Homo sapiens [14,19]. Alignment of the amino acid sequences of the fungal enzyme and these three structurally solved enzymes reveals both areas of extensive amino acid similarity and segments that appear unique to the fungal enzyme (Fig. 1). The shown alignment is the result of a combination of a structural alignment of the three crystal structures and the sequence alignment generated with ClustalX, to exclude misalignment in sequence variable regions, as especially the chitobiase from Serratia marcescens has additional domains. Starting from this alignment we created a structural model of the catalytic subunit including the small N-terminal zincin-like domain using a restraint-based comparative modeling approach. In this model we grouped the six cysteins into three pairs according to the closest distance. The three cysteine pairs Cys^{290}-Cys^{351}, Cys^{448}-Cys^{483} and Cys^{583}-Cys^{590} were adjusted into the model by repeating the modeling procedure with the additional restraints as an input for Modeller. The final model had 82.8% of residues in the most favored regions of the Ramachandran plot and an acceptable overall geometry, both determined with the ProCheck program. The overall g-factor of the structure obtained showed a value of -0.22. The g-factor tries to quantify the overall geometry and its value should be above -0.5; values below -1.0 may indicate a wrong structure. With respect to the general shortcomings of homology modelling especially in the loop and sequence variable regions, we decided to solvate the homology structure in SPC water to refine it by 20 ns of molecular dynamics in a NPT ensemble. According to the root mean square deviation of the Cα atoms the structure gets after 7 ns into an equilibrium state oscillating around a fixed value with a deviation of ± 0.3 Å. The resulting catalytic subunit has a kidney-shaped structure with approximate dimensions of 6.8 × 5.8 × 5.6 nm (Fig. 2A). All amino acids involved in catalysis are concentrated in the central TIM barrel, with the catalytic glutamic acid on the upper border (Fig. 2B, center). An overlay with the corresponding amino acids of the three template structures with our final model shows that their position is conserved among the available structures and that the active site in our model structure was stable during the molecular dynamics refinement of the overall structure (Fig. 3).

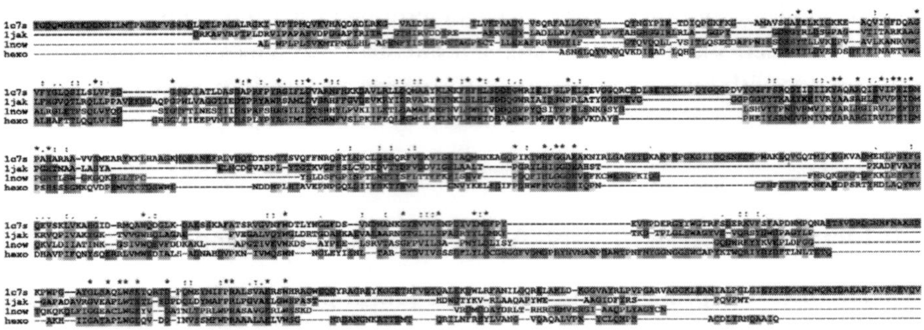

Figure 1. Primary sequence alignment. Alignment of the amino acid sequence of β-N-acetylhexosaminidase from Aspergillus oryzae with the three hexosaminidases having the solved three-dimensional structure (1c7s: Serratia marcenscens, 1jak: Streptomyces plicates, 1now: Homo sapiens).

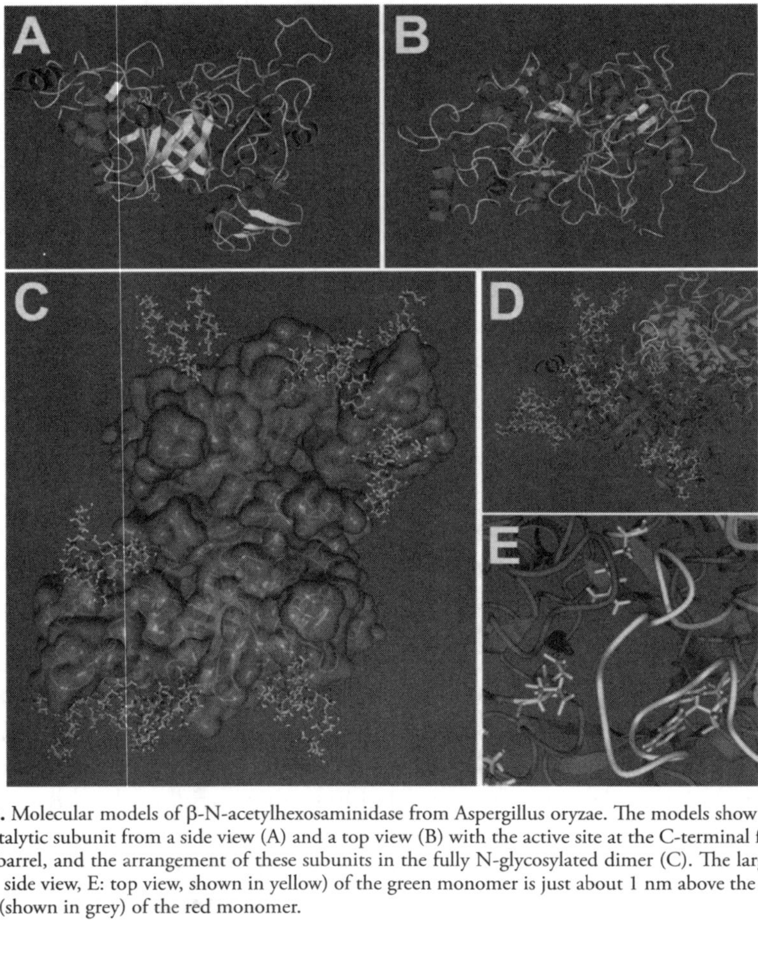

Figure 2. Molecular models of β-N-acetylhexosaminidase from Aspergillus oryzae. The models show the shape of the catalytic subunit from a side view (A) and a top view (B) with the active site at the C-terminal face of the (β, α)8-barrel, and the arrangement of these subunits in the fully N-glycosylated dimer (C). The large flexible loop (D: side view, E: top view, shown in yellow) of the green monomer is just about 1 nm above the active site residues (shown in grey) of the red monomer.

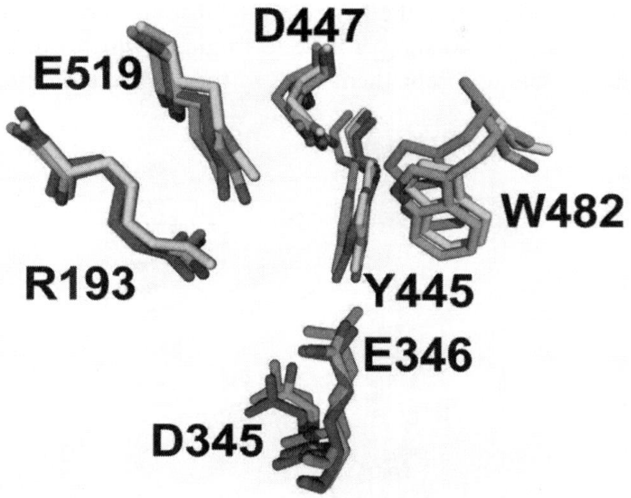

Figure 3. Overlay of the active site residues of the refined homology model with the crystal structures. Active site residues of the homology model of the complete monomer refined by 20 ns of molecular dynamics in a NPT ensemble, are shown in an overlay with PDB entry: 1c7s: Serratia marcenscens (yellow), 1jak: Streptomyces plicates (green), 1now: Homo sapiens (magenta). In the human structure the tryptophane residue is not conserved and thus missing in the structure, and in the Serratia marcenens structure the aspartic acid is mutated to alanine. The overlay shows clearly a spatial alignment with our calculated and refined structure.

Docked into the binding site, the natural substrate N, N'-diacetylchitobiose, participates in the hydrogen bonding between Asp345 and Tyr445 with the acetamido group of the non-reducing GlcNAc moiety. The non-reducing end of the disaccharide is locked into the active site owing to the hydrogen bonds of Arg193, Asp447, Trp482 and Glu519 with the C4 and C5 OH groups. The reducing end of the disaccharide is stabilized by a π-π interaction with the aromatic ring of Trp482. After the substrate being docked and the complex being solvated it shows an initial equilibration phase with rather high fluctuations of the observed binding energies and quite large changes in root mean square deviation of the binding site residues (Fig. 4). However, after 1.6 ns of MD simulation the system seems to have established the binding energy gets stable with only minor fluctuation and the root mean square deviation of the binding site residues arrives back at a stable value for each of the residues that corresponds to the concrete amino acids rigidity (Fig. 5). The mean value of the binding energy over the time interval from 1.6 ns to 3 ns is 447.1 kJ/mol. The root mean square fluctuation of the protein is measured during the last 10 ns of the 20 ns MD simulation, as the structure is at that time already several nanoseconds in a equilibrium state. Figure 6 shows all active site residues belonging to the less flexible part of the protein with minimum fluctuation in time. The root mean square fluctuation of Arg193, Tyr445, Asp447 and Trp482 describes these residues as the most rigid part of the protein and thus these amino acids should play the key role for substrate specificity. Asp345 and the catalytic residue Glu346 show a little less rigidity which can be interpreted that they must be able to orient themselves to the O-glycosidic bond after substrate binding.

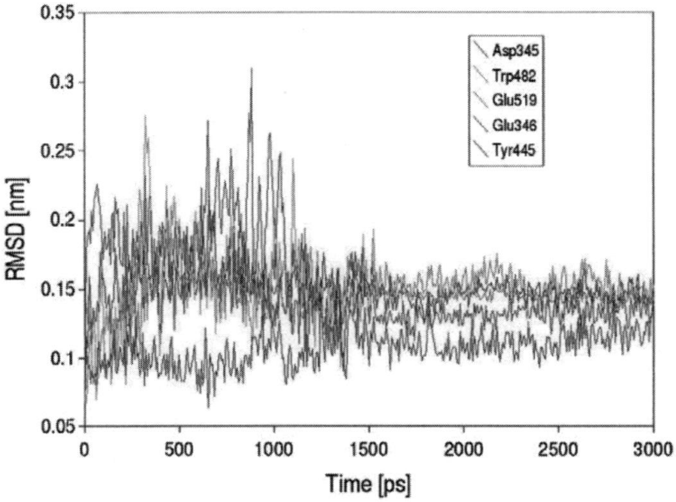

Figure 4. Root mean square deviation of the active site residues after substrate docking. The root mean square deviation of the main amino acids in the active site was monitored for 3 ns.

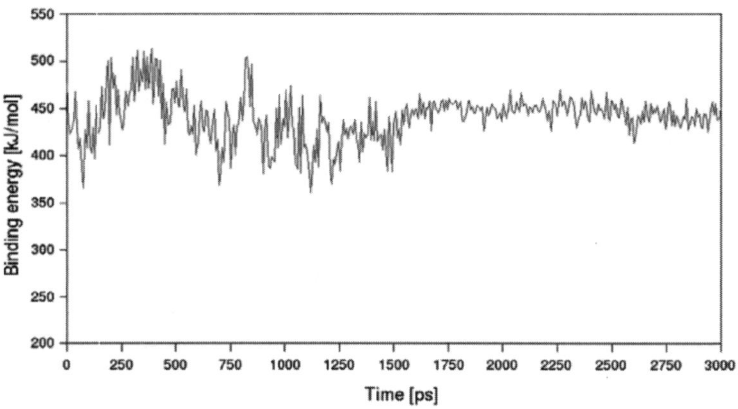

Figure 5. Binding energy of chitobiose during molecular dynamics. Chitobiose was docked into the active site of the homology model of the complete monomer, that was solvated in SPC water and refined by 20 ns of molecular dynamics in a NPT ensemble to equilibrate the homology structure. Behaviour of the substrate in the active site was monitored for 3 ns and the binding energy showed an average value of 447.1 kJ/mol in the time period of 1.6 ns to 3 ns.

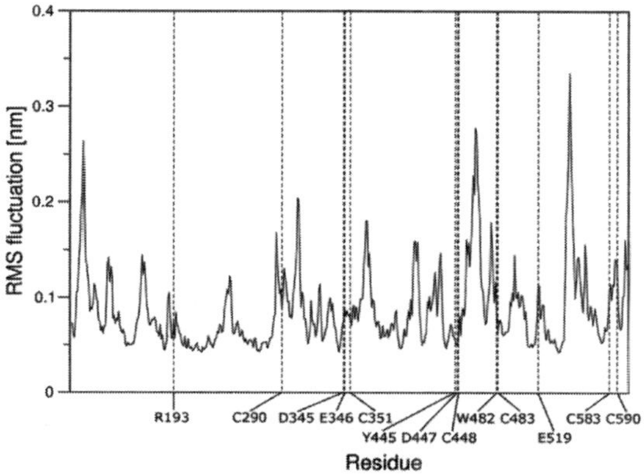

Figure 6. Protein flexibility. Root mean square fluctuation of β-N-acetylhexosaminidase from Aspergillus oryzae during the last 10 ns of MD simulation at 300 K. Amino acids in the active site and cysteins envolved in the formation of disulphide bridges are labeled.

Since initial biochemical characterization of β-N-acetylhexosaminidase from Aspergillus oryzae indicated it to be a dimer, we used the physiologically relevant dimeric crystal structure of human β-N-acetylhexosaminidase [14] (pdb ID: 1o7a) to model the dimeric fungal enzyme according to the physiolocially valid dimerization interface I found in the crystal structure (Fig. 2C). Similar like in crystal structure of human β-N-acetylhexosaminidase, the proposed dimer interface is formed mainly by loop regions at the C-terminal face of the (β, α)8-barrel, as is the active site. Furthermore, the large flexible loop residues anchored by the disulphide bridge of Cys448 with Cys483, an exclusive feature of the fungal enzymes, contributes to the dimer interface.

Detailed examination of the dimer contact surface reveals a buried surface area per monomer of 2373 Å2 and 64 residues containing atoms closer than 3.6 Å to atoms of the interface partner with 19 residues being hydrophobic (Ile, Leu, Val, Met, Tyr, Phe and Trp). We were able to identify 39 hydrogen bonds between both monomers, with the first monomer accepting 22 and donating 17. The presence of two Arg, three His, three Glu and 4 Asp explains the importance of ionic interactions between of Arg and His residues of one subunit and Asp and Glu residues of the opposite subunit. The model thus indicates that the β-N-acetylhexosaminidase dimer may be a reversible function of pH, with acidic environments favoring dissociation into subunits. Indeed, down to pH3.5 the enzyme exists exclusively in the dimeric form, but at pH2.5 it exists only as monomeric, but enzymatically active, subunits as revealed by gel filtration and enzyme assay [20]. When the dissociated enzyme is titrated back to pH5.0, after a short incubation the dimer forms to the original extent [20]. This result also indicates contact areas that are easily re-established depending on the ionization state of residues with pKa values near pH3. The role of the ionic pairs of the above type in the formation the dimeric structure of hexosaminidases is well established. In human hexosaminidase B there is Asp494-Arg533 at the interface between the catalytic subunits in the dimer [14] (Asp522 and His586 in our structure). Notably, since in human hexosaminidases the dimerization is essential for the full catalytic activitity, mutations in the above "interface" amino acids are among the well documented human mutations that have been found in the less severe (late onset) forms of Sandhoff [14] and Tay-Sachs [18,21] diseases. Interestingly, in the dimeric structure the large flexible loop of the opposite monomer is about 1 nm above the C-terminal face of the (β, α)8-barrel, and the active site giving the impression of a substrate binding site lid (Fig. 2D, E).

Six putative sites of N-glycosylation are present in the primary structure of the fungal enzyme [20]. Digestion of β-N-acetylhexosaminidase with N-glycanase results in significant mobility shift on SDS-PAGE (Figure 7A), indicating that one or more glycosylation sites are used on the secreted enzyme. In order to reveal the details of these important structural modifications, we have performed a detailed analysis of the actual occupaccy of all these sites using the standard N-glycanase/ H218O technique [22,23] and mass spectrometry which revealed that all six sites of N-glycosylation were indeed used in the actual hexosaminidase preparation. This conclusion was further supported by complete proteolytic digestion of hexosaminidase followed by isolation of the glycopeptides on immobilized plant lectin Concanavalin A, and identification by these glycopeptides by mass spectrometry. Glycopeptides that covered all six sites of N-glycosylation could be identified on the mass spectra (results not shown).

The fact that all the glycopeptides described above could be efficiently recovered on the immobilized Concanavalin A provided a strong indication for the high-mannose type oligosaccharides being present at the individual N-glycosylation sequences. In order to support this indication, the oligosaccharides were released by N-glycanase and analyzed using a combination of Dionex oligosaccharide profiling, exoglycosidase digestion, and mass spectrometry as described in the Experimental section. When we performed this analysis on oligosaccharides released from the entire hexosaminidase, we obtained the overall glycosylation of the entire enzyme indicating the hexamannosyl oligosaccharide M6 as the predominant structure (Table 1, lane Asnall sites). This overall glycosylation profile was very similar to the one we found on the two particular sites of glycosylation Asn427 and Asn499 (Table 1). As the experimental results show a M6 structure as the predominant oligosaccharide, we choose this structure as the oligosaccharide to be attached to the molecular model of β-N-acetylhexosaminidase. In order to get a preliminary insight into the structural arrangement, and possible biological role(s) of N-glycans in the fungal β-N-acetylhexosaminidase, we have sterically fitted and covalently linked the most prevalent (M6) oligosaccharide to each site of glycosylation on our model structure. This gave rise to the model of a fully glycosylated enzyme shown as the most prevalent glycoform in Fig. 2C. Although the analysis of the actual glycoforms would be a much more demanding task, the model shown in Fig. 2C illustrates nicely the surface exposure of the individual oligosaccharides protecting the enzyme dimer from the effects of the extracellular environment.

Table 1. Oligosaccharide composition at the individual sites of N-glycosylation

Site	Oligosaccharides containing the indicated number of mannoses (% of total)							
	M4	M5	M6	M7	M8	M9	M10	M11
Asn^{427}	--	19	30	21	19	11	--	--
Asn^{499}	19	26	14	14	24	3	--	--
$Asn^{all\ sites}$	9	14	26	23	15	7	4	2

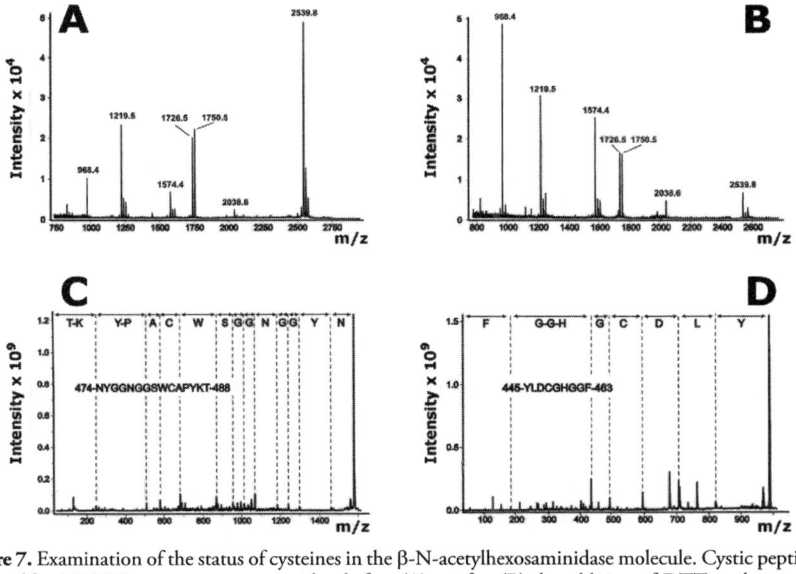

Figure 7. Examination of the status of cysteines in the β-N-acetylhexosaminidase molecule. Cystic peptides were analyzed by MALDI mass spectrometry either before (A) or after (B) the addition of DTT to the sample. The individual assignments were further confirmed by PSD MALDI mass spectrometry of peaks with m/z 1574.4 (C) and 968.4 (D).

Evidence for the Status of Cysteines Using Protein Chemistry and Mass Spectrometry

The fungal β-N-acetylhexosaminidase is secreted into the extracellular environment, indicating the possibility for forming disulfide bonds. Since there are six cysteines in the primary structure of β-N-acetylhexosaminidase, the status of their sulfhydryl groups was determined by mass spectrometry [24]. It has been found that the experimentally verified arrangement of the disulfide bridges corresponds exactly to that anticipated by the molecular model (see above). Moreover, these results were further confirmed by differential mapping of pepsin peptides with and without DTT reduction. Using this technique reverse phase HPLC identified only one peak with significant shift in the retention time. One of the components of this chromatographic peak was the peptide with m/z 2538.8 (Fig. 7A). Significant decrease in its intensity with concomitant intensity increase of peaks at m/z 1574.0 and 968.2 (Fig. 7B), observed upon addition of DTT, indicates that this peptide consists of two disulfide linked peptides. These were identified by PSD measurement as peptides Tyr^{445}-Phe^{453} and Asn^{474}-Thr^{488} containing the cysteines 448 and ^{483}, respectively. This confirms the identification of the disulfide bond between Cys^{448} and Cys^{483}.

In the model structure without disulfide bridges, Cys^{483} is 6.2 Å away from Cys^{448}. The formation of the disulfide bond in the model is leading to a somewhat different position of one of the loops, easily accommodating the experimentally determined disulfide bridge with only a relatively minor change in the overall structure. As Asp^{447} and Trp^{482} are both amino acids forming the active site, the flexible loop between Cys^{448} and Cys^{483} is right beside the substrate-binding site. The behavior of this loop measured by root mean square fluctuation during the last 10 ns MD simulation shows that it is highly flexible (Fig. 6), which may be functionally important. The nearby disulfide bond thus stabilizes not only this loop but also the active site.

The Effect of Deglycosylation on Enzyme Activity and Stability

Preliminary deglycosylation experiments indicated the possibility to deglycosylate the enzyme under native conditions (Fig. 8A, cf. lane 1 and 11). We evaluated several protocols for deglycosylation using N-glycanase or endoglycosidase H under conditions compatible with optimal stability of the deglycosylated hexosaminidase (Fig. 8A). The deglycosylated enzyme remains fully active (Fig. 8B), and displays kinetic parameters indistinguishable from those of the native enzyme (Km = 0.71 mM and Vmax = 2.16 nmol/min vs. Km = 0.45 mM and Vmax = 1.55 nmol/min for the native form).

To probe the effects of deglycosylation on enzyme stability, native and deglycosylated enzymes were incubated at a range of pH values. No significant activity differences were found at alkaline or neutral pH. However, the deglycosylated enzyme is significant less stable than the native enzyme at pH values below 4. The activity of the deglycosylated enzyme after incubation at pH3.5 and 3.0 is about 75% of the native enzyme, and at pH2.5 the deglycosylated enzyme is only half as active as the native enzyme (Fig. 8C). This result suggests that N-linked oligosaccharides may be important for the stabilization of β-N-acetylhexosaminidase at low pH.

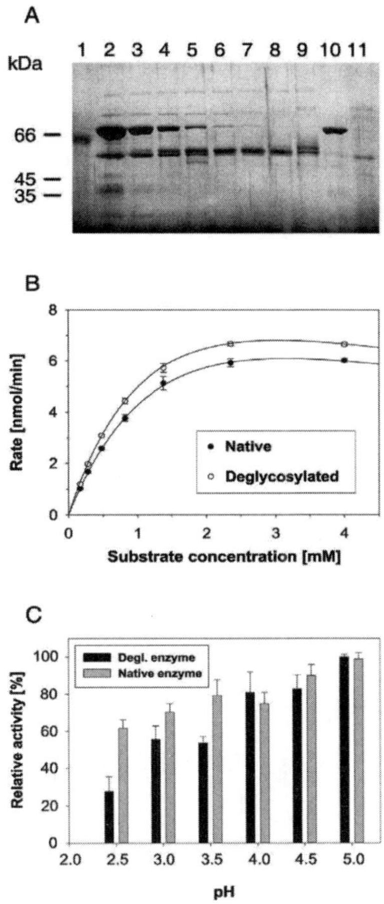

Figure 8. Effect of glycosylation on enzymatic activity and stability of β-N-acetylhexosaminidase. (A) deglycosylation of the enzyme by endoglycosidase H and N-glycanase under native conditions. β-N-acetylhexosaminidase (0.1 µg, lane 1) was deglycosylated using 10 U (lane 2), 5 U (lane 3), 2 U (lane 4), 1 U (lane 5), 0.5 U (lane 6), 0.2 U (lane 7), 0.1 U (lane 8), or 0.05 U (lane 9) of endoglycosidase H (Endo Hf). Lane 10 contains Endo Hf control, and lane 11 β-N-acetylhexosaminidase (0.1 µg) deglycosylated by 0.1 U of N-glycanase. (B) comparison of the enzymatic parameters of native and deglycsylated β-N-acetylhexosaminidase. (C) effect of β-N-acetylhexosaminidase deglycosylation on the stability of the enzyme under various pH values. The average values from triplicate determinations with the standard error indicated by the error bars are shown.

Structure and Stability Determined by Vibrational Spectroscopy

Infrared spectra of β-N-acetylhexosaminidase are characterized by two major bands at 1655 cm^{-1}, and 1543 cm^{-1} (Fig. 9A), associated with the amide I and II vibrations, respectively. The position and shape of these bands are sensitive to protein conformation and secondary structure content. The second derivative, which can identify overlapping components, reveals three major bands in the amide I region. The largest component at 1655 cm^{-1} belongs to α-helical and disordered conformations [25,26]. Two other components at 1642 cm^{-1} and 1626 cm^{-1} are assigned to β-sheets [25,26]. The unresolved band at 1684 cm^{-1} indicates the presence of antiparallel β-strands [27], and the second part of the unresolved band at 1675 cm^{-1} corresponds to β-turns [25,26]. Characteristic side chain absorption of Tyr and Phe is observable at 1517 cm^{-1} and 1493 cm^{-1}, respectively [26]. The FTIR spectrum of β-N-acetylhexosaminidase after deglycosylation (Fig. 9B) shows no significant shifts with respect to the native protein.

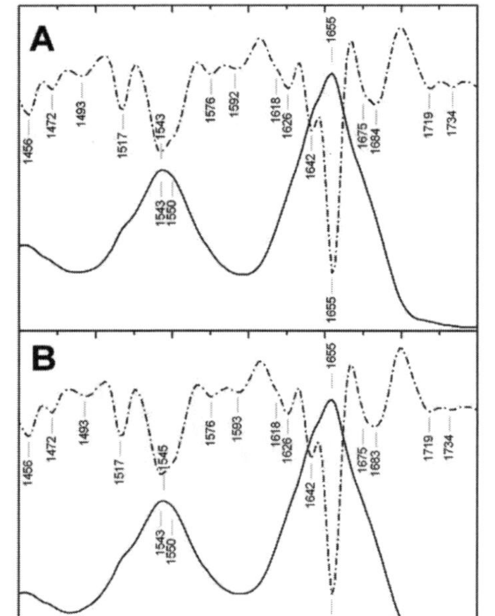

Figure 9. Infrared spectra of β-N-acetylhexosaminidase. Comparison of the infrared spectra in the amide I and II regions of β-N-acetylhexosaminidase (A) and β-N-acetylhexosaminidase deglycosylated (B). The solid curves represent the original spectra while the dashed curves are associated with the second derivative (15 pts) of the spectra.

Raman spectroscopy (Fig. 10) confirms the presence of disulfide bonds by the band at 520 cm^{-1} due to the S-S stretching vibrations. The frequency of this band is sensitive to the local conformation of a disulfide bridge. The band at 520 cm^{-1} corresponds to the sulfide bridges in a conformation close to gauche-gauche-trans (GGT) [28]. The bands in the region of stretching CS vibrations 700–745 cm^{-1} can be hardly resolved due to presence of three disulfide bridges and their probable flexibility with respect to ν CS vibrations. The intensity ratio of the tyrosine Fermi resonance doublet (828 cm^{-1} and 852 cm^{-1}; I852/I828 = 1.6) indicates that the tyrosine OH group acts as an acceptor of H-bonds suggesting that some of the tyrosines are solvent exposed [29]. The Raman spectrum after deglycosylation (Fig. 10B) shows significant changes with respect to native protein (Fig. 10A) that are clearly revealed by the difference spectrum (A-B). Almost all marker bands of aromatic side chains are affected by deglycosylation, although the secondary structure was not affected, as discussed below. Thus, changes in the Raman spectrum can be attributed to changes in environment of the residues, presumably due to unmasking of large surface areas of the protein by deglycosylation.

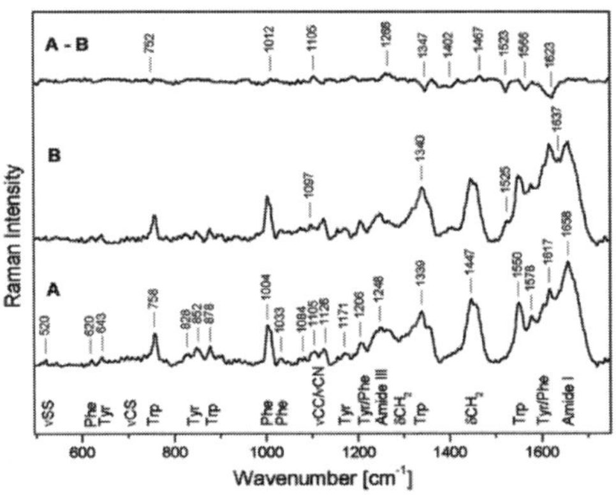

Figure 10. Raman spectra of β-N-acetylhexosaminidase. Raman spectrum of the native β-N-acetylhexosaminidase (A), of the enzyme deglycosylated by endoglycosidase H (B), and the differential spectrum of the two enzymes (A-B). The assignment of the bands is discussed in the text (ν corresponds to stretching and δ to bending vibrations).

The FTIR amide I and II bands and the Raman amide I band were analyzed by LSA (Table 2) to quantify the secondary structure content of β-N-acetylhexosaminidase and its deglycosylated form. Deglycosylation of the protein leads to small changes in secondary structure, as estimated by LSA, connected with slight decrease of β-sheets (about 3%) and slight increase of α-helices—probably on the periphery of the protein. The LSA methods employs two types of α-helical structure for Raman spectra: short helical segment up to eight residues called monohydrogen-bonded or disordered, and helical segments bracketed by at least four residues on each side named bihydrogen-bounded or ordered. Recently we proposed that the vibrations could be affected by oligomerization or complex formation [30], therefore the amount of disordered α-helix (about 15%) was included to other structure in Table 2. If we agree with interpretation that the signal from disordered α-helix is caused mostly by dimerization of β-N-acetylhexosaminidase then carbohydrate moiety plays important role in the dimer formation. The lower stability of the deglycosylated enzyme upon prolonged incubations at acidic pH (see Fig. 8C) may then be interpreted as the gradual decomposition of the enzyme dimer into the monomeric units that is fully reversible for the native but not for the deglycosylated enzyme.

Table 2. Secondary structure estimation (in %) of β-N-acetylhexosaminidase by FTIR and Raman spectroscopy compared with the model structures

Method	Secondary structure estimation				
	α-Helix	β-Sheet	β-Turn	Bend	Other
Model	32	15	14	14	25
FTIR-LSA[a]	31	19	14	14	22
Raman-LSA[b]	31	26	15	--	26

[a] Least-squares analysis of FTIR amide I and II bands according to [55].
[b] Least-squares analysis of Raman amide I band according to [54].

Discussion

We describe here results of structural analysis of dimeric, fully N-glycosylated, fungal β-N-acetylhexosaminidase isolated from the culture medium of the public collection strain of Aspergillus oryzae. The unique biology [15] and important biotechnological applications of this enzyme [8,9,17] stimulated this work as a prerequisite for structure-function studies and enzyme engineering. To create a conceptual basis for the design of structural experiments, the amino acid sequence of the Aspergillus β-N-acetylhexosaminidase determined in our laboratory [16] was used to generate a molecular model of the catalytic subunit based on sequence homology to three solved β-N-acetylhexosaminidase structures [10,12,14]. Our major experimental efforts then have been directed towards understanding the principal differences between the intracellular bacterial enzymes and the eukaryotic enzymes that have to undergo posttranslational modifications to survive better under the more aggressive conditions in lysosomes and the extracellular environment in which they function. In particular, we investigated how enzyme dimerization, disulfide bonding, and glycosylation participate in enzyme stabilization.

Multimerization of enzymes represents one strategy for catalytic subunit stabilization that is often used, for example, in thermophilic organisms. Consistent with this trend, the intracellular bacterial β-N-acetylhexosaminidases are monomeric whereas the extracellular secreted enzymes of fungi and humans occur as dimers. Dimerization of the fungal β-N-acetylhexosaminidase appears to be a reversible process that is pH dependent. This result provides strong support for the participation of titratable functional groups at the subunit interface as predicted by the constructed dimeric structure. The experimental finding that oligosaccharide moieties may also participate in the dimerization process may represent a unique feature of the exclusively extracellular enzymes.

Formation of disulfide bridges is another common way of stabilizing proteins that are secreted into the oxidative extracellular environment. The disulfide bridges detected experimentally are in complete accordance with the molecular model. One disulfide may stabilize a hinge region in a small flexible loop close to the substrate binding site. This loop is an exclusive feature of the fungal enzymes, neither present in bacterial nor mammalian structures. We hypothesize that this loop may play the role of a substrate binding site lid, as has been found previously in several enzymes including other TIM barrels [31-34], anchored by a disulfide bridge that prevents the substrate binding site from being influenced by the flexible motion of the loop. This substrate binding site lid is formed in the dimeric structure by the large flexible loop of the opposite monomer and is close above the C-terminal face of the (β, α)8-barrel, and the active site thus indicating the necessity of a functional dimer for full enzyme function.

Yet another posttranslational modification known to stabilize proteins is N-glycosylation. However, experimental investigation of its role in enzyme function and stabilization is technically difficult, and only a few studies report detailed experimental analysis [22,35]. In the case studied here we had the rare possibility to fully de-N-glycosylate the enzyme under native conditions. To probe the role of extensive glycosylation in the fungal β-N-acetylhexosaminidase experimentally we used two deglycosylation methods, which are based on the use of either N-glycanase or endoglycosidase H. The former enzyme is an asparagines amidase that cleaves the N-glycosidic (amide) bond between the asparagines and the oligosaccharide chains. This cleavage gives rise to the free β-glycosylamine, and aspartic acid that remains in the protein in the original position of the asparagines. On the other hand, the latter enzyme is a typical endoglycosidase that cleaves a glycosidic linkage between the two GlcNAc residues forming the chitobiose core of the N-linked oligosaccharide. Such a cleavage leaves the N-glycosidic bond on asparagines, and the proximal GlcNAc intact.

We employed endoglycosidase H digestion which allows to efficiently remove the mannose branches while retaining the core GlcNAc residues to avoid the potential dangers of complete protein deglycosylation that we have previously reported [36]. Deglycosylation had only limited effect on enzyme activity, but it significantly affected enzyme stability in acidic conditions. This role of N-glycosylation is consistent with the observation that the enzyme completely deglycosylated by N-glycanase has low stability and tends to precipitate.

Conclusion

In summary, we have shown that complementary biochemical and biophysical methods provided structural insight into critical features of fungal β-N-acetylhexosaminidase, even in the absence of data from protein crystallography or nuclear magnetic resonance. The results reported here also contain useful information for designing experiments that will use high resolution methods of protein structure analysis. Dimerization of the fungal β-N-acetylhexosaminidase appears to be a reversible process that is strictly pH dependent. Oligosaccharide moieties may also participate in the dimerization process which might represent a unique feature of the exclusively extracellular enzymes. Deglycosylation had only limited effect on enzyme activity, but it significantly affected enzyme stability in acidic conditions. Dimerization and N-glycosylation are the enzyme's strategy for catalytic subunit stabilization. One disulfide bridge, neither present in bacterial nor mammalian structures, anchors a flexible loop close to the active site that may play the role of a substrate binding site lid in the physiologically relevant dimer.

Methods

Molecular Modeling

The molecular model of β-N-acetylhexosaminidase was generated by a combination of energetic and homology modeling. In a first step three homologs from the glycohydrolase family 20 were extracted from the Protein Databank [37]—PDB entry: 1c7s: Serratia marcenscens [11], 1jak: Streptomyces plicates [18], 1now: Homo sapiens [19]. These proteins show a primary sequence identity about 30% and a homology around 45% (Serratia marcenscens 32/48%, Streptomyces plicates 27/42%, Homo sapiens 31/49%) making homology modeling possible. A structural alignment was generated in SwissPDBViewer [38] showing a root mean square deviation ~1.4 Å between the three structures. Then the complete primary sequence of β-N-acetylhexosaminidase from Aspergillus oryzae [16] was aligned with the three homologs in ClustalX [39] keeping the positions of the structural alignment conserved (see Fig. 1). The three-dimensional model constituted by all non-hydrogen atoms was built and examined by the Modeller6 package [40]. All three disulfide bridges were created and refinement was achieved through algorithmic analysis and minimization with the Tripos force field in the Sybyl/Maximin2 (Tripos). Hydrogen atoms were added and the model of the generated structure was minimized to convergence of the energy gradient less then 0.01 kcal/mol using the Powell minimiser. The minimization included electrostatic interactions based on Gasteiger-Hückel partial charge distributions using a dielectric constant with a distance dependent function $\varepsilon = 4r$ and a non-bonded interaction cut-off of 8 Å. The tertiary structure model was checked with Procheck [41].

The final proposed model was solvated in simple point charge water and four chloride counter-ions were added. The production runs were preceded by short equilibration runs of altogether 250 ps with positional restraints applied on the protein atoms to allow the solvent to relax and 250 ps of protein relaxation without positional contraints and a timestep of 2 fs. The simulation box was sized 1 nm in each direction from the protein surface and filled with 18950 water molecules to give a system of 62049 atoms. In a first step the system is minimized by 500 steps of steepest descent minimization, followed by 20 ps of solvent relaxation with a 1 fs timestep. Chloride ions were added by replacing four water molecules to neutralize the systems, followed by 30 ps of ion and solvent relaxation with a 1 fs timestep and 200 ps with a 2 fs timestep. The following production MD simulations, without any restraints, were 20 ns long and were run with GROMACS 3.2 [42,43] using the gmx force field, with a 5 fs time step (which is possible because dummy hydrogens are used). SETTLE (for water) and LINCS were used to constrain covalent bond lengths, and long-range electrostatic interactions were computed with the Particle-Mesh Ewald method. The temperature was kept at 300 K by separately coupling the protein and solvent to an external temperature bath ($\tau = 0.1$ ps) [44]. The pressure was kept constant at 1 bar by weak coupling ($\tau = 1.0$ ps) to a pressure bath. The protein proved to be stable during simulation.

The three dimensional structure of the N-linked complex glycan (M6 oligosaccharide structure) was calculated with Sweet [45] and the carbohydrates were sterically fitted by visual analysis and covalently attached to the molecular model of β-N-acetylhexosaminidase in Sybyl. The dimeric structure was created by fitting two monomeric structures onto the dimeric structure from Homo sapiens within YASARA [46]. Then the second monomer was placed in a distance of about 1 nm away from its final position to avoid sterical conflicts of the large flexible loop with the first monomer, followed by 2000 steps of simulated annealing minimization. The second monomer was such moved in 0.1 nm steps to its final position, each step followed by another 2000 steps of simulated annealing minimization.

Chitobiose was build in YASARA, forcefield parameters were assigned using the AutoSMILES approach [44], in a first step YASARA calculated semi-empirical AM1 Mulliken point charges that were corrected by assignment of AM1BCC atom types and improved AM1BCC charges by fragments of molecules with known RESP charges, to closer resemble RESP charges. Corresponding bond, angle and torsion potential parameters are taken from the General AMBER force field. For the docking experiments our model structure was fitted onto the crystal structure of 1qbb [10], a bacterial chitobiase complexed with N, N'-diacetylchitobiose. Chitobiose was placed in an arbitrary position according to the ligand co-ordinates in the bacterial chitobiase complex. Exact positioning of the ligand was done by a two-step procedure, energy minimization followed by a molecular dynamics. The ligand-protein system was minimized by 2000 steps followed by a 3 ns MD simulation in aqueous solution using the YAMBER2 force field [46]. The protein structure was placed into a box, which was 1 nm larger than the protein along all three axes. The box was filled with TIP3P water, sodium ions were iteratively placed at the coordinates with the lowest electrostatic potential until the cell was neutral. Molecular dynamics simulations were run with YASARA, using a multiple time step of 1 fs for intra-molecular and 2 fs for intermolecular forces. A 1.2 nm cut-off was taken for Lennard Jones forces and the direct space portion of the electrostatic forces, which were calculated using the Particle Mesh Ewald method [47] with a grid spacing 0.1 nm, 4th order B-splines, and a tolerance of 10-4 for the direct space sum. The simulation of interaction was then run at 298 K and constant pressure (NPT ensemble) to account for volume changes due to fluctuations of homology models in solution.

Interaction energies were calculated considering the internal energy obtained with the specified force field, as well as the electrostatic and Van der Waals solvation energy obtained. The electrostatic solvation energy estimates the interaction energy between the solvent and the solute by treating the solvent as a continuum without explicit solvent molecules. A first-order boundary element approximation to the solvation energy was used. Van der Waals solvation energy was calculated as a function of the solute's solvent accessible surface area [48]. The entropic cost of fixing the ligand in the binding site is almost impossible to calculate accurately, but fortunately not needed since it mainly depends on characteristics that are constant during the simulation (ligand and protein size, side-chains on the surface etc.). The entropic component is thus a constant factor that can be omitted. The more positive the interaction energy, the more favorable is the interaction in the context of the chosen force field.

Enzyme Isolation and Characterization

Aspergillus oryzae strain CCF1066 (Czech Collection of Fungi, Faculty of Science, Charles University in Prague) was grown as described in [17]. Ammonium sulfate enzyme precipitate was purified by hydrophobic chromatography on Phenyl-Sepharose 6 Fast Flow (Amersham) using elution with the reversed ammonium sulfate gradient (0.6 M to 0 M in 20 mM sodium phosphate buffer pH6.8). Partially purified enzyme was concentrated and dialyzed against 20 mM sodium citrate buffer pH3.5. The enzyme was then purified on SP-Sepharose Fast Flow (Amersham) using elution with sodium chloride gradient (0 to 1 M). The enzyme was concentrated, dialyzed against 20 mM piperazine-HCl pH5.4, and purified on MonoQ HR 10/10 column (Amersham) eluted with sodium chloride gradient (0 to 0.3 M). The enzyme was concentrated to approx. 10 mg/mL, and stored in the stabilization buffer composed of 0.5 M ammonium sulphate in 50 mM citrate buffer pH5.0. Immediately before the spectroscopic measurements, the enzyme was transferred to 50 mM bis-Tris buffer pH5.0. The purity of the enzyme was checked by SDS-PAGE, and the concentration was determined by [49].

Determination of Enzymatic Activity

Enzymatic activity was measured using 4-nitrophenyl-2-acetamido-2-deoxyglucopyranoside at either saturating concentration (5 mM, used for most determinations), or at concentrations around Km (for the evaluation of enzymatic parameters) according to the procedure of [50].

Analysis of Individual Sites of N-Glycosylation

Occupancy of sites of N-glycosylation was determined as described previously [23] by comparison of measured masses of peptides corresponding to the sites of glycosylation. The peptides were generated after SDS-PAGE separation by means of in gel digestion with sequencing grade trypsin (Promega), or sequencing grade Asp-A protease (Roche) either in normal or isotopic water H218O (Fluka). The peptides were extracted from the gel, desalted and concentrated with C-18 microcolumn (ZipTip C18, Millipore) and analyzed by MS. To analyze individual sites of N-glycosylation, the glycoprotein was digested with trypsin, and glycopeptides were captured on concanavalin A—Sepharose resin (Amersham), washed, and eluted with 0.01 M Tris-HCl, pH8.0, 0.15 M NaCl and 0.3 M D-mannose (Sigma). Individual glycopeptides were separated by reverse-phase chromatography on the Vydac C-18 column (Dionex), equilibrated in 0.1% trifluoroacetic acid, and eluted by acetonitrile gradient to 70% over 120 min. The glycopeptides were identified by MALDI MS measurements in their native and or deglycosylated state. Glycans released from glycopeptides or from the intact protein with PNGase F (New England BioLabs) in 50 mM sodium bicarbonate pH8.0 were desalted on mini-columns (10 μL of bead volume) filled with non-porous graphitized carbon

[51], and separated by HPAEC-PAD (DX500, Dionex) on Carbopac PA100 column. Elution was performed by 5 min isocratic step with 0.1 M NaOH (solvent A) followed by linear gradient from 0 to 35% of 0.1 M NaOH, 0.6 M NaOAc (solvent B) over 48 min and another linear gradient to 100% solvent B within 10 min at the flow rate 1 mL/min. The collected peaks were desalted (as above) and analysed by MALDI MS. For further characterization of the isolated oligosaccharides, the digestion with α-mannosidase was performed using 5 U/mL of the enzyme in 50 mM sodium citrate buffer pH5.0 [52].

Enzymatic Deglycosylation Under Native Conditions

100 μg of β-N-acetylhexosaminidase was fully deglycosylated either in 100 μl of 50 mM ammonium bicarbonate pH8.0 using 500 units of N-glycanase (New England BioLabs), or in 100 μL of 50 mM citrate buffer pH5.0 using 500 units of EndoH (New England BioLabs). The reaction was carried out for 16 h at 37°C.

pH Stability Test

pH stability test of the native and deglycosylated enzyme was performed as follows: 1 μL of the stock solution of enzyme was mixed with 9 μL of 50 mM sodium citrate buffer pH2.5, 3.0, 3.5, 4.0, or 4.5, or with 9 μL of 50 mM sodium phosphate buffer pH5.5, 6.0, 6.5, or 7.0, or 50 mM Tris-HCl buffer pH7.5, 8.0, 8.5, or 9.0. These mixtures were incubated for 16 h at 4°C, diluted with 40 μL of the substrate solution (50 mM sodium citrate pH5.0 with 5 mM substrate), and the enzymatic activity was determined as described in [50].

FTIR Spectroscopy

Infrared spectra of protein samples (9.5 mg/mL native and 8.5 mg/mL deglycosylated in 50 mM bis-Tris, pH5.0) were recorded in CaF2 10 μm BioCell™ (BioTools) at room temperature with a Bruker IFS-66/S FTIR spectrometer using a standard source, a KBr beamsplitter and an MCT detector. 4000 scans were collected with 4 cm^{-1} spectral resolution and Happ-Genzel apodization function. Spectral contribution of a buffer in carbonyl stretching region was corrected following the standard algorithm [53].

Raman Spectroscopy

Raman spectra were recorded in a standard 90° geometry on a multichannel instrument based on a 600-mm single spectrograph (Monospec 600, Hilger &

Watts) with a 1200-grooves/mm grating and a liquid N2-cooled CCD detection system (Princeton Instruments) having 1024 pixels along dispersion axis. A holographic notch-plus filter (Kaiser Optical Systems) was used to remove elastically scattered light. The effective spectral slit width was set to ~5 cm^{-1}. Samples were excited with 514.5 nm/50 mW line of an Ar+ laser Innova 300 (Coherent). Calibrated wavenumber scale by Ar+ plasma lines was accurate to ± 1 cm^{-1}.

Measurements were made on protein samples (8–13 mg/mL in 50 mM bis-Tris, pH5.0) in a 10-μL capillary microcell. Spectra, measured at 4°C, were averaged from 150 exposures of 120 s. Spectra were treated according to [54], then they were smoothed using 9-point Savitsky-Golay algorithm and normalized to the 1447 cm^{-1} δCH_2 band as an internal standard.

Abbreviations

DTT, dithiothreitol; EDC, 1-ethyl-3-(3'-dimethylaminopropyl)-carbodiimide hydrochloride; FTIR, Fourier transform infrared; GalNAc, N-Acetyl-D-galactosamine; GlcNAc, N-Acetyl-D-glucosamine; HPAEC-PAD, high-performance anion exchange chromatography with pulsed amperometric detection; HPLC, high performace liquid chromatography; LSA, least-squares analysis; MALDI, matrix-assisted laser desorption/ionization; ManNAc, N-Acetyl-D-mannosamine; MD, molecular dynamics; MS, mass spectrometry; PNGase F, peptide-N4-(N-acetyl-β-D-glucosaminyl) asparagine amidase F; PSD, post-source decay; RESP, restrained electrostatic potential; SDS, sodium dodecyl sulfate; SDS-PAGE, SDS polyacrylamide gel electrophoresis.

Authors' Contributions

RE was responsible for all the work on sequence alignments, molecular modeling, and molecular dynamics studies and performed most of it. MK built up the ligand structure and supervised the docking, VK jr., KH and VB performed all the IR and Raman spectroscopy measurements and the data interpretation. KH was also responsible for the hexosaminidase production. PN, PP and PM were responsible for mass spectrometry measurements, and solved the disulfide bonds arrangement, NK performed the computational glycosylation and the binding energy calculations and OP and JS were responsible for the determination of hexosaminidase glycosylation and performed the deglycosylation studies. HR took part in the enzyme purification and primary structure determination, VK conceived the study and provided the strains and advice on hexosaminidase production, KB was responsible for the overall coordination of the project and produced the first draft of the manuscript, and was responsible for hexosaminidase purification. All authors read and approved the final manuscript.

Acknowledgements

Supports from the Ministry of Education of the Czech Republic (Nos. LC06010, MSM0021620835, MSM6007665808, 1M 4635608802), the Institutional Research Concept of the Academy of Sciences of the Czech Republic (Nos. AVOZ60870520, AVOZ50200510), and the Grant Agency of the Czech Republic (Nos. 203/04/1045 and 204/06/0771) are gratefully acknowledged. The authors also would like to thank Elmar Krieger, Yasara.org, for his helping hand and Janette Carey, Princeton University, for a critical reading of this manuscript.

References

1. Gooday GW, Zhu WY, O'Donell RW: What are the roles of chitinases in the growing fungus? FEMS Microbiol Lett 1992, 100:387–392.
2. Bulawa CE: Genetics and molecular biology of chitin synthesis in fungi. Annu Rev Microbiol 1993, 47:505–534.
3. Cheng Q, Li H, Merdek K, Park JT: Molecular characterization of the β-N-acetylglucosaminidase of Escherichia coli and its role in cell wall recycling. J Bacteriol 2000, 182:4836–4840.
4. Cohen E: Chitin synthesis and inhibition: a revisit. Pest Manag Sci 2001, 57:946–950.
5. Mahuran DJ: Biochemical consequences of mutations causing the GM2 gangliosidoses. Biochim Biophys Acta 1999, 1455:105–138.
6. Křen V, Ščigelová M, Přikrylová V, Havlíček V, Sedmera P: Enzymatic-synthesis of β-N-acetylhexosaminides of ergot alkaloids. Biocatalysis 1994, 10:118–193.
7. Rajnochová E, Dvořáková J, Huňková Z, Křen V: Reverse hydrolysis catalysed by β-N-acetylhexosaminidase from Aspergillus oryzae. Biotechnol Lett 1997, 19:869–872.
8. Krist P, Herkommerová-Rajnochová E, Rauvolfová J, Semeňuk T, Vavrušková P, Pavlíček J, Bezouška K, Petruš L, Křen V: Toward an optimal oligosaccharide ligand for rat natural killer cell activation receptor NKR-P1. Biochem Biophys Res Commun 2001, 287:11–20.

9. Weignerová L, Vavrušková P, Pišvejcová A, Thiem J, Křen V: Fungal β-N-acetylhexosaminidases with high β-N-acetylgalactosaminidase activity and their use for synthesis of β-GalNAc-containing oligosaccharides. Carbohydr Res 2003, 338:1003–1008.

10. Tews I, Perrakis A, Oppenheimer A, Dauter Z, Wilson KS, Vorgias CE: Bacterial chitobiase structure provides insight into catalytic mechanism and the basis of Tay-Sachs disease. Nat Struct Biol 1996, 3:638–648.

11. Prag G, Papanikolau Y, Tavlas G, Vorgaris CE, Petratos K, Oppenheim AB: Structures of chitobiase mutants complexed with the substrate Di-N-acetyl-d-glucosamine: the catalytic role of the conserved acidic pair, aspartate 539 and glutamate 540. J Mol Biol 2000, 300:611–617.

12. Mark BL, Vocadlo DJ, Zhao D, Knapp S, Withers SG, James MNG: Crystallographic evidence for substrate-assisted catalysis in a bacterial β-hexosaminidase. J Biol Chem 2001, 276:10330–10337.

13. Williams SJ, Mark BL, Vocadlo DJ, James MNG, Withers SG: Aspartate 313 in the Streptomyces plicatus hexosaminidase plays a critical role in substrate-assisted catalysis by orienting the 2-acetamido group and stabilizing the transition state. J Biol Chem 2002, 277:40055–40065.

14. Maier T, Strater N, Schuette CG, Klingenstein R, Sandhoff K, Saenger W: The X-ray crystal structure of human β-hexosaminidase B provides new insights into Sandhoff disease. J Mol Biol 2003, 328:669–681.

15. Huňková Z, Křen V, Ščigelová M, Weignerová L, Scheel O, Thiem J: Induction of β-N-acetylhexosaminidase in Aspergillus oryzae. Biotechnol Lett 1996, 18:725–730.

16. Aspergillus oryzae beta-N-acetylhexosaminidase precursor (hexA) gene, complete cds [http://www.ncbi.nlm.nih.gov/entrez/viewer.fcgi?db=nuccore&id=29242776].

17. Hušáková L, Herkommerová-Rajnochová E, Semeňuk T, Kuzma M, Rauvolfová J, Přikrylová V, Ettrich R, Plíhal O, Bezouška K, Křen V: Enzymatic discrimination of 2-acetamido-2-deoxy-D-mannopyranose-containing disaccharides using β-N-acetylhexosaminidases. Adv Synth Catal 2003, 345:735–742.

18. Mark BL, Vocadlo DJ, Zhao D, Knapp S, Withers SG, James MNG: Biochemical and structural assessment of the 1-N-azasugar GalNAc-isofagomine as a potent family 20 β-N-acetylhexosaminidase inhibitor. J Biol Chem 2001, 276:42131–42137.

19. Mark BL, Mahuran DJ, Cherney MM, Zhao D, Knapp S, James MNG: Crystal structure of human β-hexosaminidase B: understanding the molecular basis of Sandhoff and Tay-Sachs disease. J Mol Biol 2003, 327:1093–1109.

20. Plíhal O, Sklenář J, Kmoníčková J, Man P, Pompach P, Havlíček V, Křen V, Bezouška K: N-glycosylated catalytic unit meets O-glycosylated propeptide: complex protein architecture in a fungal hexosaminidase. Biochem Soc Trans 2004, 32:764–765.

21. Lemieux MJ, Mark BL, Cherney MM, Withers SG, Mahuran DJ, James MNG: Crystallographic structure of human β-hexosaminidase A: interpretation of Tay-Sachs mutations and loss of GM2 ganglioside hydrolysis. J Mol Biol 2006, 359:913–929.

22. Bařinka C, Šácha P, Sklenář J, Man P, Bezouška K, Slusher BS, Konvalinka J: Identification of the N-glycosylation sites on glutamate carboxypeptidase II necessary for proteolytic activity. Protein Sci 2004, 13:1627–1635.

23. Gonzalez J, Takao T, Hori H, Besada V, Rodriguez R, Padron G, Shimonishi Y: A method for determination of N-glycosylation sites in glycoproteins by collision-induced dissociation analysis in fast atom bombardment mass spectrometry: identification of the positions of carbohydrate-linked asparagine in recombinant alpha-amylase by treatment with peptide-N-glycosidase F in 18O-labeled water. Anal Biochem 1992, 205:151–158.

24. Novák P, Man P, Pompach P, Hofbauerová K, Bezouška K: Straightforward Determination of Disulfide Linkages in Proteins: The Case of β-N-acetyl-Hexosaminidase from Aspergillus oryzae. Proceedings of the ASMS Conference on Mass Spectrometry and Allied Topics 2006, 54:540.

25. Arrondo JLR, Goñi FM: Structure and dynamics of membrane proteins as studied by infrared spectroscopy. Prog Biophys Mol Biol 1999, 72:367–405.

26. Fabian H, Mäntele W: Handbook of Vibrational Spectroscopy. Edited by: Chalmers JM, Griffiths PR. Chichester: John Wiley & Sons Ltd; 2002:3399–3425.

27. Yamada N, Ariga K, Naito M, Matsubara K, Koyama E: Regulation of β-sheet structures within amyloid-like β-sheet assemblage from tripeptide derivatives. J Am Chem Soc 1998, 120:12192–12199.

28. Van Wart HE, Scheraga HA: Agreement with the disulfide stretching frequency-conformation correlation of Sugeta, Go, and Miyazawa. Proc Natl Acad Sci USA 1986, 83:3064–3067.

29. Siamwiza MN, Lord RC, Chen MC, Takamatsu T, Harada I, Matsura H, Shimanouchi T: Interpretation of the doublet at 850 and 830 cm-1 in the Raman spectra of tyrosyl residues in proteins and certain model compounds. Biochemistry 1975, 14:4870–4876.

30. Ettrich R, Brandt W, Kopecký V Jr, Baumruk V, Hofbauerová K, Pavlíček Z: Study of chaperone-like activity of human haptoglobin: conformational

changes under heat shock conditions and localization of interaction sites. Biol Chem 2002, 383:1667–1676.

31. Joseph D, Petsko GA, Karplus M: Anatomy of a conformational change: hinged "lid" motion of the triosephosphate isomerase loop. Science 1990, 249:1425–1428.

32. Pakhomova S, Kobayashi M, Buck J, Newcomer ME: A helical lid converts a sulfotransferase to a dehydratase. Nat Struct Biol 2001, 8:447–451.

33. Bustos-Jaimes I, Sosa-Peinado A, Rudino-Pinera E, Horjales E, Calcagno ML: On the role of the conformational flexibility of the active-site lid on the allosteric kinetics of glucosamine-6-phosphate deaminase. J Mol Biol 2002, 319:183–189.

34. Brocca S, Secundo F, Ossola M, Alberghina L, Carrera G, Lotti M: Sequence of the lid affects activity and specificity of Candida rugosa lipase isoenzymes. Protein Sci 2003, 12:2312–2319.

35. Pfeiffer G, Strube KH, Schmidt M, Geyer R: Glycosylation of two recombinant human uterine tissue plasminogen activator variants carrying an additional N-glycosylation site in the epidermal-growth-factor-like domain. Eur J Biochem 1994, 219:331–348.

36. Hogg T, Kutá-Smatanová I, Bezouška K, Ulbrich N, Hilgenfeld R: Sugar-mediated lattice contacts in crystals of a plant glycoprotein. Acta Crystallogr D Biol Crystallogr 2002, 58:1734–1739.

37. Berman HM, Westbrook J, Feng Z, Gilliland G, Bhat TN, Weissig H, Shindyalov IN, Bourne PE: The Protein Data Bank. Nucl Acids Res 2000, 28:235–242.

38. Guex N, Peitsch MC: SWISS-MODEL and the Swiss-PdbViewer: an environment for comparative protein modeling. Electrophoresis 1997, 18:2714–2723.

39. Thompson JD, Gibson TJ, Plewniak F, Jeanmougin F, Higgins DG: The CLUSTAL_X windows interface: flexible strategies for multiple sequence alignment aided by quality analysis tools. Nucl Acids Res 1997, 25:4876–4882.

40. Sali A, Blundell TL: Comparative protein modelling by satisfaction of spatial restraints. J Mol Biol 1993, 234:779–815.

41. Laskowski RA, McArthur MW, Moss DS, Thornton JM: PROCHECK—a program to check the stereochemical quality of protein structures. J Appl Crystallog 1993, 26:283–291.

42. Berendsen HJC, van der Spoel D, van Drunen R: GROMACS: a message-passing parallel molecular dynamics implementation. Comput Phys Commun 1995, 91:43–56.

43. Lindahl E, Hess B, van der Spoel D: GROMACS 3.0: A package for molecular simulation and trajectory analysis. J Mol Modell 2001, 7:306–317.

44. Berendsen HJC, Postma JPM, van Gunsteren WF, DiNola A, Haak JR: Molecular-dynamics with coupling to an external bath. J Chem Phys 1984, 81:3684–3690.

45. Bohne A, Lang E, von der Lieth CW: W3-SWEET: Carbohydrate modeling by Internet. J Mol Model 1998, 4:33–43.

46. Krieger E, Darden T, Nabuurs SB, Finkelstein A, Vriend G: Making optimal use of empirical energy functions: force-field parameterization in crystal space. Proteins 2004, 57:678–683.

47. Essman U, Perera L, Berkowitz ML, Darden T, Lee H, Pedersen LG: A smooth particle mesh Ewald method. J Chem Phys 1995, 103:8577–8593.

48. Bultinck P, De Winter H, Langenaeker W, Tollenare J: Computational medicinal chemistry for drug discovery. CRC Press; 2003.

49. Bradford MM: A rapid and sensitive method for the quantitation of microgram quantities of protein utilizing the principle of protein-dye binding. Anal Biochem 1976, 72:248–254.

50. Li SC, Li YT: Studies on the glycosidases of jack bean meal. 3. Crystallization and properties of β-N-acetylhexosaminidase. J Biol Chem 1970, 245:5153–5160.

51. Packer NH, Lawson MA, Jardine DR, Redmond JW: A general approach to desalting oligosaccharides released from glycoproteins. Glycoconj J 1998, 15:737–747.

52. Harvey DJ: Matrix-assisted laser desorption/ionization mass spectrometry of carbohydrates. Mass Spectrom Rev 1999, 18:349–450.

53. Dousseau F, Therrien M, Pézolet M: On the spectral substraction of water from the FT-IR spectra of aqueous-solutions of proteins. Appl Spectrosc 1989, 43:538–542.

54. Williams RW: Protein secondary structure analysis using Raman amide I and amide III spectra. Methods Enzymol 1986, 130:311–331.

55. Dousseau F, Pézolet M: Determination of the secondary structure content of proteins in aqueous solutions from their amide I and amide II infrared bands. Comparison between classical and partial least-squares methods. Biochemistry 1990, 29:8771–8779.

SAM Domain-Based Protein Oligomerization Observed by Live-Cell Fluorescence Fluctuation Spectroscopy

Brian D. Slaughter, Joseph M. Huff, Winfried Wiegraebe, Joel W. Schwartz and Rong Li

ABSTRACT

Sterile-alpha-motif (SAM) domains are common protein interaction motifs observed in organisms as diverse as yeast and human. They play a role in protein homo- and hetero-interactions in processes ranging from signal transduction to RNA binding. In addition, mutations in SAM domain and SAM-mediated oligomers have been linked to several diseases. To date, the observation of heterogeneous SAM-mediated oligomers in vivo has been elusive, which represents a common challenge in dissecting cellular biochemistry in live-cell systems. In this study, we report the oligomerization and binding stoichiometry of high-order, multi-component complexes of (SAM) domain proteins Ste11 and Ste50 in live yeast cells using fluorescence fluctuation methods. Fluorescence cross-correlation spectroscopy (FCCS) and 1-dimensional photon counting histogram (1dPCH) confirm the SAM-mediated interaction and oligomerization of Ste11 and Ste50. Two-dimensional PCH (2dPCH), with endogenously expressed proteins tagged with GFP or mCherry, uniquely indicates that Ste11 and Ste50 form a heterogeneous complex in the yeast cytosol comprised of a dimer of Ste11 and a monomer of Ste50. In addition, Ste50 also exists as a high order oligomer that does not interact with Ste11, and the size of this oligomer decreases in response to signals that activate the MAP kinase cascade. Surprisingly, a SAM domain mutant of Ste50 disrupted not only the Ste50 oligomers but also Ste11 dimerization. These results establish an in vivo model of Ste50 and Ste11 homo- and hetero-oligomerization and highlight the usefulness of 2dPCH for quantitative dissection of complex molecular interactions in genetic model organisms such as yeast.

Introduction

Determining the state of protein complex formation is critical for understanding many signaling and structural pathways. Often protein interactions are mediated through conserved domains, such as the well-studied Src homology 3 (SH3) and PDZ domains [1]. The Sterile-alpha-motif (SAM) domain is another commonly occurring motif, facilitating diverse interactions including protein homodimerization, hetero-dimerization, and even RNA binding [2], [3]. Defects in the SAM domains of proteins have been observed in a number of human diseases [2], [4]–[8]. Notably, chromosomal translocation of the ETS family transcriptional regulator TEL (translocation Ets leukemia), a SAM-domain containing protein, has been frequently linked to human leukemias and it is thought that the diseases arise because SAM-mediated oligomerization constitutively activates mitogenic proteins [6], [9]–[11]. The importance of this domain has led to numerous studies determining the structure and stoichiometry of SAM-domain complexes [2]. Although in vitro SAM domains are capable of forming both homo- and hetero-oligomers, it remains unclear how SAM domains mediate protein interactions under in vivo settings, where most proteins are expressed at levels much lower than those often used in biochemical and structural analyses.

In yeast, Ste11 and Ste50 are SAM domain-containing signaling proteins involved in multiple morphogenetic pathways, including mating, invasive growth, and high-osmolarity response [12]–[14]. The interaction of these proteins through their SAM domains is thought to play a role in the delivery of Ste11, a MAP kinase kinase kinase, to the cell cortex to activate MAP kinase in response to environmental signals [15], [16]. Several groups have employed structural and biochemical methods to examine homo and hetero interactions of purified SAM domains of Ste11 and Ste50 in solution [17]–[21]. A consensus of their study is that Ste11 SAM domains form tight homodimers or high-order oligomers, whereas the Ste50 SAM domain, with a slightly different sequence, is largely monomeric in solution but can mediate strong heterodimerization with the Ste11 SAM domain [17], [18], [20], [21]. It is unknown, however, how these domains might be engaged in hetero or homo-oligomer formation in vivo, in the presence of the full length proteins expressed at their endogenous levels and in the presence of additional interacting partners.

Emerging fluorescence-based technologies probe in vivo binding equilibrium and stoichiometry of protein complexes. Fluorescence correlation spectroscopy (FCS) and fluorescence cross-correlation spectroscopy (FCCS) [22]–[26] are fluctuation techniques that analyze protein mobility, concentration, and protein-protein association (Figure 1A, 1B), and have recently been applied to live yeast cells expressing autofluorescent proteins (AFP) at the endogenous levels [27], [28]. While FCCS measures co-diffusion of two particles with different fluorescent tags, extraction of binding stoichiometry is not easily accomplished with this technique. The photon counting histogram (PCH) and similar techniques, such as fluorescence intensity distribution analysis (FIDA), are fluctuation techniques designed to analyze the oligomeric status of fluorescent species and have been applied to both in vitro and in vivo systems. These techniques determine the state of molecular homo-oligomerization [29]–[34] (Figure 1C); however, PCH does not resolve stoichiometry of heterogeneous complexes that result from dynamic protein interactions between different molecular species.

Two-dimensional PCH (2dPCH), where two different proteins are tagged with spectrally distinct probes, is a recently developed technique that can be used for simultaneous measurement of stoichiometry and interaction [35], [36]. In 2dPCH, a two-dimensional histogram of fluorescence counts in red and green channels is generated (Figure 1A, 1D) from fluctuation data. The surface of the two-dimensional histogram can be fit to yield a two-dimensional map of the brightness in each channel of diffusing species. For example, a hypothetical monomeric green probe diffusing alone with a brightness of 3000 counts per second per molecule would surface in the plot with a brightness in the green channel without a contribution in the red channel (Figure 1E). A red probe diffusing with no green particle would likewise only have a contribution in the red channel. However, if monomeric red and green probes are co-diffusing, the corresponding two-dimensional histogram would be best fit by a diffusing species with brightness contributions in both channels. In this way, both co-diffusion and stoichiometry of heterogeneous complexes may be observed. As a new technique, its in vivo applications have been limited to this point [35]. However, the combination of GFP with the improved red AFP, mCherry [37], [38], and the ease of introducing these tags to chromosomal loci through homologous recombination in yeast make it feasible to apply 2dPCH to assess high-order heterogeneous protein complexes in live yeast cells.

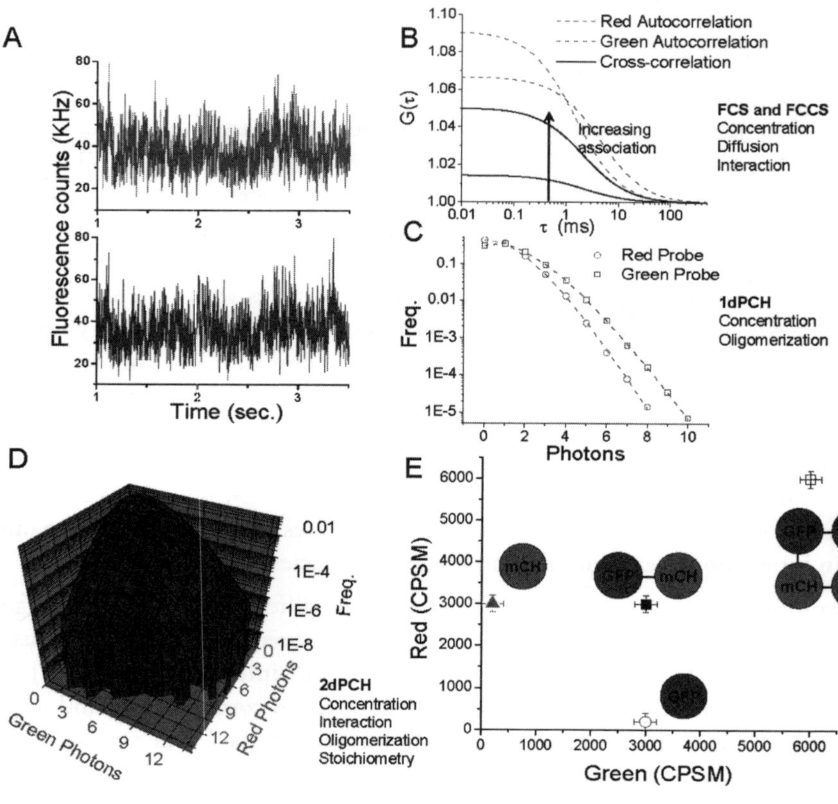

Figure 1. Fluctuation data can probe protein-protein interactions.
A. Example traces of fluctuation data for dual-color experiments. B. Data can be analyzed by correlation analysis to examine concentration, diffusion, and co-diffusion of red and green particles. C. 1dPCH examines the distribution of photon events per time interval, and reports concentration and 'brightness,' or oligomeric status. D. 2dPCH reports simultaneously concentration, interaction, oligomerization, and binding stoichiometry of heterogeneous complexes. An example two-dimensional PCH histogram is shown, with frequency versus number of green photons and number of red photons per time bin. E. Example, two-dimensional plot of a fit of modeled 2dPCH data. If a monomer red or green probe has a brightness of 3000 CPSM, for example, the plot demonstrates points one would expect to find values for with non-interacting monomeric species, or interacting monomeric species, or interacting dimeric species, as labeled.

In this study, we apply 2dPCH, FCCS, and 1d PCH in live budding yeast cells to examine the interaction between SAM domain-containing proteins Ste11 and Ste50 and the effect of mutations in Ste50's SAM domain on homotypic and heterotypic protein interactions. The data allows establishment of a dynamic model depicting homo and hetero-oligomeric complex formation among these SAM domain proteins, and represents the first application of 2dPCH to extract stoichiometry of high order heterogeneous complexes using endogenously expressed proteins in live cells.

Results

For examining Ste11 and Ste50 interaction, we constructed a yeast strain expressing Ste11-GFP and Ste50-mCherry from their respective chromosomal loci (Table 1). Fluorescence cross-correlation spectroscopy measurements were first performed as previously described [27] to confirm the expected heterotypic interaction (Figure 2). All of the fluorescence fluctuation measurements described in this study were made on the cytosolic pool by appropriately targeting the laser beam (Figure 2A and Materials and Methods). As expected, a high degree of cross-correlation was observed (Figure 2A, 2B), as demonstrated by the high amplitude of the cross-correlation curve relative to the autocorrelation curves of the individual channels. Results were quantified [39] and showed over 60% of Ste50 bound with Ste11 (Figure 2B). The interaction of these proteins remained strong after activation of the pheromone response pathway or the osmotic stress pathway (Figure 2B) (see Materials and Methods). To test if the observed interaction relies on the SAM domain, two mutations, L73A and L75A, were introduced into the Ste50 SAM domain (Table 1). These mutations were previously shown to abolish the binding between Ste11 and Ste50 SAM domains in vitro [17]. Ste50L73A-L75A-GFP was driven by the STE50 promoter from a centromeric plasmid and expressed in the ste50Δ background. The cross correlation was diminished (Figure 2A, 2B), demonstrating as expected that the Ste50 SAM domain plays a role in Ste50's interaction with Ste11 in vivo.

In vitro, the Ste11 and Ste50 SAM domains have been shown to mediate homo-oligomerization, but this has not been demonstrated in vivo when proteins are expressed at the endogenous levels. PCH analyzes the probability distribution of detected photons from a small confocal volume to calculate particle concentration and brightness, usually reported as the average number of molecules in the focal volume (N) and counts per second per molecule (CPSM), respectively (Figure 1C) [29], [40]. As a comparison technique, molecular brightness reflects the oligomeric state of the fluorescent species when compared to the brightness of a standard, for example, a known monomer or dimer of the same fluorescent molecule. For controls, yeast strains expressing monomeric, dimeric, and trimeric cytosolic GFP under the control of the BZZ1 promoter were constructed, as previously described (Table 1) [27]. The distribution of brightness values for 1dPCH measurements in live yeast cells were recorded using the Zeiss confocor 3 with 488 nm excitation and BP 505–540 nm emission collection (Materials and Methods). Box plots, as well as example curves, are shown in Figure 3A and 3B. The brightness distributions of these GFP species were easily distinguishable, providing the basis of comparison for oligomeric status of mobile GFP-tagged proteins in yeast. In addition, the BAT2 locus was replaced by cytosolic mCherry and the BZZ1 locus was replaced by mCherry-mCherry (Table 1). At the low excitation powers that are necessary to minimize photobleaching in our experiments, the brightness of mCherry was less than that of eGFP (Figure 3B) but, at over 2000 CPSM, still presented an improvement from other monomeric, red autofluorescent protein options [35], as expected based on improvements in photostability, and quantum yield [37], [38].

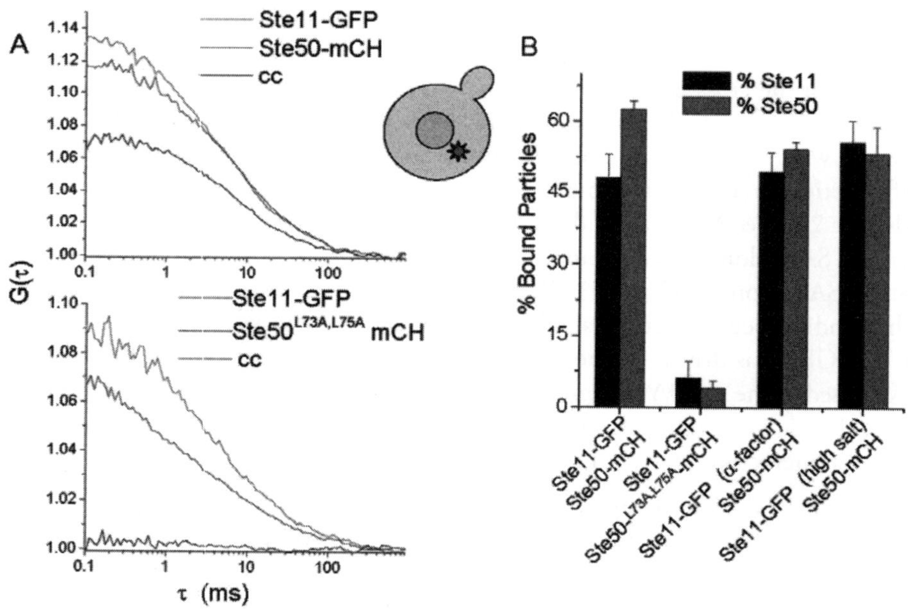

Figure 2. Cross-correlation analysis determines protein co-diffusion.
A. Autocorrelation and cross-correlation curves of Ste11-GFP and Ste50-mCherry and Ste11-GFP and Ste50-L73A-L75A-mCherry. Curves are the averages of multiple cells. B. Results for G1 cycling cells, and cells upon pheromone or osmotic stress pathway activation in live yeast cells (see Materials and Methods). Results were quantified as previously described, as the percentage of bound particles relative to total [27], [39].

Table 1. Yeast strains used in this study.

Strain	Genotype	Source
rly		[27]
2667	MATA *BAT2-GFP-mCherry::URA3* (6 Ala linker)	[27]
2748	MATA p*BZZ1::GFP:HIS5*	[27]
3118	MATA *;STE11-GFP::HIS5*	[27]
3120	MATA *;STE50-GFP::HIS5*	this study
3126	MATA p*BZZ1::GFP-GFP-GFP:URA3*	[27]
3165	MATA p*BZZ1::GFP-GFP:URA3*	[27]
3232	MATA *;STE11-GFP::HIS5 STE50mCherry::URA3*	this study
3282	MATA *;ste50Δ; STE11-GFP::URA3;* CEN *HIS5* pISTE50^{L73AL75A}-*mCherry*	this study
3283	MATA *;ste50Δ;* CEN *HIS5* pISTE50^{L73AL75A}-*GFP*	this study
3291	MATA *;pBZZ1::GFP:HIS5 pBAT2::mCHERRY::URA3*	this study
3489	MATA p*BZZ1::mCherry-mCherry::URA3*	this study

all strains are S288C background, his3Δ1;leu2Δ0;met15Δ0;ura3Δ0.

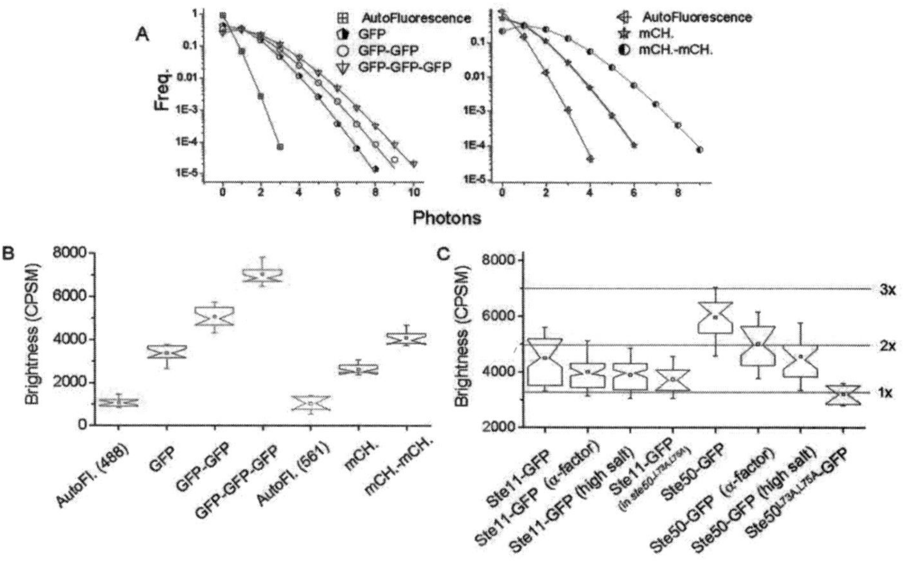

Figure 3. 1dPCH analysis of Ste50-GFP and Ste11-GFP probes homo-oligomerization.
A. Example curves for GFP and mCherry (mCH.) controls in live yeast. B. Notched box plots of PCH fits, ranging from 18 to 30 individual, 7 second data traces from 5 to 10 cells. For auto-fluorescence measurements, data represents 7 measurements for mCherry and 15 measurements for GFP. 50 μs bins were used to generate the PCH distributions. C. Notched box plots of 1dPCH fits of GFP tagged species, with lines (same color scheme as in B) representing average brightness values of monomer, dimer, and trimer controls for a basis of comparison.

Box plots of average brightness for individual 1dPCH measurements of Ste11 and Ste50 are presented in Figure 3C, with lines representing average brightness values of monomer, dimer, and trimer controls for a basis of comparison. Ste11 exhibited an average brightness close to that of dimeric cytosolic GFP, whereas Ste50 showed an average brightness in-between dimeric and trimeric cytosolic GFP. Interestingly, the SAM domain mutant, Ste50L73A-L75A-GFP, revealed a brightness much reduced relative to Ste50-GFP, near that of monomeric GFP, suggesting that these mutations also affect homo-oligomerization of Ste50. Surprisingly, PCH of Ste11-GFP in a yeast strain where the only form of Ste50 was untagged Ste50L73A-L75A revealed a decreased brightness, distinct from the distribution of Ste11-GFP in wild-type cells ($p<0.05$), suggesting that the SAM domain of Ste50 is also required for stabilization of Ste11 homo-oligomers.

To examine the effect of signals that normally activate MAP kinase cascade on Ste11-GFP and Ste50-GFP complexes, we activated the yeast mating response pathway by treatment of yeast cells with 50 μM α-factor or the osmotic stress pathway by treating the cells with 0.4 mM NaCl for 30 minutes [14] (see Materials and Methods). Average brightness values for PCH measurements are shown in Figure 3C. The average brightness of Ste11-GFP and Ste50-GFP were slightly decreased in response to both conditions.

The average brightness values of Ste11-GFP and Ste50-GFP, which were above the brightness of monomeric GFP, suggested an ability of these proteins to form homo-oligomeric structures, but the composition of the complexes was unclear. This represents a difficulty with 1dPCH. For example, the average brightness of Ste50-GFP could be explained by a distribution of dimers and trimers, but could also be explained by a distribution of monomers and high-order oligomers, or any other combination. In an ideal case, such as a solution measurement, sufficient statistics can be obtained to accurately distinguish a distribution of species freely from PCH data without a priori knowledge. However, this is not the case in live yeast cells due to limits of laser exposure time to minimize photobleaching. Similarly, other live cell studies have also found it necessary to make certain reasonable assumptions and/or fix brightness values to fit fluctuation data to distributions to extract additional information [32], [40].

To better examine the stoichiometry of the Ste11 and Ste50 complexes, 2dPCH was performed. As a proof of principle, we first applied 2dPCH to a yeast strain expressing GFP and mCherry physically linked to the cytosolic protein Bat2 [27] (Table 1). The 2dPCH histograms of Bat2-GFP-mCherry fit well to a 1 species model, with diffusing particles having coincidence brightness in the GFP and mCherry channels with values consistent with monomeric GFP and mCherry (Figure 4, compare to Figure 3B). As a negative control, 2dPCH was conducted for a yeast strain expressing unlinked, cytosolic GFP and cytosolic mCherry (Table 1). As expected, the data fit well with a two-species model (average chi^2 = 0.9) but not with a one-specie model (average chi^2 = 4.5). The resulting brightness values for the 2dPCH data sets were consistent with those expected for monomeric GFP and mCherry. In all two-species 2dPCH fits, an F test was used to validate the necessity of the second-component (F>97%).

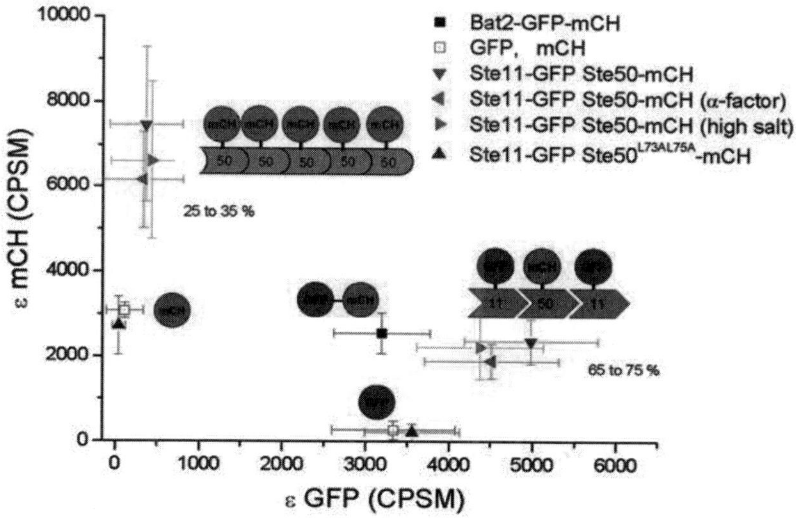

Figure 4. 2dPCH analysis of Ste50-mCherry and Ste11-GFP detects binding stoichiometry.
50 μs bins were used. Data were fit to a one-component model or two-component model, as explained in the text. Symbols and bars represent the averages and standard deviations, respectively. Schematic representations of average stoichiometry observed; possible geometries of the interactions (see main text) are displayed next to the corresponding regions of the graph.

2dPCH was applied to the Ste11-GFP, Ste50-mCherry fluctuation data, and revealed the in vivo binding stoichiometry of the complexes (Figure 4). Again, a one-specie model did not adequately fit the data (average chi^2 = 4.1); a two-component fit was necessary (average chi^2 = 1.2). The data reveal a dominant specie (N comprises approximately 65 to 75% of the total particles from the fit) with a mCherry average brightness (2400±150) consistent with that of a monomer ($p = 0.12$) and a GFP average brightness of 4980±230, which is indistinguishable from the 1dPCH GFP-GFP dimer brightness ($p = 0.7$). Thus, the data suggests a dominant complex in the yeast cytosol consisting of monomeric Ste50 and dimeric Ste11. A second abundant specie revealed by the 2dPCH consisted of a high order oligomer of Ste50 that is not associated with Ste11. This data reveals a mutual exclusiveness between Ste50 homo-oligomerization and Ste50 forming a complex that contains two molecules of Ste11. Based on 1dPCH data, we expected a small fraction of a third specie, consisting of monomeric Ste11, but a three-component fit cannot be confirmed with statistics provided by the live cell measurements. Consistent with the lack of cross-correlation reported in Figure 2, 2dPCH of Ste11-GFP, Ste50L73A-L75A-mCherry did not fit with a one-specie model (average chi^2 = 6.8), but rather a two-species model (average chi^2 = 1.2), with non-interacting, monomeric species. This result confirms the FCCS and 1dPCH results that the Ste50 SAM domain is required for homo-oligomerization of Ste50, interaction of Ste50 with Ste11, and it plays a role in stabilization of the Ste11 dimer.

The effects of activation of the mating pathway and osmotic stress pathway were subtle, with the dominant specie in either case still consisting of monomeric Ste50 interacting with Ste11. The average GFP brightness value of this dominant complex under these conditions were lower than that in cycling cells; this decrease was statistically significant at the 95% level for high salt conditions relative to wt ($p = 0.05$) but not at the 95% confidence limit for α-factor treated cells ($p = 0.11$). The trend is consistent with a lower average brightness of the Ste11 component of this complex, and perhaps a distribution of interacting species that varies between 2:1 and 1:1 Ste11:Ste50. The average brightness of the Ste50 high-order oligomer observed by 2dPCH was significantly reduced upon activation of the pheromone pathway ($p = 0.02$), while the average brightness of the oligomer upon activation of the osmotic stress pathway was not reduced at a statistically high confidence level ($p = 0.24$).

Thus, a possible effect of activation of these pathways is a trend toward a reduction in the size of the high-order Ste50 oligomer. This is consistent with change observed by 1dPCH as the decreased average brightness of Ste50-GFP upon activation of the signaling pathways. However, at this point we are uncertain how brightness of mCherry containing complexes scales with number of mCherry subunits at high stoichiometry. The brightness of GFP complexes scales well with GFP subunits, as demonstrated by the fit of the average brightness of the monomeric, dimer, and trimeric controls (Figure 5). We revisited the 1dPCH data of Ste50-GFP to attempt to better quantify the stoichiometry of the high-order, Ste50 oligomer. The result that the dominant specie of Ste50 was a monomer, as shown by 2dPCH, provided an important constraint for fitting the 1dPCH data to a distribution. Therefore the 1dPCH data for Ste50-GFP was fit to a distribution consisting of a fixed monomer brightness and freely varied oligomer brightness. N for each species was also freely varied. The Ste50 1dPCH data was well fit (average chi^2 = 0.9) to a distribution that consists of a large percentage of monomer and a small percentage of high order oligomer, in percentages roughly equivalent with those found using 2dPCH. The constrained fits revealed that the treatment with α-factor or high salt led to a slightly decreased percentage of oligomer, and also a decreased oligomer brightness (Figure 5). Using this analysis, average brightness values of the Ste50 oligomer were approximately 12,500, 8900, and 7700 CPSM for cycling cells, α-factor treated cells, and cells at high salt, respectively (Figure 5). Assuming that the linearity of GFP brightness continues to hold at a high number of subunits, we estimate the number of Ste50 subunits in this complex to be ~5 to 6 in cycling cells, ~4 in α-factor treated cells, and ~3 to 4 in high salt condition. While the slight decrease in brightness of the Ste50 oligomer was also observed with mCherry in the 2dPCH data, it was not nearly as pronounced as that observed with 1dPCH (Figure 3D) or the global fit (Figure 5), raising the possibility that mCherry brightness may not scale linearly with subunit number at high stoichiometry, perhaps due to a self-quenching mechanism or increased propensity for photobleaching or photoblinking relative to GFP.

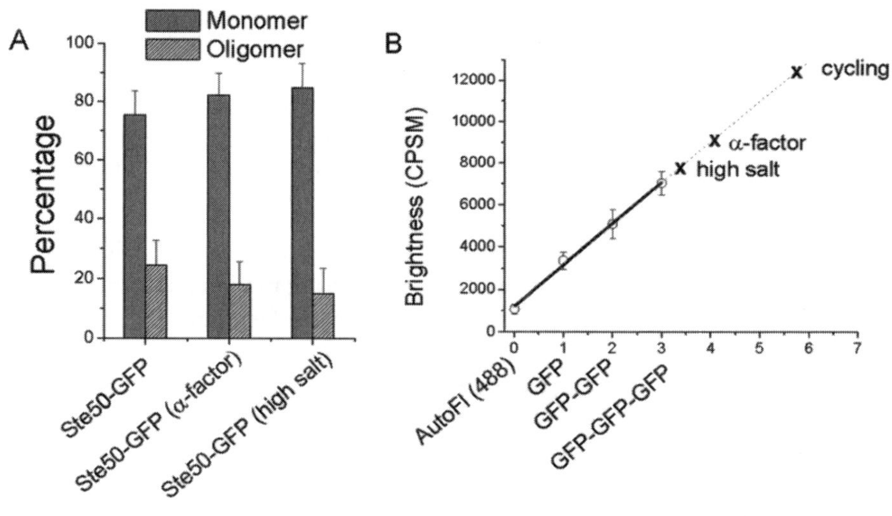

Figure 5. Constrained fits of the 1dPCH data to two-species allows for the examination of monomer and oligomer populations.
A. A model was assumed where Ste50 could exist as either a monomer with fixed brightness, or oligomer with unconstrained brightness and number (see Materials and Methods, and main text). Results are displayed to show the percentage of each component. Error bars are the standard error of the mean. B. Average brightness values for autofluorescence, GFP, GFP-GFP, and GFP-GFP-GFP from the data in Figure 3, fit to a line. Error bars are the standard deviation. The line represents the best fit of the data to a linear model with a slope of 1959 and intercept 1193, which was then extrapolated toward higher brightness. Average brightness observed for the Ste50 oligomer from the analysis described above are marked on the extrapolated part of the plot.

Discussion

The analyses presented above demonstrate that whereas FCCS and 1dPCH provide valuable information on the strength of the interaction between two molecular species and average oligomerization status of individual molecular species, respectively, 2dPCH is better suited for revealing the binding stoichiometry of protein complexes in vivo. Specifically in this study, 2dPCH revealed a predominant complex of Ste11 and Ste50 with a 2:1 binding stoichiometry, and a pool of large, Ste11-free Ste50 oligomers. The heterotypic complex is consistent with previous biochemical data [17], [18], [21]. However, the requirement of Ste50 SAM domain for the integrity of this complex in vivo suggests that either Ste11 molecules in this complex do not directly dimerize through their SAM domains but rather they each bind a common Ste50 SAM domain at two different surfaces (Figure 4); or, a strong direct interaction between two Ste11 molecules is stabilized by the Ste50 SAM domain. The former possibility is supported by the structural study demonstrating that Ste50 binds Ste11 in a head to tail fashion [19], which could also explain the ability of Ste50 to form homo-oligomers (Figure 4). The most unexpected finding was the presence of large Ste50, but not Ste11, homo-oligomers, because in vitro the Ste11 SAM domain, but not the Ste50 SAM domain, has the strong propensity to form oligomers [2], [6], [20], [21]. Even though we do not have an explanation for the difference between the in vivo and in vitro observations, this result highlights the need to directly probe protein complex stoichiometry using techniques such as 2dPCH in live-cell settings.

The SAM domain-mediated interaction between Ste50 and Ste11 is known to be important for efficient MAP kinase signaling during mating and osmotic stress responses [12], [15]. It is thought that this interaction facilitates the targeting of Ste11 to the site of receptor signaling at the plasma membrane where Ste20, the upstream kinase for Ste11, is activated by the small GTPase Cdc42 [12], [14], [15], [41] Ste50 itself is recruited to the membrane through an interaction of the C-terminal RA domain with Cdc42 [16]. An ability of one Ste50 to simultaneously bring two molecules of Ste11 to the site of active Ste20 on the cortex may significantly enhance the efficiency of Ste11 phosphorylation by Ste20, while the large Ste50 homo-oligomers may be a dynamic reservoir for Ste50 that could buffer the concentration of Ste50 monomers available for interaction with Ste11. A recent study showed that the mobile Ste11 concentration increases following pheromone stimulation [27], and this would be consistent with a need to mobilize some of the Ste50 reserved in the homo-oligomers to the monomer pool, leading to the reduced average size of the Ste50 oligomer.

The homo-oligomers of Ste50 observed in yeast may be similar to the high affinity, high-order oligomers proteins observed for SAM domains from proteins such as translocation ETS leukemia (TEL) or Eph receptor tyrosine kinase (EphB2) [6], [42]. As mentioned above, the constrained two-component fit of the 1dPCH data for Ste50-GFP reported a stoichiometry of the Ste50-oligomer in cycling cells of 5 to 6 subunits. Given the caveat that the brightness of GFP might not rise linearly with high-order oligomers, it is still intriguing that the estimated stoichiometry of Ste50 in vivo compares well to the stoichiometry proposed for oligomers of TEL or EphB2 [2], [6], [42]. While these models generally predict a linear or head-to-tail model, if this applies to the Ste50 oligomer it would be difficult to explain the observed lack of binding to Ste11. One possibility is that formation of the Ste50 oligomer induces a conformational change in the Ste50 SAM domain that makes binding to Ste11 less likely, as depicted in Figure 4. It may be interesting to test if the TEL SAM domain homo-oligomers, which are thought to account for certain types of human leukemia [2], [6], [9]–[11], also modulate the interaction with other regulatory molecules and respond dynamically to morphogenetic signals.

Materials and Methods

Yeast Culture

One-step COOH-terminus genomic tagging was used for generating yeast strains expressing both GFP and mCherry labeled genes [43], [44] unless otherwise specified. Correct tagging was verified by PCR. For correlation analysis, yeast cells were grown in synthetic complete media to mid-log phase. For examination of Ste50 SAM mutants, Ste50-GFP or Ste50-mCherry was subcloned into a centromeric plasmid, with its promoter region. Mutations were made and verified with sequencing. The centromeric plasmids were transformed into Δste50 strains for analysis. For examination of the mating pathway, pheromone (α-factor) was incubated with yeast at a concentration of 50 μM for 2 hours. For activation of the HOG pathway, yeast cells were treated with 0.4 mM NaCl for 30 minutes as described [14]. Yeast cells were immobilized on glass for analysis.

FCS

The experimental set-up for FCS, FCCS, and 1dPCH was used as previously described [27]. Briefly, for cross-correlation studies, the 488 nm and 561 nm laser lines were used with the HFT 488/561 dichroic to excite GFP and mCherry, respectively. Emission was split with an HFT565 dichroic, and a 505–540 BP emission filter was used for the green emission channel and a LP 580 nm filter for the red channel. The pinhole was set to 1.0 airy unit in the red channel. Autocorrelation curves for the individual channels and a cross-correlation curve between the channels was calculated by the Zeiss Confocor-3 software according to equations 1 and 2, respectively:

$$G_{ac}(\tau) = \frac{\langle \delta F(t) \cdot \delta F(t+\tau) \rangle}{\langle F(t) \rangle^2} \quad (1)$$

$$G_{cc}(\tau) = \frac{\langle \delta P1(t) \cdot \delta P2(t+\tau) \rangle}{\langle P1(t) P2(t) \rangle} \quad (2)$$

where $\delta F(t) = F(t) - \langle F \rangle$ and P1 and P2 represent photon counts in channel 1 and 2, respectively.

For cross-correlation experiments, the number of bound particles was calculated from Eq. 3 [39], where the inverse amplitude of the autocorrelation curves was used to calculate the number of red (NRT) or green particles (NGT) (Eq. 4). Ncc is the inverse amplitude of the cross-correlation curve, and the cross-talk between channels, Q, was estimated to be approximately 5% for the GFP and mCherry probes using the filter sets described above. This is discussed in more detail elsewhere [27]. A volume correction was applied to take into account small differences in red, green, and cross-correlation volumes [45].

$$N_{bound} = \frac{N_{GT}(N_{RT} + Q \cdot N_{GT})}{N_{cc}} - N_{GT} \cdot Q \quad (3)$$

$$N = \gamma / (G_0 - 1) \cdot \quad (4)$$

PCH

1dPCH was conducted as previously described [27]. The 488 nm laser line was used with a HFT 488/561 main dichroic, an HFT565 secondary emission dichroic, and a 505–540 BP emission filter. Fluorescence traces were collected in 7 second increments, and 4 to 5 measurements were collected per cell. Data was arranged as a histogram of number of photon events per unit time using a bin time of 50 µs. Data was fit with the PCH algorithm [29], [30] to extract an average brightness and particle number per measurement. A 3-dimensional Gaussian focal volume (1-photon) was used. Control strains expressing GFP, GFP-GFP and GFP-GFP-GFP linked proteins (under the BZZ1 promoter) were used for monomer, dimer, and trimer brightness controls (Table 1). Importantly, to verify the reliability of 1-photon PCH in live yeast, the brightness values for the control strains were linearly spaced, taking auto-fluorescence into account.

For 2dPCH [35], [36], the experimental set-up was identical to the cross-correlation set-up described above and previously [27]. Data were taken in ten-second increments, and arranged in a two-dimensional histogram of counts in each channel as a function of frequency for 50 μs bins. Data were fit to one or two components, and brightness values and particle number in each channel were allowed to freely vary. An F-test was used to validate the necessity of two-component fits (F>97%). The plots in Figure 4 represent averages and standard deviations for N between 6 and 15 cells.

For fitting Ste50-GFP, 1dPCH data to a two-component model, as shown in Figure 5, the brightness of a monomer specie was fixed. The number of monomer, brightness of oligomer, and number of oligomer were freely varied.

Acknowledgements

We are grateful to Giulia Rancati, Stowers Institute for Medical Research (SIMR) for providing useful yeast strains, Boris Rubinstein (SIMR) for insightful discussions on data analysis, Jay Unruh (University of California, Irvine) for homebuilt software, and SIMR Molecular Biology facility for mutant generation.

Authors' Contributions

Conceived and designed the experiments: RL BS JS JH. Performed the experiments: BS. Analyzed the data: BS. Contributed reagents/materials/analysis tools: JS JH WW. Wrote the paper: RL BS.

References

1. Pawson T, Nash P (2003) Assembly of cell regulatory systems through protein interaction domains. Science 300: 445–452.
2. Qiao F, Bowie JU (2005) The many faces of SAM. Sci STKE 2005: re7.
3. Kim CA, Bowie JU (2003) SAM domains: uniform structure, diversity of function. Trends Biochem Sci 28: 625–628.
4. McGrath JA, Duijf PH, Doetsch V, Irvine AD, de Waal R, et al. (2001) Hay-Wells syndrome is caused by heterozygous missense mutations in the SAM domain of p63. Hum Mol Genet 10: 221–229.
5. Sahin MT, Turel-Ermertcan A, Chan I, McGrath JA, Ozturkcan S (2004) Ectodermal dysplasia showing clinical overlap between AEC, Rapp-Hodgkin and CHAND syndromes. Clin Exp Dermatol 29: 486–488.

6. Kim CA, Phillips ML, Kim W, Gingery M, Tran HH, et al. (2001) Polymerization of the SAM domain of TEL in leukemogenesis and transcriptional repression. Embo J 20: 4173–4182.
7. Cicero DO, Falconi M, Candi E, Mele S, Cadot B, et al. (2006) NMR structure of the p63 SAM domain and dynamical properties of G534V and T537P pathological mutants, identified in the AEC syndrome. Cell Biochem Biophys 44: 475–489.
8. Kantaputra PN, Hamada T, Kumchai T, McGrath JA (2003) Heterozygous mutation in the SAM domain of p63 underlies Rapp-Hodgkin ectodermal dysplasia. J Dent Res 82: 433–437.
9. Boccuni P, MacGrogan D, Scandura JM, Nimer SD (2003) The human L(3) MBT polycomb group protein is a transcriptional repressor and interacts physically and functionally with TEL (ETV6). J Biol Chem 278: 15412–15420.
10. Jousset C, Carron C, Boureux A, Quang CT, Oury C, et al. (1997) A domain of TEL conserved in a subset of ETS proteins defines a specific oligomerization interface essential to the mitogenic properties of the TEL-PDGFR beta oncoprotein. Embo J 16: 69–82.
11. Tognon CE, Mackereth CD, Somasiri AM, McIntosh LP, Sorensen PH (2004) Mutations in the SAM domain of the ETV6-NTRK3 chimeric tyrosine kinase block polymerization and transformation activity. Mol Cell Biol 24: 4636–4650.
12. Ramezani-Rad M (2003) The role of adaptor protein Ste50-dependent regulation of the MAPKKK Ste11 in multiple signalling pathways of yeast. Curr Genet 43: 161–170.
13. Posas F, Witten EA, Saito H (1998) Requirement of STE50 for osmostress-induced activation of the STE11 mitogen-activated protein kinase kinase kinase in the high-osmolarity glycerol response pathway. Mol Cell Biol 18: 5788–5796.
14. Tatebayashi K, Yamamoto K, Tanaka K, Tomida T, Maruoka T, et al. (2006) Adaptor functions of Cdc42, Ste50, and Sho1 in the yeast osmoregulatory HOG MAPK pathway. Embo J 25: 3033–3044.
15. Wu C, Jansen G, Zhang J, Thomas DY, Whiteway M (2006) Adaptor protein Ste50p links the Ste11p MEKK to the HOG pathway through plasma membrane association. Genes Dev 20: 734–746.
16. Truckses DM, Bloomekatz JE, Thorner J (2006) The RA domain of Ste50 adaptor protein is required for delivery of Ste11 to the plasma membrane in the filamentous growth signaling pathway of the yeast Saccharomyces cerevisiae. Mol Cell Biol 26: 912–928.

17. Grimshaw SJ, Mott HR, Stott KM, Nielsen PR, Evetts KA, et al. (2004) Structure of the sterile alpha motif (SAM) domain of the Saccharomyces cerevisiae mitogen-activated protein kinase pathway-modulating protein STE50 and analysis of its interaction with the STE11 SAM. J Biol Chem 279: 2192–2201.

18. Kwan JJ, Warner N, Pawson T, Donaldson LW (2004) The solution structure of the S.cerevisiae Ste11 MAPKKK SAM domain and its partnership with Ste50. J Mol Biol 342: 681–693.

19. Kwan JJ, Warner N, Maini J, Chan Tung KW, Zakaria H, et al. (2006) Saccharomyces cerevisiae Ste50 binds the MAPKKK Ste11 through a head-to-tail SAM domain interaction. J Mol Biol 356: 142–154.

20. Bhattacharjya S, Xu P, Chakrapani M, Johnston L, Ni F (2005) Polymerization of the SAM domain of MAPKKK Ste11 from the budding yeast: implications for efficient signaling through the MAPK cascades. Protein Sci 14: 828–835.

21. Bhattacharjya S, Xu P, Gingras R, Shaykhutdinov R, Wu C, et al. (2004) Solution structure of the dimeric SAM domain of MAPKKK Ste11 and its interactions with the adaptor protein Ste50 from the budding yeast: implications for Ste11 activation and signal transmission through the Ste50-Ste11 complex. J Mol Biol 344: 1071–1087.

22. Bacia K, Kim SA, Schwille P (2006) Fluorescence cross-correlation spectroscopy in living cells. Nat Methods 3: 83–89.

23. Bacia K, Schwille P (2003) A dynamic view of cellular processes by in vivo fluorescence auto- and cross-correlation spectroscopy. Methods 29: 74–85.

24. Thompson NL, Lieto AM, Allen NW (2002) Recent advances in fluorescence correlation spectroscopy. Curr Opin Struct Biol 12: 634–641.

25. Elson EL, Magde D (1974) Fluorescence correlation spectroscopy. I. Conceptual basis and theory. Biopolymers 13: 1–27.

26. Haustein E, Schwille P (2003) Ultrasensitive investigations of biological systems by fluorescence correlation spectroscopy. Methods 29: 153–166.

27. Slaughter BD, Schwartz JW, Li R (2007) Mapping dynamic protein interactions in MAP kinase signaling using live-cell fluorescence fluctuation spectroscopy and imaging. Proc Natl Acad Sci USA 104: 20320–20325.

28. Maeder CI, Hink MA, Kinkhabwala A, Mayr R, Bastiaens PI, et al. (2007) Spatial regulation of Fus3 MAP kinase activity through a reaction-diffusion mechanism in yeast pheromone signalling. Nat Cell Biol 9: 1319–1326.

29. Chen Y, Muller JD, So PT, Gratton E (1999) The photon counting histogram in fluorescence fluctuation spectroscopy. Biophys J 77: 553–567.

30. Chen Y, Muller JD, Ruan Q, Gratton E (2002) Molecular brightness characterization of EGFP in vivo by fluorescence fluctuation spectroscopy. Biophys J 82: 133–144.

31. Hink MA, Shah K, Russinova E, de Vries SC, Visser AJ (2007) Fluorescence Fluctuation Analysis of AtSERK and BRI1 Oligomerization. Biophys J.

32. Saffarian S, Li Y, Elson EL, Pike LJ (2007) Oligomerization of the EGF receptor investigated by live cell fluorescence intensity distribution analysis. Biophys J 93: 1021–1031.

33. Digman MA, Brown CM, Horwitz AF, Mantulin WW, Gratton E (2007) Paxillin dynamics measured during adhesion assembly and disassembly by correlation spectroscopy. Biophys J.

34. Kask P, Palo K, Ullmann D, Gall K (1999) Fluorescence-intensity distribution analysis and its application in biomolecular detection technology. Proc Natl Acad Sci USA 96: 13756–13761.

35. Hillesheim LN, Chen Y, Muller JD (2006) Dual-color photon counting histogram analysis of mRFP1 and EGFP in living cells. Biophys J 91: 4273–4284.

36. Chen Y, Tekmen M, Hillesheim L, Skinner J, Wu B, et al. (2005) Dual-color photon-counting histogram. Biophys J 88: 2177–2192.

37. Shaner NC, Campbell RE, Steinbach PA, Giepmans BN, Palmer AE, et al. (2004) Improved monomeric red, orange and yellow fluorescent proteins derived from Discosoma sp. red fluorescent protein. Nat Biotechnol 22: 1567–1572.

38. Shaner NC, Steinbach PA, Tsien RY (2005) A guide to choosing fluorescent proteins. Nat Methods 2: 905–909.

39. Rigler R, Foldes-Papp Z, Meyer-Almes FJ, Sammet C, Volcker M, et al. (1998) Fluorescence cross-correlation: a new concept for polymerase chain reaction. J Biotechnol 63: 97–109.

40. Chen Y, Wei LN, Muller JD (2003) Probing protein oligomerization in living cells with fluorescence fluctuation spectroscopy. Proc Natl Acad Sci USA 100: 15492–15497.

41. Lamson RE, Winters MJ, Pryciak PM (2002) Cdc42 regulation of kinase activity and signaling by the yeast p21-activated kinase Ste20. Mol Cell Biol 22: 2939–2951.

42. Thanos CD, Goodwill KE, Bowie JU (1999) Oligomeric structure of the human EphB2 receptor SAM domain. Science 283: 833–836.

43. Huh WK, Falvo JV, Gerke LC, Carroll AS, Howson RW, et al. (2003) Global analysis of protein localization in budding yeast. Nature 425: 686–691.

44. Sheff MA, Thorn KS (2004) Optimized cassettes for fluorescent protein tagging in Saccharomyces cerevisiae. Yeast 21: 661–670.

45. Schwille P, Meyer-Almes F-J, Rigler R (1997) Dual-color fluorescence cross-correlation spectroscopy for multicomponent diffusional analysis in solution. Biophys J 72: 1878–1886.

In-Vivo Optical Detection of Cancer Using Chlorin E6—Polyvinylpyrrolidone Induced Fluorescence Imaging and Spectroscopy

William W. L. Chin, Patricia S. P. Thong, Ramaswamy Bhuvaneswari, Khee Chee Soo, Paul W. S. Heng and Malini Olivo

ABSTRACT

Background

Photosensitizer based fluorescence imaging and spectroscopy is fast becoming a promising approach for cancer detection. The purpose of this study was to examine the use of the photosensitizer chlorin e6 (Ce6) formulated in polyvinylpyrrolidone (PVP) as a potential exogenous fluorophore for fluorescence imaging and spectroscopic detection of human cancer tissue xenografted in preclinical models as well as in a patient.

Methods

Fluorescence imaging was performed on MGH human bladder tumor xenografted on both the chick chorioallantoic membrane (CAM) and the murine model using a fluorescence endoscopy imaging system. In addition, fiber optic based fluorescence spectroscopy was performed on tumors and various normal organs in the same mice to validate the macroscopic images. In one patient, fluorescence imaging was performed on angiosarcoma lesions and normal skin in conjunction with fluorescence spectroscopy to validate Ce6-PVP induced fluorescence visual assessment of the lesions.

Results

Margins of tumor xenografts in the CAM model were clearly outlined under fluorescence imaging. Ce6-PVP-induced fluorescence imaging yielded a specificity of 83% on the CAM model. In mice, fluorescence intensity of Ce6-PVP was higher in bladder tumor compared to adjacent muscle and normal bladder. Clinical results confirmed that fluorescence imaging clearly captured the fluorescence of Ce6-PVP in angiosarcoma lesions and good correlation was found between fluorescence imaging and spectral measurement in the patient.

Conclusion

Combination of Ce6-PVP induced fluorescence imaging and spectroscopy could allow for optical detection and discrimination between cancer and the surrounding normal tissues. Ce6-PVP seems to be a promising fluorophore for fluorescence diagnosis of cancer.

Background

As with most cancers, early diagnosis is critical to achieve favorable prognosis. Currently, random surveillance biopsies are the existing gold standard for the identification of lesions in pre-neoplastic conditions. However this method is prone to sampling error, time-consuming, subjective and cost-inefficient. A diagnostic method that could provide rapid, automated classification of cancer lesions would increase the efficiency and comprehensiveness of malignancy screening and surveillance procedures. A variety of optical techniques have recently been utilized for the diagnostic study of cancerous tissue. These include fluorescence spectroscopy [1], Raman spectroscopy [2], light scattering spectroscopy [3], and Fourier-transform infrared spectroscopy [4]. These optical spectroscopic techniques are capable of providing biochemical and morphological information in short integration times, which can be used for automated diagnosis of intact tissue. However, in order to be useful as a comprehensive screening procedure, the optical technique must allow rapid real time imaging of a large area of tissue rather than point by point measurement, such that suspicious regions could be identified accurately and biopsied for histopathologic correlation [5].

With the advent of molecular probes, imaging methods such as ultrasound, microCT (Computed Tomography), microMRI (Magnetic Resonance Imaging), and microPET (Positron Emission Tomography) can be conducted not only to visualize gross anatomical structures, but also to visualize substructures of cells and monitor molecule dynamics [6]. Imaging of endogenous or exogenous fluorochromes has several important advantages over other optical approaches for tumor imaging. This imaging technique relies on fluorochrome induced fluorescence, reflectance, absorption or bioluminescence as the source of contrast, while imaging systems can be based on diffuse optical tomography, surface-weighted imaging, phase-array detection, intensified matrix detector and charged-coupled device camera detection, confocal endomicroscopy, multiphoton imaging, or microscopic imaging with intravital microscopy [7,8]. Fluorescence ratio imaging is a method widely used for optical diagnosis of cancer after administration of a photosensitizer [9]. Enhanced contrast between tumor and adjacent normal tissue can be obtained based on calculating the ratio between red intensity of the photosensitizer (600–700 nm) over the blue/green intensity of the back-scattered excitation light or tissue autofluorescence (450–550 nm). Many investigations have confirmed good agreement with the histopathological extent of the tumor, implying that this technique can be applied as a useful tool for indicating tumor boundary [10].

A number of fluorochromes such as fluorescein, toluidine blue, cyanine dyes and indocyanine green have been described with variable stabilities, quantum efficiencies, and ease of synthesis. However, most of the fluorochromes are not tumor specific and are rapidly eliminated from the organism. Chemically and endogenously synthesized fluorochromes such as porphyrin based photosensitizers have properties that may be utilized both experimentally and clinically. Porphyrins have been known to naturally localize in malignant tissue where they emit light when irradiated at certain wavelengths, providing a means to detect tumor by the location of its fluorescence. However, one of the major limitation is its slow clearance from tissues and long period of skin phototoxicity. Moreover, the porphyrin's core absorbs wavelengths of light too short for optimal penetration in tissue. As such, by reducing a pyrrole double bond on the porphyrin periphery, a chlorin core compound can be generated with a high absorption at longer wavelengths of 660—670 nm that can penetrate deeper in human tissue than those of porphyrins. Of particular interest among the evaluated chlorins is the naturally occurring chlorin e6 (Ce6) [11]. Ce6 has improved efficacy and has decreased side effects compared to first generation photosensitizers from hematoporphyrin derivatives. Due to the importance of Ce6's characteristic fluorescence properties, there is a need to identify new formulations that are stable, exhibit ease in manufacturing and selectively deliver the photosensitizer to target tissue in an efficient manner. Hence, we have investigated the use of Ce6 in combination with the polymer polyvinylpyrrolidone (Ce6-PVP). Polyvinylpyrrolidone is one of the most important excipient used in modern pharmaceutical technology. We have previously described the selective localization and photodynamic activity of Ce6-PVP in nasopharyngeal and lung carcinoma models that provided rationale for its use as a therapeutic agent for photodynamic therapy [12,13]. By employing a chick chorioallantoic membrane model, Ce6-PVP was shown to selective accumulate in bladder tumors xenografts and had a faster clearance from normal CAM when administered topically compared to systematic administration [14]. The uptake ratio of Ce6-PVP was found to have a 2-fold increase across the CAM when compared to that of Ce6, indicating that PVP was able to facilitate diffusion of Ce6 across the membrane [15]. Furthermore, Ce6-PVP had less in vivo systemic phototoxic effect compared to Ce6 alone after light irradiation in photodynamic therapy in mice bearing tumors [16]. Using a chemical fluorescence extraction technique and cuvette-based spectrofluorimetry, our data demonstrated that the distribution of Ce6-PVP drug were much lower in normal organs such liver, spleen, kidney, brain, heart and lung compared to Ce6 delivered using dimethylsulfoxide (DMSO) [17]. We also postulated that the extent of tumor necrosis post Ce6-PVP mediated photodynamic therapy (PDT) was dependent on the plasma concentration of Ce6-PVP, implying a vascular mediated cell death mechanism [18].

In this study, we have evaluated the usefulness of Ce6-PVP to accurately define the margin of the tumor from its normal adjacent tissue in the chick chorialllantoic membrane (CAM) tumor model. We also presented visual information on Ce6-PVP induced fluorescence in tumor and gross anatomical structure of normal organ of a murine model. Fluorescence spectroscopy measurements were also performed to characterize emission spectra from these tissue samples as well as to corroborate results from fluorescence images. Finally, a pilot trial was carried out to validate the use of Ce6-PVP as a clinically relevant diagnostic photosensitizer using both imaging and spectroscopy modality for differentiation of normal and tumor tissue in a patient.

Methods

Photosensitizer

The formulation of Ce6-PVP, also known as Fotolon or Photolon was supplied by HAEMATO-science GmbH, Germany. It is a co-lyophilisate of Ce6 sodium salt and PVP (a pharmaceutical grade polymer, molecular mass ≈ 12,000 g/mol) in a 1:1 mass ratio [19].

Cell Culture

MGH (European Collection of Cell Cultures), a poorly differentiated human bladder carcinoma cells were grown as a monolayer in RPMI-1640 medium supplemented with 10% fetal bovine serum, 1% non-essential amino acids (Gibco, USA), 1% sodium pyruvate (Gibco, USA), 100 units mL-1 penicillin-streptomycin (Gibco, USA) and incubated at 37°C, 95% humidity and 5% CO_2.

Chick Choriallantoic Membrane Tumor Model

Fertilized chicken eggs were incubated at 37°C in a humidified atmosphere inside a hatching incubator equipped with an automatic rotator (Octagon 20, Brinsea, Somerset, UK). At embryo age (EA) 7, a window of about 1.5 cm was opened in the eggshell to detach the shell membrane from the developing CAM. Then, the window was sealed with sterilized parafilm to avoid contamination and the eggs were returned to the static incubator for further incubation until the day of experiment. On EA 9, approximately 5–10 × 10^6 MGH cells were inoculated on the CAM. The window of the eggs were resealed with sterile parafilm and returned to the static incubator. Grafted cells were allowed to grow on the CAM for up to 5 days. On EA 14, Ce6-PVP was dissolved in 0.9% sodium chloride to constitute a stock solution of 1 mg/mL. The stock solution was further diluted to obtain a volume of 500 µL containing a dose of 1 mg/kg body weight of the chick's embryo. The photosensitizer was applied on the entire surface of the CAM and left to incubate for 30 min. The window was resealed to avoid evaporation of the drug solution from the CAM. After 30 min incubation, macroscopic fluorescence imaging was performed at 0.5, 1, 2, 3, 4, and 5 h post drug administration using a commercially available fluorescence endoscopic system (Karl Storz, Tuttlingen, Germany). A modified xenon short arc lamp (D—Light system in blue light mode, Karl Storz) filtered by a band pass filter (380—450 nm) was used for excitation of photosensitizer in tissue. Fluorescence was captured via a sensitive CCD camera (Tricam SL PAL, Karl Storz) attached to an endoscope integrated with a long pass filter (cut-off wavelength 470 nm). This observation LP filter of the endoscope only minimally transmits the diffuse back-scattering excitation light with a peak at 450 nm (blue light), while has a transmission of over 98% in the 470—800 nm range. The red channel registered the photosensitizer's fluorescence and the blue channel captured the diffusely back-scattered excitation light. A short exposure of the surface of tissue to the excitation light (10 s) was performed to avoid excessive photobleaching effects. White light imaging was used to correlate the boundaries of tumors and organs. All procedures involving preparation and administration of the photosensitizer were conducted under low ambient lighting.

Murine Tumor Model

A total of 10 Balb/c athymic nude mice and C57 mice, 6–8 weeks of age, weighing an average of 24 g were obtained from the Animal Resource Centre, Western Australia and Centre for Animal Resources, National University of Singapore respectively. Before inoculation, the cell layer was washed with phosphate-buffered saline, trypsinized, and counted using a hemocytometer. Approximately 3.0×10^6 MGH cells suspended in 150 μl of Hanks' Balanced Salt Solution (Gibco, USA) were injected subcutaneously into the lower flanks of Balb/c athymic nude mice. The animals were used for experiments when the tumors measured around 7—10 mm in diameter. This ensured that the tumor sizes were kept consistent to minimize variations due to the degree of vascularization of the implants. Mice were injected with a dose of 5 mg/kg of Ce6-PVP via tail vein injection. At 1, 3 and 6 h, mice were sacrificed and the skin overlaying the tumor was carefully removed to expose the tumor and normal peritumoral muscle for fluorescence imaging. C57 mice were used for imaging and spectroscopy of normal organs. All procedures were approved by the Institutional Animal Care and Use Committee, SingHealth, Singapore, in accordance with international standards.

Spectroscopic Measurement Using Fiber Optics-Based Fluorescence Spectrometer

The spectral measurement was performed on mice sacrificed at 1 and 3 h post Ce6-PVP administration. A fiber optics-based fluorescence spectrometer (Spex SkinSkan, JY Inc., Edison, NJ, USA) was used for the measurement of fluorescence intensity of Ce6-PVP. A monochromator with a 150-W Xenon lamp was used as the excitation light source. The excitation light (400 nm) was guided to illuminate samples by one arm of a Y-type quartz fiber bundle, and the emission fluorescence was collected by another arm of the fiber bundle, guided to another motor-controlled monochromator. The resulting emission spectra were recorded from 650 to 750 nm, in 1 nm increments, collected using the DataMax version

2.20 (Instruments SA, Inc.) software package. The optical fiber tip was placed on the measuring sites and fluorescence intensity spectra were measured. After each measurement, the optical fiber tip was carefully cleaned to remove the possible remaining drug on the tip.

Human Subject

After informed consent, 1 patient with histologically proven angiosarcoma was recruited in this pilot case study. One tumor was located on the scalp, and 2 at the temperomandibular joint. These tumors had been previously treated with Ce6-PVP photodynamic therapy. The patient was intravenously administered with Ce6-PVP with a dose of 2.0 mg/kg for repeated photodynamic therapy on existing and a new angiosarcoma lesion on the scalp and face. Before light irradiation, fluorescence imaging and spectroscopy were performed on 3 angiosarcoma lesions, normal scalp and skin at 1 and 3 h post drug administration. The patient had to remain in subdued light throughout the imaging period. This study was approved by the National Cancer Centre Singapore's Institutional Review Board.

Data Analysis

To evaluate the quality of discrimination between healthy and tumor tissues of the fluorescence images in the CAM model, the red to blue ratio algorithm was applied. Such algorithm is independent of the geometries of excitation/collection of signals and the power of excitation during the fluorescence imaging process. The sensitivity and the specificity of the classifier were calculated using the receiver-operator characteristics (ROC) curves by plotting the fluorescence intensity of tumor against the fluorescence intensity of normal CAM tissue using the GraphPad software (GraphPad Prism™ Version 4.03, San Diego, USA). The ROC curve illustrates the trade-off between sensitivity and specificity for the different threshold of red to blue ratios to distinguish healthy from tumor tissue. Next, the cut-off value corresponding to the highest combined sensitivity and specificity was chosen and evaluated. For fluorescence spectroscopy, the emission spectra data presented in the results are the absolute fluorescence intensities of the tissues after pre-processed using Fourier transformation to decrease noise levels and normalizing the spectra to baseline at 700 nm. The normalization procedure significantly reduced the within-class variances. Spectral data from the various organs and tissues were analyzed to determine spectral line shape and the peak fluorescence intensities at region of 660––690 nm.

Results

Fluorescence Imaging of Bladder Tumor Xenografts on the CAM Model

Fluorescence was not observed from the tumors under blue light illumination before drug administration. After topical administration of Ce6-PVP, an intense red fluorescence in the bladder tumor xenografts was observed, suggesting selective localization in the malignant cells. Fluorescence in the normal CAM tissue was lower compared to fluorescence in the tumor tissue, suggesting either a lower uptake or faster clearance rate from normal tissue of the CAM (Figure 1). The

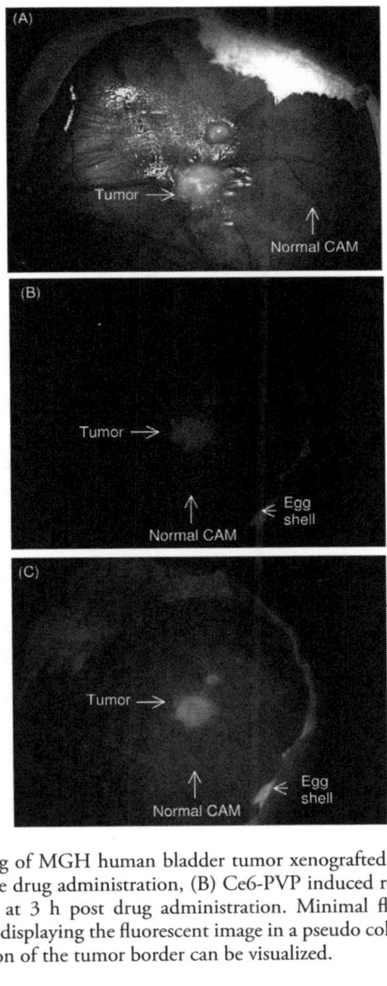

Figure 1. Fluorescence imaging of MGH human bladder tumor xenografted on the CAM model. (A) White light image of the tumor before drug administration, (B) Ce6-PVP induced red fluorescence in tumor imaged under blue light illumination at 3 h post drug administration. Minimal fluorescence was observed in the adjacent normal CAM. (C) By displaying the fluorescent image in a pseudo color using simple image processing technique, a clear discrimination of the tumor border can be visualized.

fluorescence retention from 1 to 5 h post topical administration of Ce6-PVP in bladder tumor xenografts on CAM was tabulated using the red to blue ratio algorithm and fitted into a ROC curve to validate the ability of Ce6-PVP to discriminate tumor from adjacent normal CAM membrane. By applying a cut-off value to these ratios as a diagnostic criterion, it allows the generation of sensitivity and specificity values to distinguish tumor from healthy CAM. A cut-off red to blue ratio of > 1.08 gives the highest combined sensitivity and specificity were 70.8% (95% CI 48.9% to 87.4%) and 83.3% (95% CI 62.6% to 95.3%) respectively (Figure 2). Raising the value to > 1.33 gives the sensitivity and specificity values of 62.5% (95% CI 40.59% to 81.20%) and 91.2% (95% CI 73.0% to 99.0%) respectively.

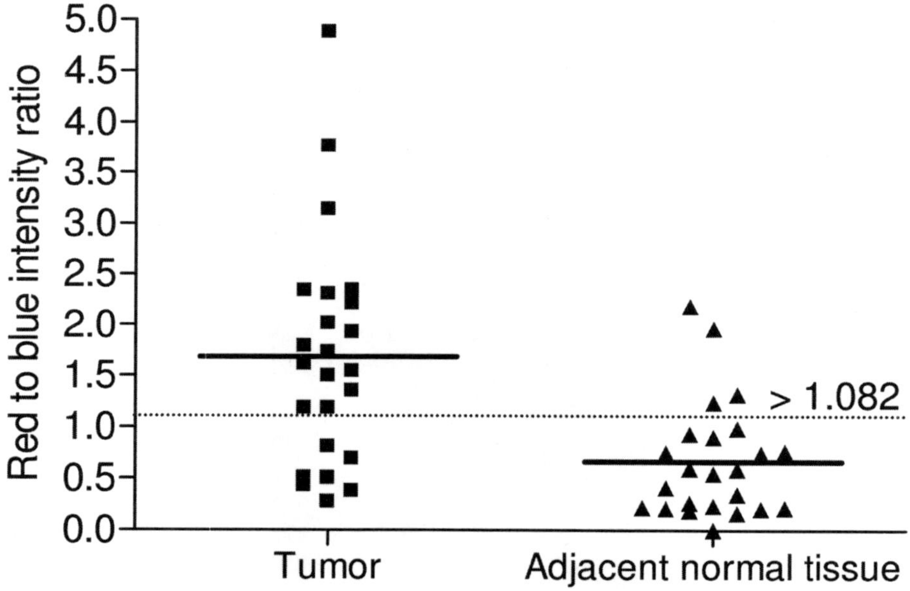

Figure 2. A scatter plot comparing the fluorescence intensity in tumor and the adjacent normal CAM tissue was compiled from 1—5 h post topical drug administration. The points on the scatter plot are normalized individual measurements from 24 eggs. The dotted line is the cut-off fluorescence intensity threshold derived from the ROC curve to classify tumor from normal tissue with a sensitivity and specificity of 70.8% (95% CI 48.9% to 87.4%) and 83.3% (95% CI 62.6% to 95.3%) respectively.

Fluorescence Imaging of Tumor and Normal Organs in Mice Models

Fluorescence imaging was performed on the bladder tumor xenograft, peritumoral muscle, and normal bladder at 1 and 3 h post intravenous injection of

Ce6-PVP in mice (Figure 3). Representative fluorescence images of skin and various internal organs taken at 1 h post Ce6-PVP administration are presented in Figure 4. Overall, tumor fluorescence was observed to be more intense compared to the adjacent peritumoral muscle. Fluorescence intensity in bladder tumor was also higher compare to fluorescence of normal bladder tissues. The internal organs were also found to yield substantial fluorescence especially the gall bladder, liver, stomach, small and gastrointestinal tracts. Minimal fluorescence was observed in the heart and spleen. The fluorescence intensity in muscle, skin, lung, liver, and bladder dropped at 3 h post drug administration. There was little or no fluorescence remaining in heart and spleen. In contrast, the tumors showed sustained fluorescence intensity at 3 h post drug administration. By 6 h post drug administration, minimal fluorescence was detected in the gastrointestinal tract, liver and bladder (Figure 5).

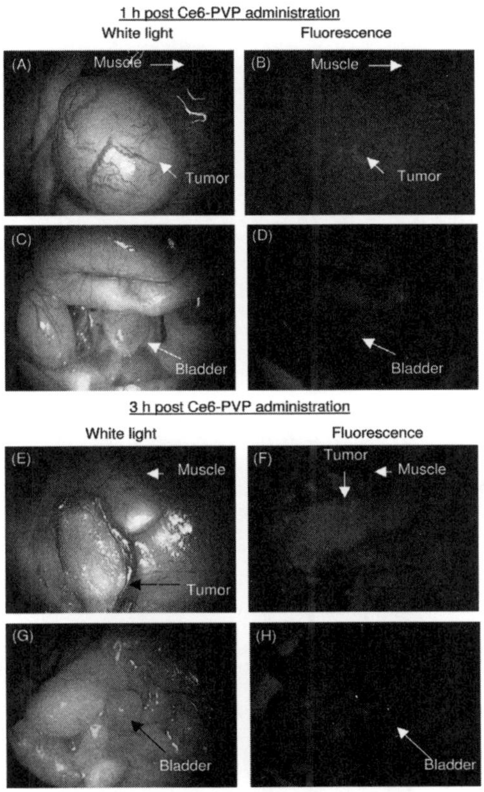

Figure 3. Macroscopic fluorescence imaging in bladder tumor xenografts and normal bladder at 1 h and 3 h post-intravenous administration of 5 mg/kg Ce6-PVP. Ce6-PVP induced fluorescence can be characterized by red fluorescence. At both time points, higher fluorescence intensity was observed in the tumor compared to the adjacent muscle and normal bladder of the mice.

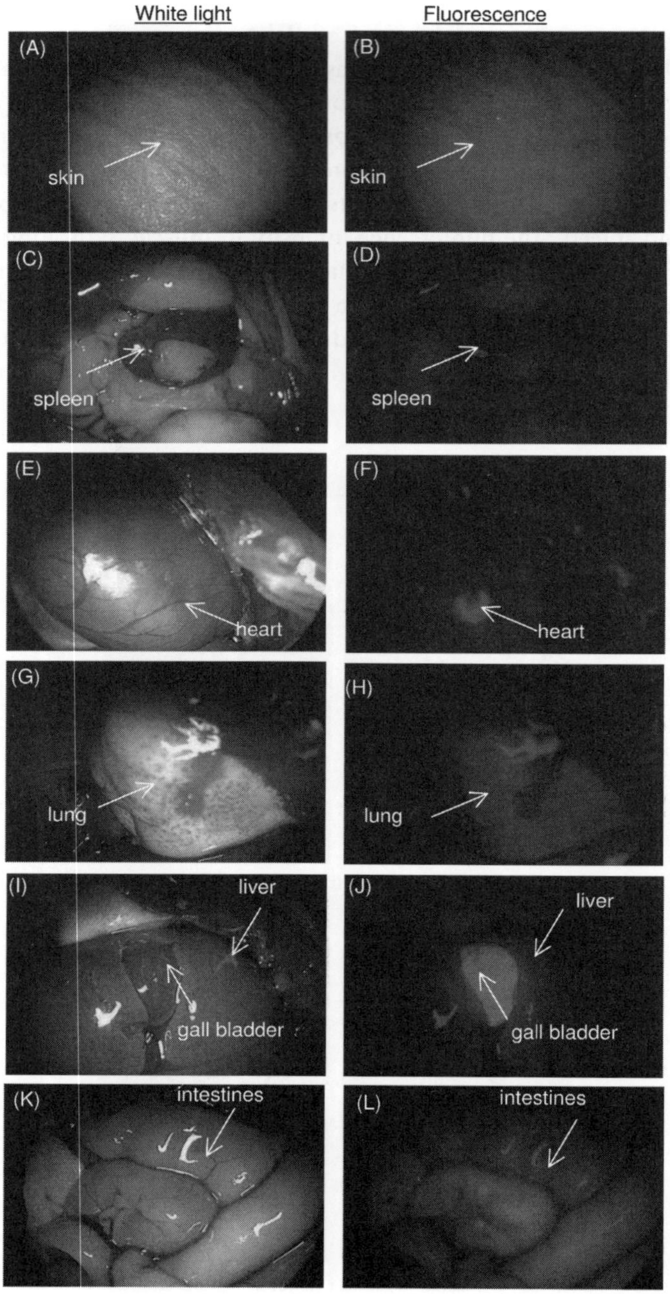

Figure 4. Macroscopic white light and fluorescence imaging in skin, heart, lung, gall bladder, liver, spleen, kidney and gastrointestinal tract at 1 h post-intravenous administration of 5.0 mg/kg Ce6-PVP.

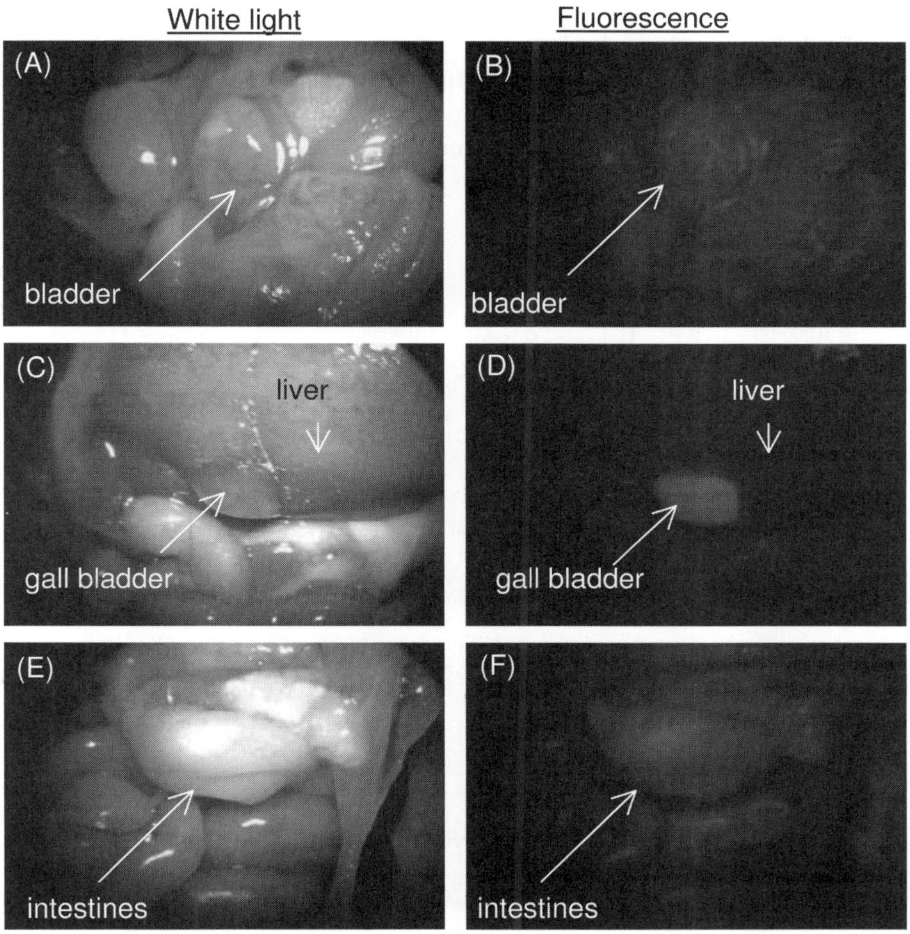

Figure 5. Macroscopic white light (A, C, E) and fluorescence (B, D, F) imaging in normal bladder, liver, gall bladder, and gastrointestinal tract at 6 h post-intravenous administration of 5.0 mg/kg Ce6-PVP.

In Vivo Fiber Optic Spectrofluorometric Measurement

Typical fluorescence emission spectra from tumor, adjacent peritumoral muscle and normal bladder after i.v. administration of Ce6-PVP are shown in Figure 6. In general, the peak fluorescence intensities of tumor were higher than those of normal sites. The greatest intensity occurred in the region between 660—670 nm. When the spectra were normalized to baseline value, changes in peak intensity became evident. The line-shape differences were predominantly due to increased Ce6-PVP accumulation of tumor in the red region (emission peak at 665 nm). Fluorescence emission spectra of skin, heart, lung, gall bladder, liver, spleen, kidney and gastrointestinal tract are shown in Figure 7. Except for the gall bladder, all other organs showed a decreased of fluorescence emission of Ce6-PVP at 3 h compared to 1 h post drug administration. Essentially, fluorescence images that showed greater red fluorescence intensity had visibly higher spectral peak, thus suggesting that that the macroscopic fluorescence imaging were reproducible.

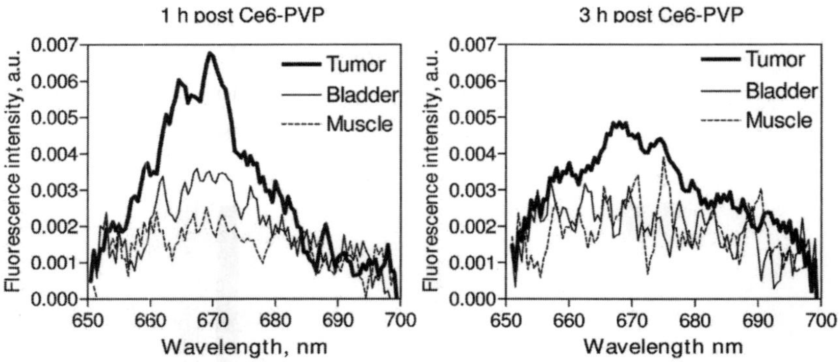

Figure 6. Comparison of emission spectra of bladder tumor xenograft, normal bladder and muscle of the murine model at 1 and 3 h post administration of Ce6-PVP using 400 nm excitation. The spectral signatures showed a peak at the wavelength 665—670 nm in tumor while the fluorescence intensity of normal bladder and muscle is weaker than that of the tumor tissue.

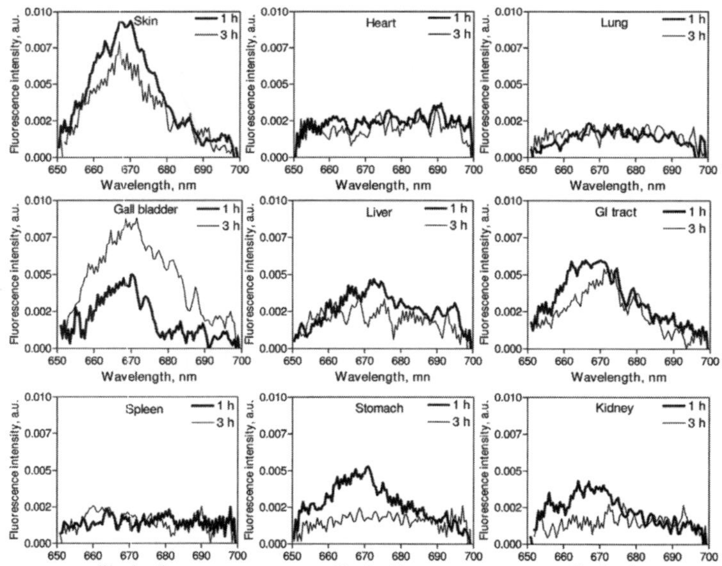

Figure 7. Comparison of emission spectra in the 650—700 nm region using 400 nm excitation in various normal organs at 1 and 3 h post Ce6-PVP administration. Except for skin and gall bladder, it is evident that the emission spectra of normal organs were lower compared to the emission spectra of tumor.

Fluorescence Imaging and Spectrofluorometric Measurement in a Patient

Fluorescence imaging and spectroscopy carried out on 3 tumors in an angiosarcoma patient showed that the tumors developed maximum fluorescence emission intensity at 3 h post Ce6-PVP administration (Figure 8). No observable variations were found for the intensity of the fluorescence between tumors. The fluorescence kinetics study findings from this consistent to those of our earlier results [17]. Spectra from tumor areas show a clear distinct Ce6-PVP induced fluorescence spectrum that discriminate between the tumor and normal skin, with tumor showing higher fluorescence emission intensity compared to normal tissue.

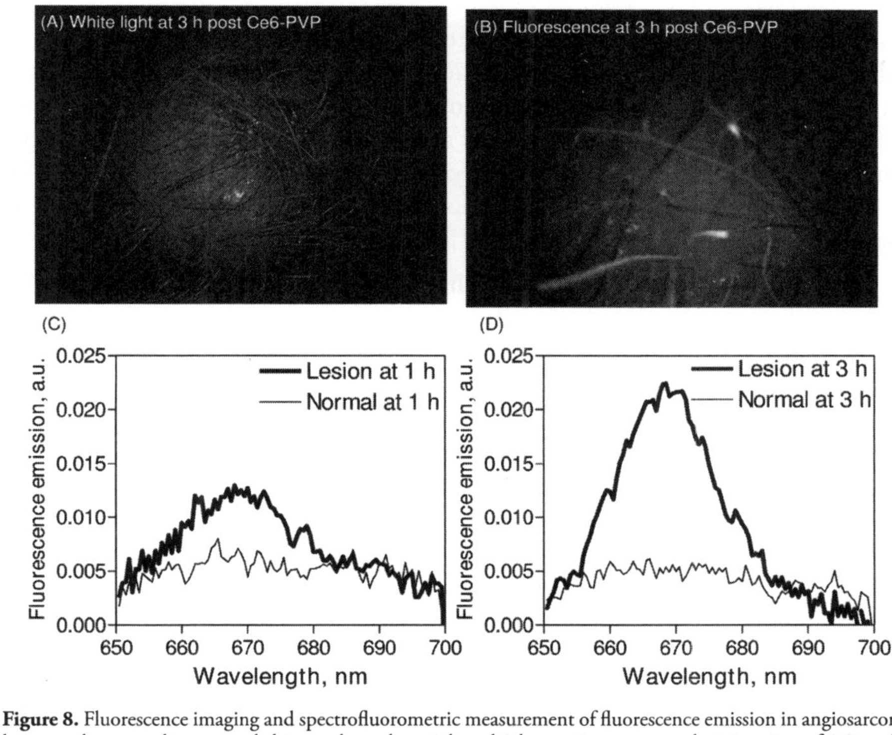

Figure 8. Fluorescence imaging and spectrofluorometric measurement of fluorescence emission in angiosarcoma lesion and surrounding normal skin on the scalp at 1 h and 3 h post-intravenous administration of 2.0 mg/kg Ce6-PVP.

Discussion

Fluorescence imaging approaches are increasingly being used as a medical diagnostic procedure to assess tissue malignancy over conventional methods because they do not use potentially harmful ionizing radiation [20]. In situations where discrimination of suspicious lesion is clinically problematic, fluorescence imaging may provide added advantage in demarcating abnormal tissue. The development of photosensitizer based fluorescence imaging is hindered by problems such as skin photosensitivity, poor selectivity of the photosensitizer, and formulation issues. For these reasons, Ce6 was formulated with PVP to address these issues. Formulations using biocompatible polymers such as PVP are increasingly being used in the pharmaceutical industry for enhancing drug solubility and bioavailability. Complexation of Ce6 with PVP was found to prevent Ce6 aggregation in aqueous media and led to an enhancement of Ce6 fluorescence quantum yield, while keeping the quantum yield of the intersystem crossing essentially unchanged [21]. In this study we have further examined the potential clinical use of Ce6-PVP in cancer imaging and diagnosis. We first tested out the feasibility of this photosensitizer as an exogenous fluorophore on the CAM tumor model. We were able to observe the red fluorescence emitted by the tumor tissue excited using a filtered xenon lamp excitation, which enabled clear determination of the tumor margin. The sensitivity of Ce6-PVP was more than 80%, however, the specificity remained low. We have recently reported that the new formulation of Ce6-PVP (> 95% purity level of active Ce6) demonstrated a higher sensitivity (98%) and specificity (82%) on the CAM model [15].

Results from the CAM experiments have provided the motivation to examine Ce6-PVP fluorescence distribution in bladder tumor xenograft as well as in various normal organs of a murine model. Macroscopic fluorescence imaging showed that there was considerable distinction in the localization of fluorescence in tumor compared to other organs that could enable discrimination between tumor and normal organs. Organs of elimination and detoxification such as skin, gall bladder and gastrointestinal tracts were characterized by high photosensitizer accumulation efficiency. In contrast, all other normal organs such as muscle and bladder had much lower photosensitizer accumulation at 1 h post drug administration. Blood vessels growing on the tumor can also be observed because they contrast with the fluorescence of the tumors. At 3—6 h post drug administration, a decrease of fluorescence intensity became evident on all normal organs, confirming that Ce6-PVP has fast clearance rate from normal organs. In some instances, we have observed variability of fluorescence intensity on the surface tissue such as stomach and lung. This is possibly attributed to the variations of the tissue optical properties of the organs given by their color, density and composition.

The key issue in fluorescence imaging is that the emitted fluorescence intensity measured from a tissue surface is not necessarily proportional to the fluorophore concentration because the light is altered by the tissue's intrinsic absorption and scattering properties. Hence we have employed the utility of spectrometric point fluorescence detection as a complementary technique. Spectra measurements were carried out at 1 and 3 h post drug administration to correlate the tumor intensity ratios obtained with fluorescence imaging to the tumor fluorescence spectral signal of the tissue. All the macroscopic fluorescence images correlated well to the spectra measurement. The Ce6-PVP induced spectra emission after normalization demonstrated a good separation to differentiate malignant tumors from normal tissues. Besides measuring physical parameters such as concentration of photosensitizer and tissue properties, this method can potentially improve the assessment of cancer location and its extent within the local-regional area. While fluorescence point spectroscopy studies are promising, it has several drawbacks as a screening tool as it can only interrogate a small volume of tissue (typically, 0.5—1 mm3) directly beneath the probe tip. Point measurements inevitably involve a degree of random sampling, which may not allow identification of early stage disease [22]. Hence, the combination techniques of fluorescence imaging and spectroscopy have been proven in good agreement with the actual tumor boundary found by histopathological mapping and early stage of disease [23,24].

Recently, we have reported photodynamic therapy—activated immune response against distant untreated tumours in recurrent angiosarcoma [25] and preferential accumulation of Ce6-PVP in angiosarcoma compared to normal skin following intravenous administration in 3 patients [17]. High dose PDT carried out at a high fluence rate resulted in local control of the disease for up to a year; however, the disease recurred and PDT had to be repeated [26]. During the repeat PDT session, we measured the fluorescence of 3 different lesions using fluorescence imaging followed by spectroscopy. The results confirmed that fluorescence imaging clearly captures the fluorescence in angiosarcoma and good correlation was found between fluorescence imaging and spectral measurement in the patient. This is in agreement with other reports, that fluorescence ratio imaging in combination with relative spectral measurement of the photosensitizer might be a viable method for the optical diagnosis of cancer [27]. Furthermore, in vivo and real-time determination of the time course of photosensitizer's fluorescence could potentially be a crucial pre-irradiation screening tool to determine the exact location and extent of the tumor before photodynamic therapy.

Conclusion

It is shown that Ce6-PVP has a rapid accumulation in the tumor, and a relatively short half-life in normal organs. When excited by blue light, Ce6-PVP accumulating cells can be visualized and located in the tissue by virtue of its fluorescence. The main advantage of Ce6-PVP induced fluorescence imaging is its increased tumor selectivity with the ability to clearly define the tumor margin. Combination of Ce6-PVP induced fluorescence imaging and spectroscopy could allow detection and discrimination between cancer and the surrounding normal tissues.

Competing Interests

The authors declare that they have no competing interests.

Authors' Contributions

WWC and PST conceived of the study and carried out all the experimental study. RB and KCS participated in the clinical study. PWH and MO participated in the coordination of the study. All authors read and approved the final manuscript

Acknowledgements

WW Chin and PS Thong are the recipients of the Singapore Millennium Foundation scholarship.

References

1. DaCosta RS, Wilson BC, Marcon NE: Fluorescence and spectral imaging. ScientificWorldJournal 2007, 7:2046–2071.
2. Wong Kee Song LM, Marcon NE: Fluorescence and Raman spectroscopy. Gastrointest Endosc Clin N Am 2003, 13(2):279–296.
3. Perelman LT: Optical diagnostic technology based on light scattering spectroscopy for early cancer detection. Expert Rev Med Devices 2006, 3(6):787–803.
4. Schultz CP: The potential role of Fourier transform infrared spectroscopy and imaging in cancer diagnosis incorporating complex mathematical methods. Technol Cancer Res Treat 2002, 1(2):95–104.
5. Ell C: Improving endoscopic resolution and sampling: fluorescence techniques. Gut 2003, 52(Suppl 4):iv30–33.
6. Kherlopian AR, Song T, Duan Q, Neimark MA, Po MJ, Gohagan JK, Laine AF: A review of imaging techniques for systems biology. BMC Syst Biol 2008, 2:74.

7. Chen Y, Gryshuk A, Achilefu S, Ohulchansky T, Potter W, Zhong T, Morgan J, Chance B, Prasad PN, Henderson BW, et al.: A novel approach to a bifunctional photosensitizer for tumor imaging and phototherapy. Bioconjug Chem 2005, 16(5):1264–1274.
8. Thong PS, Olivo M, Kho KW, Zheng W, Mancer K, Harris M, Soo KC: Laser confocal endomicroscopy as a novel technique for fluorescence diagnostic imaging of the oral cavity. J Biomed Opt 2007, 12(1):014007.
9. Andersson-Engels S, Klinteberg C, Svanberg K, Svanberg S: In vivo fluorescence imaging for tissue diagnostics. Phys Med Biol 1997, 42(5):815–824.
10. Zheng W, Olivo M, Soo KC: The use of digitized endoscopic imaging of 5-ALA-induced PPIX fluorescence to detect and diagnose oral premalignant and malignant lesions in vivo. Int J Cancer 2004, 110(2):295–300.
11. Kostenich GA, Zhuravkin IN, Furmanchuk AV, Zhavrid EA: Photodynamic therapy with chlorin e6. A morphologic study of tumor damage efficiency in experiment. J Photochem Photobiol B 1991, 11(3–4):307–318.
12. Ramaswamy B, Manivasager V, Chin WW, Soo KC, Olivo M: Photodynamic diagnosis of a human nasopharyngeal carcinoma xenograft model using the novel Chlorin e6 photosensitizer Fotolon. Int J Oncol 2005, 26(6):1501–1506.
13. Chin WW, Heng PW, Olivo M: Chlorin e6—polyvinylpyrrolidone mediated photosensitization is effective against human non-small cell lung carcinoma compared to small cell lung carcinoma xenografts. BMC Pharmacol 2007, 7:15.
14. Chin WW, Lau WK, Bhuvaneswari R, Heng PW, Olivo M: Chlorin e6-polyvinylpyrrolidone as a fluorescent marker for fluorescence diagnosis of human bladder cancer implanted on the chick chorioallantoic membrane model. Cancer Lett 2007, 245(1–2):127–133.
15. Chin WW, Heng PW, Lim PL, Lau KO, Olivo M: Membrane transport enhancement of chlorin e6-polyvinylpyrrolidone and its photodynamic efficacy on the chick chorioallantoic model. Journal of Biophotonics 2008, 1(5):395–407.
16. Chin WW, Lau WK, Heng PW, Bhuvaneswari R, Olivo M: Fluorescence imaging and phototoxicity effects of new formulation of chlorin e6-polyvinylpyrrolidone. J Photochem Photobiol B 2006, 84(2):103–110.
17. Chin WW, Heng PW, Thong PS, Bhuvaneswari R, Hirt W, Kuenzel S, Soo KC, Olivo M: Improved formulation of photosensitizer chlorin e6 polyvinylpyrrolidone for fluorescence diagnostic imaging and photodynamic therapy of human cancer. Eur J Pharm Biopharm 2008.

18. Chin WW, Heng PW, Bhuvaneswari R, Lau WK, Olivo M: The potential application of chlorin e6-polyvinylpyrrolidone formulation in photodynamic therapy. Photochem Photobiol Sci 2006, 5(11):1031–1037.
19. Isakau HA, Trukhacheva TV, Zhebentyaev AI, Petrov PT: HPLC study of chlorin e6 and its molecular complex with polyvinylpyrrolidone. Biomed Chromatogr 2007, 21(3):318–325.
20. Brindle K: New approaches for imaging tumour responses to treatment. Nat Rev Cancer 2008, 8(2):94–107.
21. Isakau HA, Parkhats MV, Knyukshto VN, Dzhagarov BM, Petrov EP, Petrov PT: Toward understanding the high PDT efficacy of chlorin e6-polyvinylpyrrolidone formulations: Photophysical and molecular aspects of photosensitizer-polymer interaction in vitro. J Photochem Photobiol B 2008, 92(3):165–174.
22. Stringer M, Moghissi K: Photodiagnosis and fluorescence imaging in clinical practice. Photodiagnosis and Photodynamic Therapy 2004, 1(2):9–12.
23. Wagnieres GA, Star WM, Wilson BC: In vivo fluorescence spectroscopy and imaging for oncological applications. Photochem Photobiol 1998, 68(5):603–632.
24. Mang T, Kost J, Sullivan M, Wilson BC: Autofluorescence and Photofrin-induced fluorescence imaging and spectroscopy in an animal model of oral cancer. Photodiagnosis and Photodynamic Therapy 2006, 3(3):168–176.
25. Thong PS, Ong KW, Goh NS, Kho KW, Manivasager V, Bhuvaneswari R, Olivo M, Soo KC: Photodynamic-therapy-activated immune response against distant untreated tumours in recurrent angiosarcoma. Lancet Oncol 2007, 8(10):950–952.
26. Thong PS, Olivo M, Kho KW, Bhuvaneswari R, Chin WW, Ong KW, Soo KC: Immune response against angiosarcoma following lower fluence rate clinical photodynamic therapy. J Environ Pathol Toxicol Oncol 2008, 27(1):35–42.
27. Kopriva I, Persin A, Zorc H, Pasic A, Lipozencic J, Kostovic K, Loncaric M: Visualization of basal cell carcinoma by fluorescence diagnosis and independent component analysis. Photodiagnosis and Photodynamic Therapy 2007, 4(3):190–196.

A Three-Dimensional Multivariate Image Processing Technique for the Analysis of FTIR Spectroscopic Images of Multiple Tissue Sections

Bayden R. Wood, Keith R. Bambery, Corey J. Evans, Michael A. Quinn and Don McNaughton

ABSTRACT

Background

Three-dimensional (3D) multivariate Fourier Transform Infrared (FTIR) image maps of tissue sections are presented. A villoglandular adenocarcinoma from a cervical biopsy with a number of interesting anatomical features was used as a model system to demonstrate the efficacy of the technique.

Methods

Four FTIR images recorded using a focal plane array detector of adjacent tissue sections were stitched together using a MATLAB® routine and placed in a single data matrix for multivariate analysis using Cytospec™. Unsupervised Hierarchical Cluster Analysis (UHCA) was performed simultaneously on all 4 sections and 4 clusters plotted. The four UHCA maps were then stacked together and interpolated with a box function using SCIRun software.

Results

The resultant 3D-images can be rotated in three-dimensions, sliced and made semi-transparent to view the internal structure of the tissue block. A number of anatomical and histopathological features including connective tissue, red blood cells, inflammatory exudate and glandular cells could be identified in the cluster maps and correlated with Hematoxylin & Eosin stained sections. The mean extracted spectra from individual clusters provide macromolecular information on tissue components.

Conclusion

3D-multivariate imaging provides a new avenue to study the shape and penetration of important anatomical and histopathological features based on the underlying macromolecular chemistry and therefore has clear potential in biology and medicine.

Background

The ability to generate and manipulate three-dimensional (3D) images of body parts or tissue sections is extremely useful in determining the extent and penetration of disease or tissue degeneration. Conventional ways of generating such 3D images are Computerized Tomography (CT), Positron Emission Tomography (PET), Magnetic Resonance Imaging (MRI) and 3D ultrasound. X-ray based techniques are becoming more useful with the increased contrast available by coupling the technique with synchrotron radiation and using phase contrast and diffraction enhanced imaging. These techniques do not supply information on the macromolecular composition in the image contrast, whereas spectroscopy based techniques do, and hence 3D IR imaging would provide a useful and novel alternative with the advantage of image contrast based directly on the underlying macromolecular composition. The lack of penetration of mid IR radiation into tissue precludes real time imaging of whole samples but an alternative is to build a composite from 2D images of adjacent sections of tissue thus providing a method to gauge the extent and penetration of disease, which may be of clinical value.

This has the advantage of not requiring a chemical or immunological staining protocol to provide biochemical information. High-speed low-cost computers, in combination with infrared imaging instruments based on Focal Plane Array (FPA) detectors, allow the image acquisition and reconstruction to be achieved within a reasonable time frame.

The adaptation of multi-channel infrared array detectors from military hardware to FTIR microscopes in the early 1990s resulted in new methodologies to investigate the macromolecular architecture of cells in tissue sections [1]. The new generation of FPA and more recently linear array detectors are capable of recording thousands of spectra in rapid time. Each pixel is essentially a digital hyper-spectral data cube containing absorbance, wavenumber and x,y spatial coordinates. Univariate or chemical maps can be plotted based on peak height, integrated areas under specific bands or band ratios. While these maps provide spatial information on the distribution and relative concentration of the major macromolecules they are not useful in correlating anatomical and histopathological features with corresponding spectral profiles [2]. Multivariate imaging techniques including Unsupervised Hierarchical Cluster Analysis (UHCA) [2-9], K-means clustering [8,10], Principal Components Analysis (PCA) [11], Linear Discriminant Analysis [12], Fuzzy C-means clustering [8,13] and neural networks [11] have proven to be invaluable in the identification of spectral groups or "clusters" which can be directly compared to stained tissue sections. In multivariate methods, the information of the entire spectrum can be utilized for the analysis. The first part of the analysis requires a distance matrix to be calculated. This can be achieved using a number of different algorithms including D-values (Pearson's correlation coefficient), Euclidean distances, normalized Euclidean distances, Euclidean squared distances and City Block all of which are available in the Cytospec™ software package [14] and appear to produce similar cluster maps although the time taken for each method can vary. We used the D-values method because this is a well-established linear regression method that is suited to relative concentration data. One disadvantage of this algorithm is that it is computationally more demanding than others; therefore more time is required for the distance matrix calculation.

In cluster analysis a measure of similarity is established for each class of related spectra and a mean characteristic spectrum can be extracted for each class. In the final step, all spectra in a cluster are assigned the same color. In the false color maps, the assigned color for each spectral cluster is displayed at the coordinates at which each data cube was collected. The mean spectrum of a cluster represents all spectra in a cluster and can be used for the interpretation of the chemical or biochemical differences between clusters. There are also a variety of algorithms to select from to perform cluster analysis, including Ward's algorithm, which we employ because it minimizes the heterogeneity of the clusters.

The high correlation of spectral clusters with anatomical and histopathological features has been conclusively demonstrated for a number of different tissue types including cervical [2,3], breast [10,15], liver [4,7], brain [5], mouth [6], intestine [8,16], skin [17], bone [18,19], cornea [20] and prostate [21]. Hitherto FTIR multivariate imaging has been mainly restricted to the generation of 2D cluster maps. The exception is Mendelsohn and coworkers [22] who constructed a 3D univariate map of cortical bone based on peak ratios from serial two-dimensional sections. By interfacing two types of software namely Cytospec [14] and SCIRun [23] and writing a simple "stitching" algorithm we are able to generate 3D multivariate cluster maps from multiple tissue sections. The ability to visualize 3D FTIR cluster maps provides a new avenue to assess variation in multiple tissue sections and to determine the penetration of histopathological structures based on the underlying macromolecular structure of the diseased tissue.

Methods

Following approval from the Royal Women's Hospital Research and Human Research Ethics Committees and the Monash University Standing Committee on Ethics in Research Involving Humans, written, informed preoperative consent was obtained from the patient and a cervical tissue sample exhibiting villoglandular adenocarcinoma was then obtained by cone biopsy. The tissue sample was then embedded in a paraffin block and sliced by microtome into 4 μm sections. One group of four sections was mounted on glass slides and stained with the routine histopathology stain Hematoxylin and Eosin (H&E) for light microscope examination. Hematoxylin has an affinity with nucleic acids and Eosin has an affinity for the cellular cytoplasm. An adjacent group of four sections was deparaffinized, mounted on Kevley™ "low e" IR reflective microscope slides and imaged with a Varian Stingray FTIR microscope system equipped with a 64 × 64 pixel HgCdTe liquid nitrogen cooled FPA with a 15× Cassegrain objective. FTIR hyper-spectral data images were recorded in the range 4000-950 cm^{-1} at 6 cm^{-1} resolution and with 16 scans co-added. For each of the four sections, step-motion control of the microscope stage was used to construct a 16 tile (4 × 4) FTIR image mosaic from FPA recordings collected as 16 pixel aggregates. Thus the spatial resolution obtained is approximately 22 μm per pixel aggregate. Each FTIR image was therefore 2.0 mm^2 in area and with the four 4 μm thick adjacent sections giving a total sampled volume of 1,400 × 1,400 × 16 μm. A spatial resolution of 22 μm per pixel was used as this provided FTIR images that covered an area of tissue large enough to encompass several examples of anatomically different tissue types.

Using a MATLAB® routine developed by our group, the four FTIR images were stitched together side by side (or "unfolded") to give a single large 2D image frame (see 1 "cyto4fs.m" a program for stitching multiple tissue sections together for use in Cytospec™ spectroscopic software). The absorbance was integrated over a large spectral region (1750-950 cm^{-1}) to assess sample thickness using a routine in Cytospec™. This avoids inaccuracies with too thin samples with low absorbance or too high absorbance that result in non-linear detector response. A spectrum is rejected if the determined integration value is higher or lower than a pre-defined threshold (1500 and 50 arbitrary units). Spectra that passed the thickness quality test were converted to second derivative spectra using a Savitsky-Golay algorithm (13 smoothing points). UHCA (D-values, Ward's algorithm) was performed to generate 4 clusters from second derivative spectra over the 1272-950 cm^{-1} spectral window. The resultant cluster map was then reorganized (or "back folded") into the four individual 2D cluster maps, each map corresponding to one of the FTIR images. The four cluster maps were saved in an image file format with a unique false color assigned to each cluster and then aligned or "registered" as separate floating layers of a single image in the GIMP [24] image-processing program. This registration step is necessary as the sample orientation was not identical in both rotation and translation on each of the four acquired 2D mosaic images. Proper pixel correspondence from one image to another was easily achieved using this manual approach given the small number of image layers. The registered layers constituted a best fit because some slightly unequal distortion of the tissue matrix was observed presumably caused by the sectioning and preparation processes. For this 3D imaging technique, special care must be taken to ensure the sections are not stretched or distorted when deposited onto the slides.

The SCIRun [23] software suite provides a graphical user interface for rapid development of "networks" of instruction routines for the stacking and rendering of the input data (see 2 "SCIRun adenocarcinoma.net" for the 3D image processing program modules and configuration parameters for SCIRun). The registered images were loaded into SCIRun as a set of indexed integer values (1 to 4 corresponding to each cluster) and then "stacked" into a scalar volume field of cluster values from which the 3D cluster maps were rendered.

3D univariate chemical maps depicting a single spectral feature were also generated. The spectra were vector normalized over the 1800-950 cm^{-1} range and then integrated under the absorbance band of interest using a trapezoidal baseline function in Cytospec™. The 3D univariate maps were rendered from a scalar volume field of absorbance values generated from 2D FTIR images stacked in SCIRun. The 3D univariate maps were plotted with a 256 rainbow color palette using Gaussian interpolation between the data grid points to produce a smoothly varying color field. The 3D cluster maps, on the other hand, were plotted in a palette of only 4 false colors and box interpolated, with one false color corresponding to each cluster. Figure 1 depicts a schematic of the overall process from spectral acquisition to 3D image reconstruction.

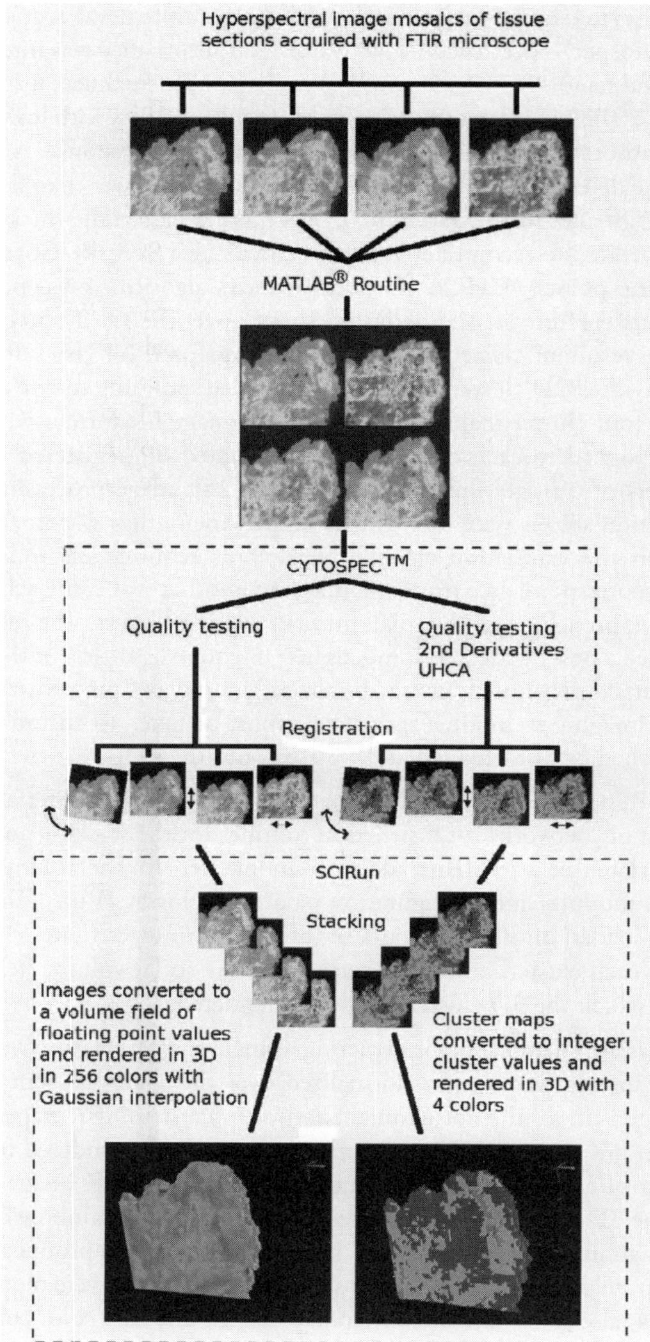

Figure 1. Schematic showing steps in the construction of 3D FTIR multivariate images.

Results and Discussion

Figure 2 shows a H&E stained 2D cervical section exhibiting a relatively rare form of neoplasm known as villoglandular adenocarcinoma. The neoplasm is characterized by the presence of long villous fronds and papillae lined by columnar cells with intact cytoplasmic borders and displays minimal atypia. [25] Spherical clusters of cells with smooth intact communal cytoplasmic rings are also associated with this condition [25]. The sample makes an ideal model for 3D unsupervised hierarchical cluster analysis because it exhibits a variety of anatomical and histopathological features, including connective tissue, red blood cells, inflammatory exudate and glandular cells. Figure 3a depicts a chemical map generated from all four sections simultaneously by integrating the area under the band in 1275-1190 cm^{-1} region associated mainly with phosphodiester contributions form nucleic acids. The chemical maps show a good correlation with morphology; however, specific correlations with anatomical and histopathological features cannot be gauged with this form of processing.

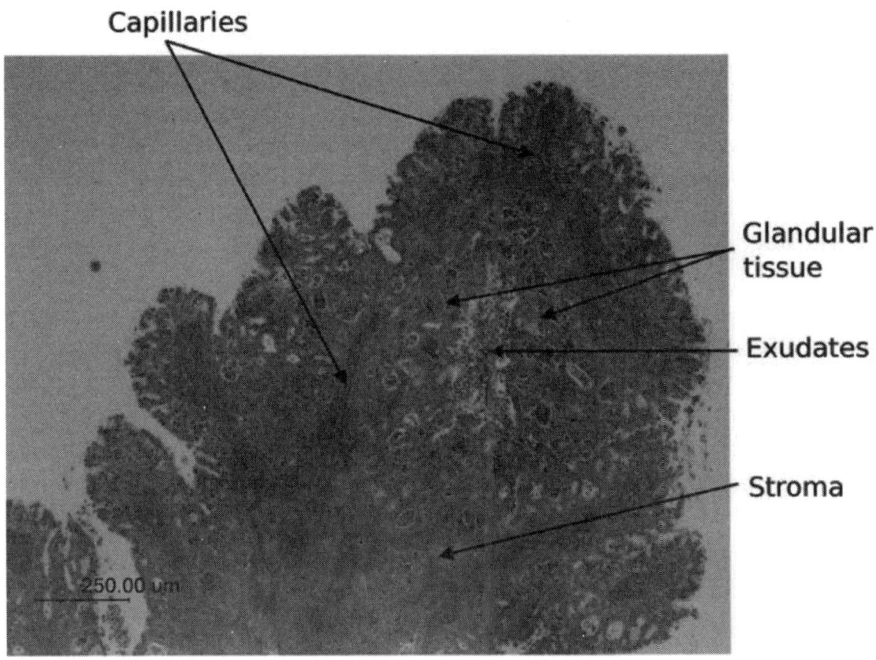

Figure 2. Light micrograph of a labeled H&E stained cervical section exhibiting villoglandular adenocarcinoma.

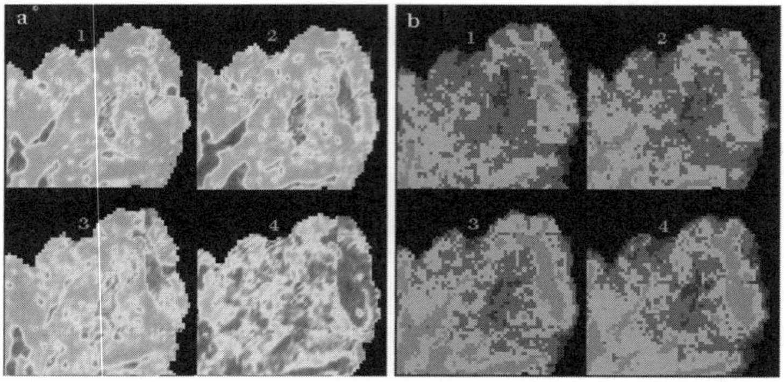

Figure 3. Hyper-spectral FTIR data processing performed simultaneously on 4 adjacent tissue sections from a cervical biopsy sample. The numbers 1 through 4 identify the individual sections in the figure. In (a), a univariate chemical image obtained from the integrating the area under the 1275-1190 cm^{-1} region after baseline subtraction and in (b), a 4 cluster map derived from analysis over the 1272-950 cm^{-1} spectral window. The cluster map false color scheme corresponds to brown for exudates, blue for inflamed glandular tissue, green for connective tissue and orange for blood filled capillaries as described in the text.

It is necessary to perform UHCA over the entire set of spectra collected to fully characterize the range of spectral variations through all the tissue sections. Performing separate UHCA on each individual tissue section would give a different clustering result due to changes (although generally small) in the biochemical composition between sections. For this reason the images were "stitched" together into a single frame to enable spectral pre-processing and UHCA to be performed in Cytospec™ on all spectra from all images simultaneously. UHCA was performed on the 1272-950 cm^{-1} region on second derivative vector normalized spectra simultaneously on four adjacent sections and the resultant cluster maps are displayed in figure 3b. The cluster maps show a general similarity and successfully highlight the major anatomical features. The orange cluster represents red blood cells embedded in the stromal matrix. The light green cluster is predominantly stroma, while the brown is mainly lymphocyte exudates. The blue cluster is predominantly glandular tissue. In tissue sections 3 and 4 there is an increase in the area of connective tissue (green cluster) relative to glandular tissue (blue cluster) when compared to sections 1 and 2 indicating penetration of the glandular tissue into the connective layer.

Figure 4 shows the raw and second derivative mean extracted spectra color coded the same as the clusters in figure 3b. The maps and corresponding spectra are very similar for each section indicating that the biochemistry between the adjacent sections is consistent. The spectra exhibit dramatic changes in the amide I mode both in terms of bandwidth and position. The peak center varies from approximately 1643 cm^{-1} to 1659 cm^{-1}. This variation is attributed to physical-chemical changes in the tissue matrix. Dramatic variation occurs in areas of thin tissue and on the periphery of tissue sections with the net result a shifting of the amide I mode along with a concomitant increase in the amide II/amide I ratio. This effect is clearly observed in the mean extracted spectrum from the brown cluster which shows the amide I mode appearing at 1643 cm^{-1} and an amide II/amide I ratio that is much greater for this spectrum when compared to the other spectra. Such strong distortions and shifts in band shape were recently addressed by Romeo et al. [9] who reported a method to correct for the "dispersion artifact." To minimize correlations with physical information the analysis was carried out using the 1272-950 cm^{-1} region, which omits the proteinaceous range (1720-1380 cm^{-1}) that may be strongly distorted by the dispersion artifact. Spectra from lymphocyte exudates and glandular tissue are dominated by a band at ~1240 cm^{-1} which is assigned to the asymmetric phosphodiester stretching vibration of nucleic acids. This band shows the most variation between all 4 mean extracted cluster spectra. The mean extracted spectrum from the stromal areas (light green) has contributions from collagen vibrations although the distinctive collagen triplet in the 1300-1200 cm^{-1} cannot be observed due to infiltration by red blood cells, lymphocyte exudates and glandular tissue into the connective layer.

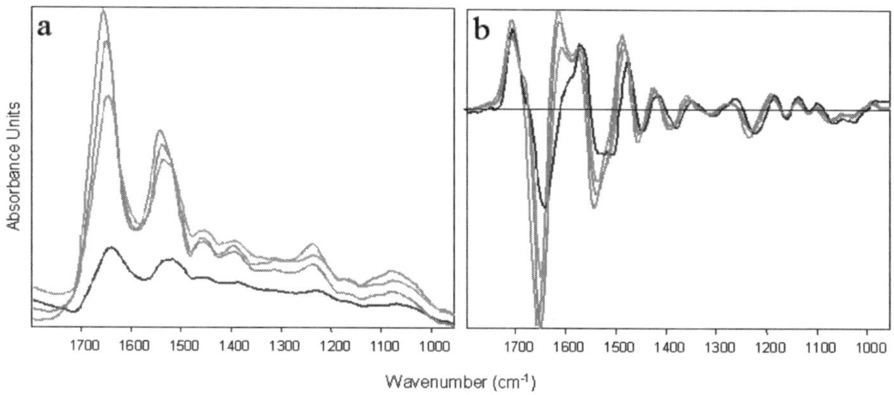

Figure 4. Mean cluster spectra (a) and mean 2nd derivative cluster spectra (b) from a four cluster analyses based on the range 1272-950 cm^{-1} and color coded to correspond to the cluster areas depicted in figures 3 (b).

The 3D chemical image constructed from 4 adjacent sections and generated by integrating the area underneath the peaks in the 1272-950 cm^{-1} region is presented in figure 5. In figure 5a the image is orientated to show the first section of the tissue block (section 1) while in figure 5b the last section (section 4) is oriented towards the viewer. The darkest orange areas in section 1 (figure 5a) correlate well with the stroma and glandular tissue while the darkest orange area shown in section 4 (figure 5b) is associated mainly with the stroma.

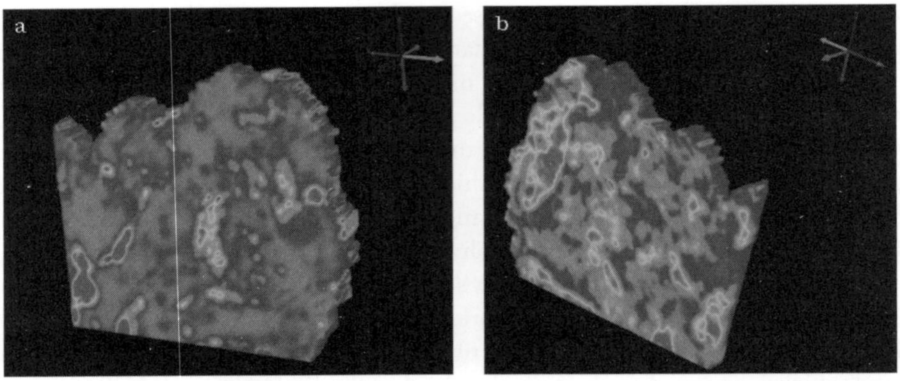

Figure 5. Two views of a 3D univariate chemical map plotting integrated absorbance over the spectral region 1275-1190 cm^{-1} after baseline correction. Red indicates areas of highest absorption and blue indicates areas of lowest absorption. The view is looking toward the section 1 side of the sampled volume in (a) and towards the section 4 side in (b).

Figure 6 shows a 3D UHCA map performed on the 4 sections simultaneously. The map shows excellent correlation with the anatomical and histopathological features indicated in figure 2. The cluster colors are the same as those used in figure 3b. The 3D UHCA map enables one to visualize the extent of penetration of the anatomical features and the degree of variation from section to section. Moreover, 3D FTIR multivariate processing enables visualization of thick tissue sections that cannot normally be analyzed using conventional mid IR spectroscopic techniques due to the limited depth penetration of IR radiation. The thin sections (4 µm) required for use with the Kevley slides are less than the thickness of a single cervical cell consequently multiple sections enable the analysis of whole cells and also minimizes the effects of orientation artifacts that can arise during tissue sectioning. Individual clusters can be studied by rendering the image in semi-transparent mode. Figure 7 is identical to figure 6 but with the stroma cluster removed from the plot and with glandular tissue now depicted in semi-transparent blue. By making the image semi transparent one can visualize clusters in the center of the 3D FTIR image and examine the shape and penetration of important anatomical and histopathological features.

The time required to acquire and compile 3D FTIR univariate (chemical) images involves the following intervals:

a) Approximately 10 minutes to acquire each 2D FTIR image from each individual tissue section.

b) Two minutes to run the MATLAB® stitching program.

c) Registration of the images was done "by hand" in this study and was consequently quite time consuming, however, it is envisioned that suitable software could be developed to automate the registration process thereby reducing the time required for this to a few minutes.

d) About 1 minute is required for SCIRun to stack, interpolate and render a single 3D image frame.

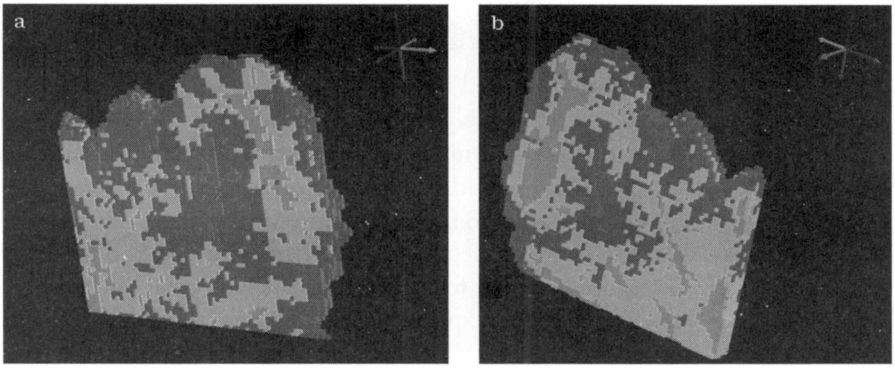

Figure 6. Two views of 3D cluster maps for 4 clusters obtained from analysis in the 1272-950 cm^{-1} spectral region. The cluster map false colors are as described in the caption of figure 3 (b). The view is looking toward the section 1 side of the sampled volume in (a) and towards the section 4 side in (b).

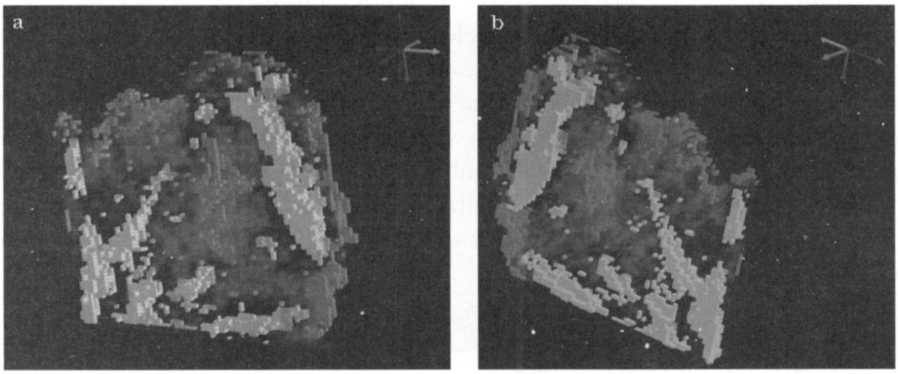

Figure 7. Two views of 3D cluster maps identical to the maps in figure 6 but with the stroma cluster removed from the plot and with glandular tissue now depicted in semi-transparent blue. The view is looking toward the section 1 side of the sampled volume in (a) and towards the section 4 side in (b).

A 3D univariate image could be obtained in less than 1 hour from commencement of FTIR scanning if a routine data-handling pipeline was incorporated. In approximately 1 hour a 3D movie, which are composed of a few hundred individual 3D image frames can be produced in SCIRun.

The production of 3D UHCA cluster images is a significantly slower process than generating univariate 3D maps because in addition to the steps delineated above for univariate maps UHCA must be performed. UHCA is computationally intensive and requires approximately 2 hours (Pentium 4, 3.4 GHz, Hyper-Threading, 2 Gb RAM) for the processing of four FTIR images stitched together. Compilation of UHCA 2D maps from large collections of tissue sections would be prohibitively slow for the current technique to have value as a rapid diagnostic tool. We are currently testing an artificial neural network alternative to UHCA.

FTIR imaging is resolution limited by diffraction to scales on the order of a few microns and hence, it is not always possible to unambiguously assign the obtained spectroscopic information as unique to particular sub-cellular structures. Nevertheless, FTIR imaging does provide valuable information on the overall biochemical composition when applied to tissue structures that are larger than a few microns in extent. The multivariate spectroscopic 3D-imaging method described in this work could be readily adapted for use with other emerging biophotonics techniques, most particularly Raman spectroscopic mapping which can provide macromolecular information on sub-cellular length scales.

Conclusion

The coupling of vibrational spectroscopy with 3D multivariate processing greatly extends the capabilities of this technology in medical diagnostics. From a biomedical perspective existing pathological and histochemical protocols depend on sample morphology and visualization. Therefore the ability to maintain spatial integrity in three dimensions while assessing precise spectroscopic data intrinsic to a tissue sample represents an ideal combination. Three-dimensional multivariate processing provides a new way of visualizing tissue blocks based on the underlying biochemistry of the tissue matrix and will therefore have significant application in biology and medicine.

Competing Interests

The author(s) declare that they have no competing interests.

Authors' Contributions

BRW conceived of the study, and participated in its design and coordination and helped to draft the manuscript. KBR took the FTIR measurements and performed the image construction and helped draft the manuscript. CE developed the MATLAB® routine for "stitching" multiple tissue sections. MQ provided the samples and did the histology. DM supervised the project and proof read the manuscript. All authors read and approved the final manuscript.

Acknowledgements

The authors thank Dr Virginia Billson (Royal Women's Hospital, Melbourne) for histopathology advice and the National Health and Medical Research Council of Australia for grant support. Dr Wood is funded by an Australian Synchrotron Research Program Fellowship Grant and a Monash Synchrotron Research Fellowship. Mr Finlay Shanks is thanked for instrumental support and Mr Clyde Riley (Royal Women's Hospital, Melbourne) for sectioning.

References

1. Lewis EN, Treado PJ, Reeder RC, Story GM, Dowrey AE, Marcott C, Levin IW: Fourier transform spectroscopic imaging using an infrared focal-plane array detector. Anal Chem 1995, 67:3377–3381.
2. Wood BR, McNaughton D, Chirboga L, Yee H, Diem M: Fourier transform infrared mapping of the cervical transformation zone, and dysplastic squamous epithelium. Gynecol Oncol 2003, 93:59–68.
3. Wood BR, Bambery K, Quinn MA, McNaughton D: Infrared imaging of normal and diseased cervical tissue sections - a comparison of FPA and synchrotron imaging. Proceedings of SPIE-Smart Materials, Nano-, and Micro-Smart Systems 2005, 5651:78–84.
4. Diem M, Chiriboga L, Yee H: Infrared spectroscopy of human cells and tissue. VII. Strategies for analysis of infrared tissue mapping data and applications to liver tissue. Biopolymers (Biospectroscopy) 2000, 57:282–290.
5. Kneipp K, Kneipp H, Corio P: Surface enhanced and normal stokes and anti-Stokes and antistokes Raman spectroscopy of single walled carbon nanotubes. Phys Rev Lett 2000, 84:3470–3473.
6. Schultz CP, Mantsch HH: Biochemical imaging and 2D classification of keratin pearl structures in oral squamous cell carcinoma. Cell Mol Biol 1998, 44:203–210.
7. Jackson M, Ramjiawan B, Hewko M, Mantsch HH: Infrared microscopic functional group mapping and spectral clustering analysis of hypercholesteramic rabbit liver. Cell Mol Biol 1998, 44:89–98.

8. Lasch P, Haensch W, Naumann D, Diem M: Imaging of colorectal adenocarcinoma using FT-IR microspectroscopy and cluster anaysis. Biochim Biophys Acta 2004, 1688:176–186.
9. Mohlenhoff B, Romeo M, Diem M, Wood BR: Mie-type scattering and non-Beer-Lambert absorption behaviour of human cells in infrared micro-spectroscopy. Biophys J 2005, 88:3635–3640.
10. Zhang L, Small GW, Haka AS, Kidder LH, Lewis EN: Classification of Fourier transform infrared microspectroscopic imaging data of human breast cells by cluster analysis and artificial neural networks. Appl Spectrosc 2003, 57:14–22.
11. Lasch P, Naumann D: FT-IR microspectroscopic imaging of human carcinoma thin sections based on pattern recognition techniques. Cell Mol Biol 1998, 44:189–202.
12. Mansfield JR, McIntosh K, Crowson AN, Mantsch HH, Jackson M: A LDA-guided search engine for the non-subjective analysis of infrared microspectroscopic maps. Appl Spectrosc 1999, 53:1323–1330.
13. Mansfield JR, Sowa MG, Scarth GB, Somorjai RL, Mantsch HH: Analysis of spectroscopic imaging data by fuzzy C-means clustering. Anal Chem 1997, 69:3370–3374.
14. Lasch P: CytospecTM. A Matlab based application for infrared imaging see http://www.cytospec.com for details.
15. Fabian H, Lasch P, Boese M, Haensch W: Infrared microspectroscopic imaging of benign breast tumour tissue sections. J Mol Struct 2003, 661–662:411–417.
16. Lasch P, Haensch W, Lewis EN, Kidder LH, Naumann D: Characterization of colorectal adenocarcinoma by spatially resolved FT-IR microspectroscopy. Appl Spectrosc 2002, 56:1–9.
17. Rerek ME, Moore DJ, Mendelsohn R, Paschalis EP: Infrared microspectroscopic imaging of skin. Biophys J 2000, 78:250A.
18. Boskey AL, Mendelsohn R: Infrared analysis of bone in health and disease. J Biomed Opt 2005, 10:31102.
19. Marcott C, Reeder RC, Paschalis EP, Tatakis DN, Boskey AL, Mendelsohn R: Infrared microspectroscopic imaging of biomineralized tissues using a mercury-cadmium-telluride focal-plane array detector. Cell Mol Biol 1998, 44:109–115.
20. German MJ, Pollock HM, Zhao B, Tobin MJ, Hammiche A, Bentley A, Cooper LJ, Martin FL, Fullwood NJ: Characterization of putative stem cell

populations in the cornea using synchrotron infrared microspectroscopy. Invest Ophthal Vis Sci 2006, 47:2417–2421.

21. German MJ, Hammiche A, Ragavan N, Tobin MJ, Cooper LJ, Matanhelia SS, Hindley AC, Nicholson CM, Fullwood NJ, Pollock HM, Martin FL: Infrared spectroscopy with multivariate analysis potentially facilitates the segregation of different types of prostate cell. Biophys J 2006, 90:3783–3795.

22. Ou-Yang H, Paschalis EP, Boskey AL, Mendelsohn R: Chemical structure-based three-dimensional reconstruction of human cortical bone from two-dimensional infrared images. Appl Spectrosc 2002, 56:419–422.

23. Scientific Computing and Imaging Institute (SCI): SCIRun: A Scientific Computing Problem Solving Environment. http://softwaresciutahedu/scirunhtml 2002.

24. http://www.gimp.org

25. Novotny DB, Ferlisi P: Villoglandular adenocarcinoma of the cervix: cytologic presentation. Diag Cytopath 1997, 17:383–387.

Molecular Mapping of Periodontal Tissues Using Infrared Microspectroscopy

Allan Hynes, David A. Scott, Angela Man, David L. Singer, Michael G. Sowa and Kan-Zhi Liu

ABSTRACT

Background

Chronic periodontitis is an inflammatory disease of the supporting structures of the teeth. Infrared microspectroscopy has the potential to simultaneously monitor multiple disease markers, including cellular infiltration and collagen catabolism, and hence differentiate diseased and healthy tissues. Therefore, our aim was to establish an infrared microspectroscopy methodology with which to analyze and interpret molecular maps defining pathogenic processes in periodontal tissues.

Methods

Specific key cellular and connective tissue components were identified by infrared microspectroscopy and using a chemical imaging method.

Results

Higher densities of DNA, total protein and lipid were revealed in epithelial tissue, compared to the lower percentage of these components in connective tissue. Collagen-specific tissue mapping by infrared microspectroscopy revealed much higher levels of collagen deposition in the connective tissues compared to that in the epithelium, as would be expected. Thus inflammatory events such as cellular infiltration and collagen deposition and catabolism can be identified by infrared microspectroscopy.

Conclusion

These results suggest that infrared microspectroscopy may represent a simple, reagent-free, multi-dimensional tool with which to examine periodontal disease etiology using entirely unprocessed tissue sections.

Background

Periodontitis is defined by the inflammatory destruction of the supporting structures of teeth, including the periodontal ligament and alveolar bone. Periodontal diseases are generally chronic in nature and usually persist in the absence of treatment [1,2]. These diseases are the result of exposure of the periodontium to dental plaque biofilms that accumulate on the teeth to form bacterial masses at or below the gingival margin [3].

Biomedical utilization of the electromagnetic spectrum of light has revolutionized the practice of medicine over the centuries, most recently through the dramatic healthcare advances afforded by the development of magnetic resonance imaging (see Figure 1). The last region of the spectrum to be applied to the practice of medicine is the infrared region. Infrared (IR) spectroscopy is now being increasingly utilized in multiple biomedical settings [4].

IR spectroscopy can distinguish differences in the characteristics of diverse molecules by probing chemical bond vibrations and use these molecular and submolecular profiles to define and differentiate "diseased" and "healthy" tissues [4]. As covalent bonds vibrate, they absorb energy in the form of IR light (see Figure 2). The wavelength of light that is absorbed depends on the nature of the covalent bond (e.g. C=O, N-H), the type of vibration (bending, stretching, etc.), and the environment of the bond. The IR spectrum of a tissue sample can be regarded as molecular fingerprint of the tissue. If this molecular fingerprint is modified by a disease process, then IR spectroscopy can be used to detect and monitor the disease process.

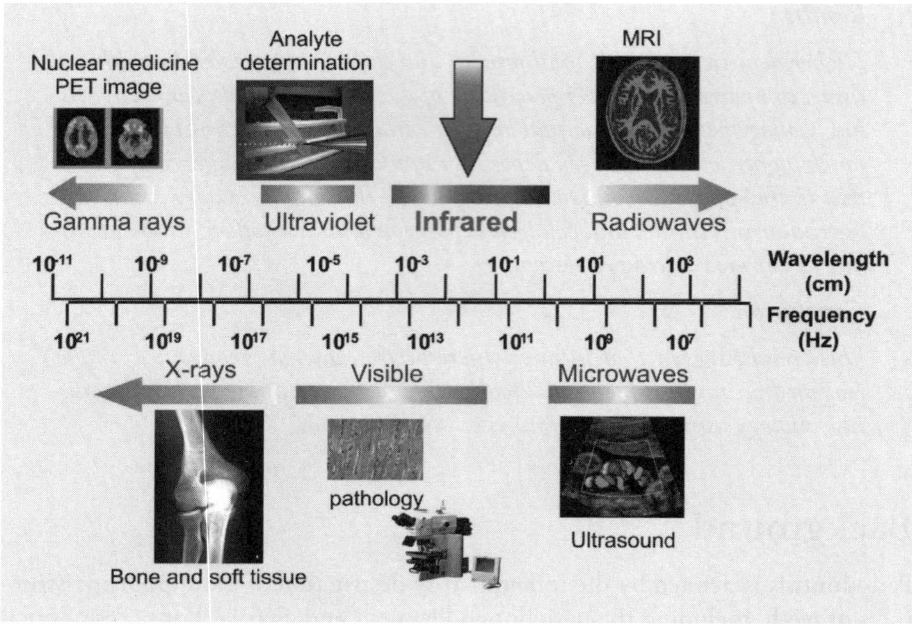

Figure 1. Biomedical applications of the electromagnetic spectrum. The classic electromagnetic spectrum is shown aligned with common, established biomedical applications. The infrared region of the electromagnetic spectrum lies between the visible and microwave regions, as indicated by the red arrow.

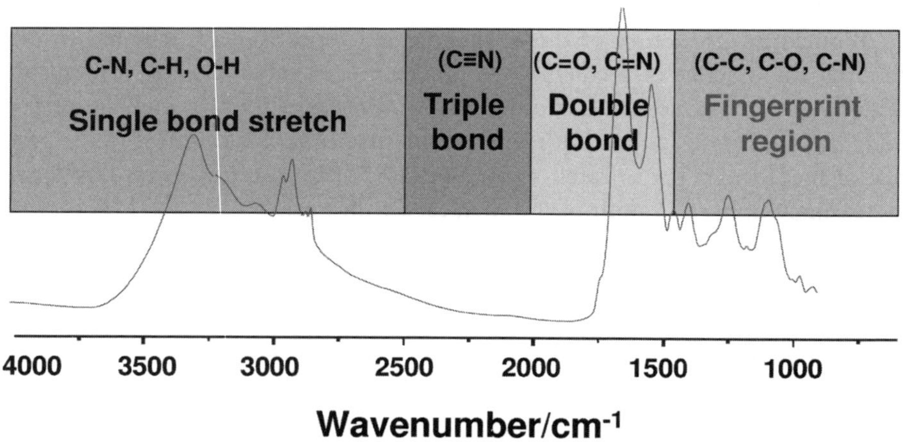

Figure 2. Representative spectrum in the mid-IR region. The region below 1500 cm^{-1} is defined as the fingerprint region. The region above 1500 cm-1 is the functional group region.

IR microspectroscopy is a relatively new technique in which infrared spectra are observed through a microscope that transmits and detects infrared radiation.

IR microspectroscopy has been previously utilized to monitor variations in the catabolism and anabolism of collagen and other components within cardiac tissues and oral cancer [5,6]. Collagens exhibit a series of unique IR absorption bands between 1000 and 1300 cm^{-1}. Specifically, the strong band at 1204 cm-1 has been identified as typical of collagen deposition [5,6]. By integrating the intensity of this absorption band, one can readily plot molecular contour maps that clearly delineate areas of collagen deposition. Inflammation-driven collagen degradation is a hallmark of periodontitis [7,8], with strong dissolution of collagen types I and III observed [9,10].

Particular absorptions are assigned to various functional groups in an attempt to extract biochemical information. Absorptions between 1620–1680 cm^{-1} are usually attributed to amide I vibration of proteins, while absorptions at 1080 and 1240 cm^{-1} are attributed to PO_2- symmetric and asymmetric stretching vibrations of DNA phosphodiester groups [4]. Using this data qualitative and semi-quantitative information can be extracted from the spectra.

Therefore, we set out to establish an IR microspectroscopic methodology that would allow the analysis and interpretation of molecular maps defining pathogenic processes in unprocessed periodontal tissues. This task is much more difficult in periodontal tissues in comparison to most other tissues due to the following factors; (1) Periodontal soft tissues are very thin (to a few mm); (2) Periodontal tissues excised at surgery are generally small (around 5 mm2); and (3) Excised periodontal tissues are particularly friable. Each of these factors present problems in embedding, sectioning, and orientating unfixed periodontal samples.

Methods

Subjects

Patients referred to the Graduate Periodontics Clinic, University of Manitoba for periodontal therapy and giving written, informed consent were recruited. Inclusion criteria were a clinical diagnosis of chronic periodontitis and a requirement for periodontal surgery at a diseased site (probing pocket depth ≥ 5 mm, bleeding on probing, and clinical attachment loss ≥ 3 mm). Exclusion criteria were tobacco smoking, pregnancy, a requirement for antibiotic prophylaxis prior to periodontal probing, prolonged anti-inflammatory medications within the past 3-months (e.g. NSAIDs, steroids, antibiotics, or immunosuppressants), any systemic condition that may interfere with the study, such as inflammatory diseases, diabetes or blood dyscrasias, and lesions of the gingiva unrelated to plaque-induced periodontal disease. All subjects had received oral hygiene instruction and root planing prior to surgery.

Expired-Air Carbon Monoxide Measurement

Self-reported non-smoking status was validated at the time of recruitment by analysis of expired-air CO concentrations (PiCO meter, Bedfont Sci., UK and associated PiCO chart software), calibrated according to the manufacturers instructions. Non-smokers were required to exhibit expired-air CO concentrations <10 ppm.

Periodontal Tissue Collection and Processing

On the day of surgery, prior to any anesthesia being administered, clinical measurements were obtained from the surgical site: probing depths, bleeding on probing and clinical attachment level (in that order). Twenty periodontal tissue samples were obtained from patients in good general health who were undergoing surgical treatment for chronic periodontitis. Mid-interproximal tissue was preferred, as these sites show increased clinical and histological signs of inflammation compared to other gingival sites [11]. Periodontal tissue samples were snap frozen in isopentane supercooled in liquid nitrogen and then stored at -80°C.

A small amount of optimal cutting tool embedding media (Miles Inc., Elkhart, IN.) was applied to the tissue samples in order to facilitate attachment to the cryotome. A series of 10 μm sections were then cut, mounted onto an IR transparent barium fluoride window, and air-dried. Adjacent 10 μm samples were mounted and processed for conventional histology including visualization of tissue integrity and inflammatory foci by H & E staining.

Infrared Microspectroscopy

Infrared microspectroscopy was performed using a Bruker FTIR spectrometer and IR microscope system equipped with a liquid nitrogen cooled mercury cadmium telluride detector. For IR spectral acquisition, the microscope aperture was closed to allow the IR beam to illuminate an area of tissue measuring 50 μm × 50 μm, thereby masking all other regions of the tissue section. For each 50 μm × 50 μm demarcation, 64 interferograms were collected. The signal was averaged against a blank area as background and Fourier-transformed to generate IR spectra with a nominal resolution of 4 cm^{-1}. After each acquisition, the stage was stepped 50 μm under computer control and the next spectrum acquired. This process was repeated until the complete area of interest was mapped. General biochemical mappings regarding the specific components of proteins, lipid, and DNA were obtained using a chemical imaging method. Specifically, the bands used for protein mapping were the amide I band (1654 cm^{-1}); lipid mapping used the lipid CH stretching vibrations at 2800–3000 cm^{-1}; and DNA mapping employed the PO_2- symmetric stretching vibration of DNA phosphodiester groups at 1080 cm^{-1}. To highlight collagen deposition, a unique collagen IR band at 1204 cm^{-1} was used to generate tissue maps, as we have described previously in cardiac tissues [5]. A digital CCD video camera was coordinated to the IR microscope to record each mapped area for future band and area correlation. Microphotography of the identical mapping position in the adjacent H and E-stained tissue section allowed the comparison of general histological features with the IR mapping data.

Data Processing for IR Microspectroscopy

Because most IR bands are broad and are composed of overlapping components, it is necessary to pre-process the original spectra by applying a band-narrowing algorithm that separates the individual bands. All data processing was performed using the Cytospec V software package http://www.cytospec.com webcite. To permit a useful comparison of the cluster analysis, uniformly pre-treated data was used. All the original IR spectra were converted into second derivative spectra using the Savitzky/Golay algorithm with a 9-point window for the multivariate statistical analysis. Derivative spectra were scaled before the cluster analysis where the sum squared deviation over the indicated wavelengths equals unity (vector normalization) [12]. The unsupervised cluster analysis used Ward's minimum variance algorithm and Euclidean distances as distance measure. The Euclidean distance between spectra is calculated and the pair of spectra with the least distance is grouped to create a cluster. Then the separation between this cluster and all other spectra is calculated and another two closest spectra/clusters are joined to form a new cluster. This procedure continues until all spectra/clusters are combined. In this unsupervised classification no information about the disease state of the samples is needed, only the similarity or dissimilarity of their infrared spectra is used for this classification.

Results

IR microspectroscopic imaging generated molecular tissue maps that provide a spectral signature of the intensity and spatial location of the chemical components of the diseased periodontal tissues examined. A representative microphotograph of periodontal tissue stained with H and E is shown in the left panel of Figure 3, while the spectra generated from defined area by IR microspectroscopy indicated by the square box are shown in the right panel. The H and E staining reveals typical histological constitution of periodontal tissue, i.e. the gingival epithelium comprising the epithelial layer that covers the external surface of the gingival and underlying connective tissue composed of gingival fibers, ground substance, and cells, including neural and vascular elements. Spectra from both the epithelium and connective tissues display the distinctive infrared spectral signatures of specific molecular structures, such as DNA, protein and lipids, as shown in the right panel.

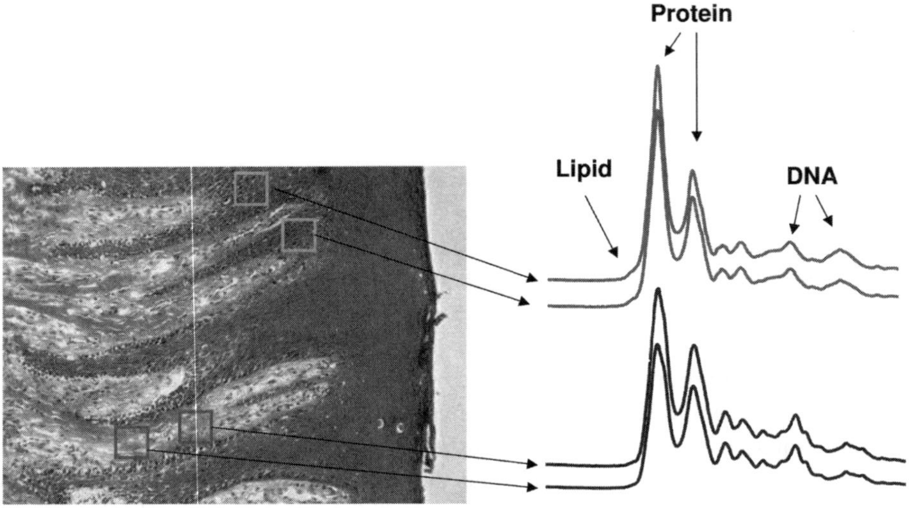

Figure 3. Comparison of periodontal tissues visualized by IR microspectroscopic mapping and by conventional H and E staining. Major periodontal tissue components (DNA, protein, and lipid), as determined by IR microspectroscopic mapping, are correlated with conventional H&E staining in a representative periodontal tissue section. IR spectra obtained from within the square boxes at different areas of the tissue section (periodontal epithelium and nearby connective tissue) are shown.

In order to reveal more subtle molecular differences in the periodontal tissue, the average spectra of these areas were generated using cluster analysis, as shown in Figure 4. As expected, both spectra show the signature of typical tissue components, i.e. total protein and collagen-specific signals, lipids (membranes), and nucleic acids. For instance, the absorptions at 1654 cm^{-1} arise from the amide I C=O stretching vibrations of the peptide groups in proteins [13]. The band at 1080 cm-1 is due to vibration of the phosphodiester groups in DNA that can be used to identify the content of cellular nucleus. The information regarding relative lipid concentration in both tissues can be found at the 1740 cm^{-1} band that originates from the ester C=O group while collagen substance in the tissue can be traced by looking at the specific band at 1204 cm^{-1}, as marked in the figure [5]. The top panel in Figure 4 is the difference spectrum, generated by subtracting the average spectrum of connective tissue from that of epithelium. The difference spectra help identify the specific molecular components that differ most between the two groups of spectra. Upwardly pointing peaks (positively shaded areas) are indicative of the presence of more of a certain molecular component in the epithelium tissue compared to that in the connective tissue, and vice-versa for the downwardly pointing peaks (negatively shaded areas). There are several important changes in the IR spectra that involve molecular vibrations of proteins, including collagen, lipids, and DNA. As expected, a higher cellular constituent content (protein, lipids and DNA) was found in the epithelial tissue, compared to the lower percentage of these components in the connective tissue. However, the collagen content was much higher in the connective tissue than in the epithelium.

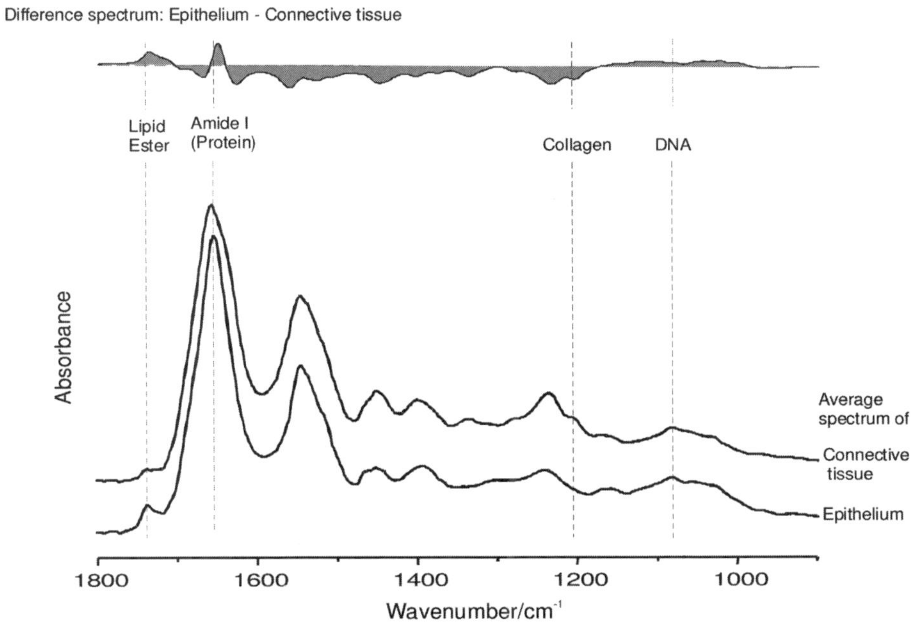

Figure 4. Comparison of the mean infrared absorption spectra* of epithelial (lower spectrum) and connective tissue (upper spectrum) and the epithelial-connective tissue difference spectrum. *Spectra are baseline corrected.

Based upon these molecular differences in periodontal tissues, IR chemical mappings were established and compared with conventional H and E-stained sections, as shown in Figure 5. Specifically, regions of interest in the periodontal tissue were identified using the H and E stained section and a high quality microscope in the visible range. Then, the corresponding features were identified by anatomical landmarks on the unstained adjacent tissue sections; subsequently, the tissue section was transferred to the IR microscope and aligned using the anatomical landmarks, as shown in Figure 5A and 5B. The numbers in the X- and Y-axes indicate the step scans employed in the acquisition of IR mapping. Therefore, a total of 962 spectra in this section were obtained and stored in the hypercube. To visualize the spectral data from the hypercube, the easiest way is to present "horizontal" slices through the hypercube, which displays intensity values at a given wavenumber for all spectral vectors as a false color representation [14]. The intensity value of each spectrum is assigned a color code and displayed against the X and Y coordinates of the spectral element. By examining different "color slices" (i.e., the intensities at different spectral elements), variations in the chemical composition can be detected for various pixels in the hypercube. As demonstrated in Figure 5C, the chemical IR mapping based on protein concentrations in the periodontal tissue revealed highest protein content in the epithelium, which was well correlated with tissue components observed by conventional H and E staining. The cell density in the epithelial tissue is significantly higher than that in the connective tissue (Figure 3). The epithelial cells provide the major source of protein noted in the IR spectra, especially at the level of functional group mapping. However, although the epithelial tissue contains more total protein than the connective tissue, the latter contains much more collagen. Other important cellular chemical mappings were also generated from the IR maps, as shown in Figure 6A–D. These false color IR molecular have the familiar appearance of traditional histological sections.

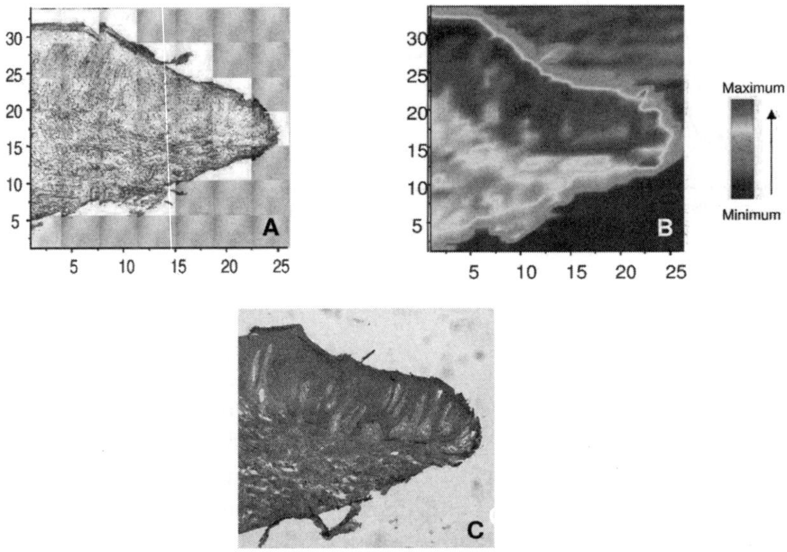

Figure 5. Comparison of unstained periodontal tissue section with H and E staining and the associated IR mapping. The original periodontal tissue used for IR mapping (A), as defined by the adjacent tissue stained with H&E (B), correlates very well with the IR chemical map based on protein band intensity (C). The color bar indicates the relative intensity of the specific IR band in the tissue.

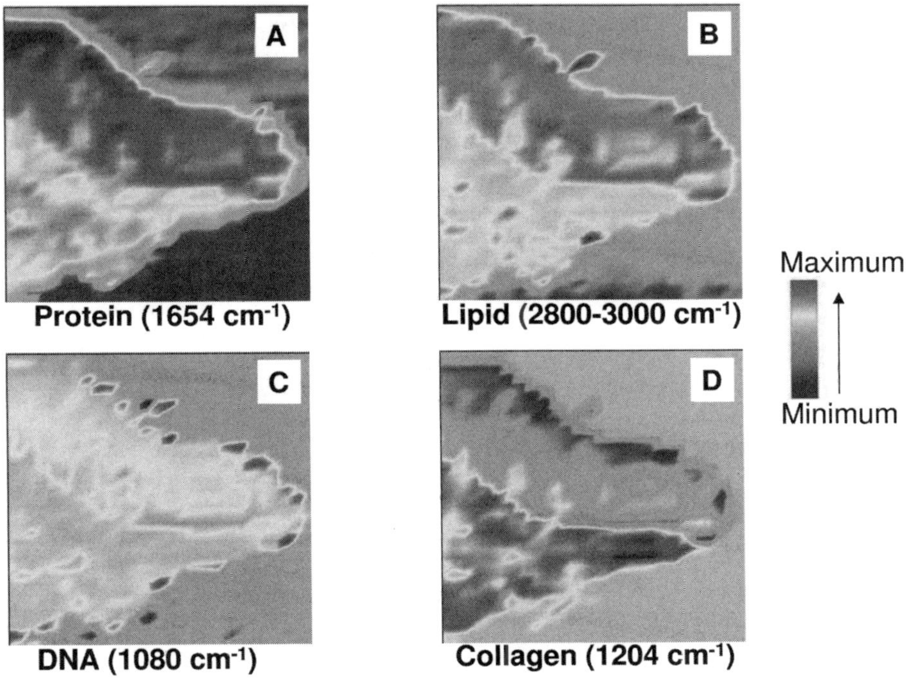

Figure 6. IR chemical mappings in periodontal tissue. Four major molecular distributions (total protein; lipid; DNA; collagen) in periodontal tissue (A-D) were generated based on the corresponding IR molecular signals. Higher concentrations of the major cellular components were observed in the periodontal epithelium (A-C) whereas higher amounts of collagen exist in the connective tissue (D), as expected.

Finally, we employed a multivariate statistical method, CLA (cluster analysis), to verify that the epithelial cells can be readily separated from connective tissues, as shown in Figure 7. The advantage of this computational approach is that it operates unsupervised: no prior knowledge of reference spectra of any tissue type is required, and characteristic spectra and membership of the families of spectra identified by the characteristic spectra are established, regardless of the number of different spectral families and the differences between spectral families [14]. This cluster analysis uses a large portion of the spectrum, rather than a few selected points, regions, or integrated regions, to determine whether spectra are related or not. Because the differences between spectra of normal and diseased tissue are generally small, this method offers the advantage of emphasizing differences in the overall shape, rather than selected intensities, for the discrimination of spectra. In this study, the entire range from 900–1800 cm^{-1} was used and nine groups based on their spectral features were generated.

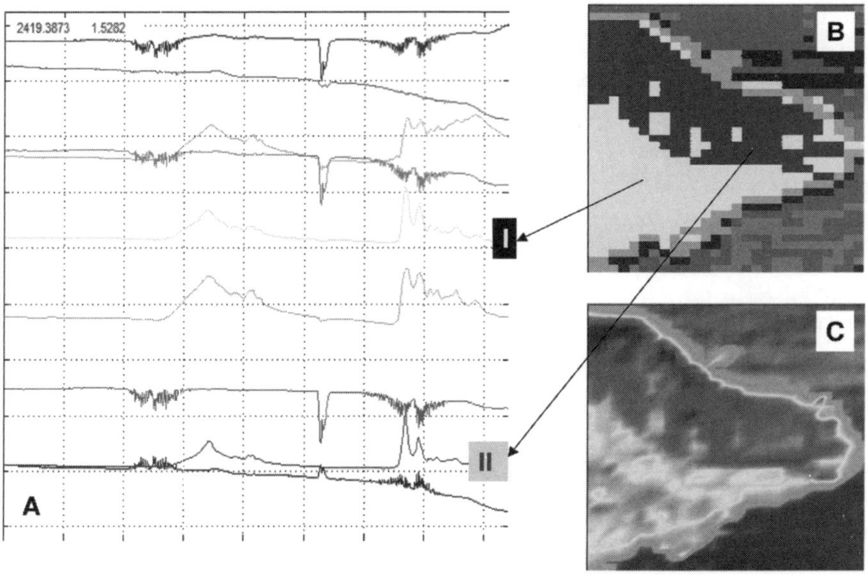

Figure 7. Cluster mapping of periodontal tissue. Average spectra were generated from nine groups based on their overall spectral features (A); each representing one class in the cluster map of periodontal tissue (B) produced by the cluster analysis. Good correlation can be observed between the cluster mapping (B) and corresponding chemical mapping (protein) (C) based on the differential spectral features of the epithelial and connective tissues.

In the final step of cluster analysis, all spectra in the same family are assigned a color code, and small, colored squares are drawn at the pixel coordinates of all spectra belonging to the same family to produce our false color maps. As shown in Figure 7B, the blue color (line II) indicates that the spectra from this group were in the same family, originating from the epithelium, like the IR protein mapping of the same tissue shown in Figure 7C. The yellow color in Figure 7B (line I) represents those similar spectra generated from connective tissue. Panel A in Figure 7 displays the average spectrum produced by the cluster analysis with color codes. In other words, the average spectrum from epithelium and connective tissues are in blue and yellow, marked II and I, respectively, and also correspond to the spectra analyzed in full detail in Figure 4.

Discussion

IR microspectroscopy possesses numerous advantages over traditional approaches to pathology. Fixation and staining of tissues are not required before histological viewing, little or no sample preparation is necessary and only minimal technical expertise is required by the operator. The method lends itself readily to rapid, high-volume repetitive measurements. IR microspectroscopy is non-destructive, meaning that the sample may be saved and passed on for further measurements if required. Furthermore, IR microspectroscopy provides information concerning the molecular structure of the tissue and multiple analytes may be measured simultaneously from a single spectrum. This combination of features is simply unavailable from visual microscopy.

Our initial data suggest that infrared microspectroscopy represents a suitable tool with which to simultaneously monitor multiple disease markers in periodontal biopsies, including cellular infiltration, collagen catabolism, and other differences in the molecular profile of diseased tissues. We have established a methodology by which IR microspectroscopy is capable of revealing several major biochemical components and specific features, including collagen content, in the studied tissue using "digital staining," without the need for any chemical reagents or probes. IR maps of inflammation-driven collagen degradation in periodontal tissue sections can therefore be constructed and analyzed. The promising preliminary results, obtained in establishing this IR microspectroscopic methodology, suggest a potential role for infrared microspectroscopy in understanding the inflammatory processes underlying the progression of periodontitis.

The major problem in adapting IR technology to the study of inflammatory processes in gingival tissues was the small size and fragility of the periodontal biopsies themselves. However, we are able to process these small tissue samples and can clearly differentiate predominantly cellular and predominantly acellular areas of tissue, and visualize areas of collagen deposition and degradation. The next steps will be to compare IR microscopic maps of diseased and healthy tissues; to correlate NIR-based definitions of pathological tissue changes with disease severity; and to correlate IR maps with classical clinical signs of periodontal diseases, such as edema, gingival bleeding, and periodontal pocket depths on probing, and clinical attachment loss.

Conclusion

As inflammatory cell infiltration and subsequent collagen degradation are hallmarks of periodontitis [7-10], and because IR spectral analyses can determine such tissue events, in addition to multiple other tissue changes at the molecular and sub-molecular level, then this IR microspectroscopic methodology can now be applied to hypothesis-driven research that aims to identify disease-related pathological changes in periodontal tissues.

List of Abbreviations

FTIR, Fourier-transform infrared; IR, infrared

Competing Interests

The authors have no conflicting interests or financial implications related to the publication of this review article. The research activities described herein are humanitarian (non-profit making) in nature. However, certain application of infrared technology may be patentable.

Authors' Contributions

The original hypothesis was developed by DAS, K-ZL, and DLS. DAS performed the literature review included in the manuscript. AGH and DLS identified suitable clinical cases. AGH performed the surgeries and prepared tissue sections for sectioning. AGH and AM performed most of the infrared microspectroscopy experiments. K-ZL, DAS and MGS performed the extrapolation of data from the IR spectra and the spectral data analysis. All authors made editorial contributions to the manuscript.

Acknowledgements

This study was supported by funding from the National Research Council, Canada and by the Health Sciences Research Foundation, Canada.

References

1. Page RC, Shroeder HE: Pathogenesis of inflammatory periodontal disease. A summary of current work. Lab Invest 1976, 34:235–265.
2. Page RC: The role of inflammatory mediators in the pathogenesis of periodontal disease. J Periodont Res 1991, 26:230–242.
3. Loesche WJ: The antimicrobial treatment of periodontal disease: changing the treatment paradigm. Crit Rev Oral Biol Med 1999, 10:245–75.
4. Jackson M, Sowa MG, Mantsch HH: Infrared spectroscopy: a new frontier in medicine. Biophys Chem 1997, 68:109–25.

5. Liu KZ, Dixon IM, Mantsch HH: Distribution of collagen deposition in cardiomyopathic hamster hearts determined by in microscopy. Cardiovasc Pathol 1999, 8:41–47.

6. Schultz CP, Liu KZ, Kerr PD, Mantsch HH: In-situ infrared histopathology of keratinization in human oral/or pharyngeal aqueous cell carcinoma. Oncol Res 1998, 10:277–286.

7. Ejeil AL, Gaultier F, Igondjo-Tchen S, Senni K, Pellat B, Godeau G, Gogly B: Are cytokines linked to collagen breakdown during periodontal disease progression? J Periodontol 2003, 74:196–201.

8. van der Zee E, Everts V, Beertsen W: Cytokines modulate routes of collagen breakdown. J Clin Periodontol 1997, 24:297–305.

9. Hillmann G, Krause S, Ozdemir A, Dogan S, Geurtsen W: Immunohistological and morphometric analysis of inflammatory cells in rapidly progressive periodontitis and adult periodontitis. Clin Oral lnvestig 2001, 5:227–35.

10. Feldner BD, Reinhardt RA, Garbin CP, Seymour GJ, Casey JH: Histological evaluation of interleukin-1 beta and collagen in gingival tissue from untreated adult periodontitis. J Periodontal Res 1994, 29:54–61.

11. Abrams K, Caton J, Polson AM: Histologic comparison of interproximal gingival tissues related to the presence or absence of bleeding. J Periodontol 1984, 55:629–632.

12. Lasch P, Haensch W, Naumann D, Diem M: Imaging of colorectal adenocarcinoma using FT-IR microspectroscopy and cluster analysis. Biochim Biophys Acta 2004, 176–186.

13. Liu KZ, Jia J, Newland AC, Kelsey SM, Mantsch HH: Quantitative determination of apoptosis on leukemic cells by infrared spectroscopy. Apoptosis 2001, 6:269–278.

14. Diem D, Chiriboga L, Yee H: Infrared spectroscopy of human cells and tissue. VIII. Strategies for analysis of infrared tissue mapping data and applications to liver tissue. Biopolymers 2000, 57:82–90.

The Use of Coumarins as Environmentally-Sensitive Fluorescent Probes of Heterogeneous Inclusion Systems

Brian D. Wagner

ABSTRACT

Coumarins, as a family of molecules, exhibit a wide range of fluorescence emission properties. In many cases, this fluorescence is extremely sensitive to the local environment of the molecule, especially the local polarity and micro-viscosity. In addition, coumarins show a wide range of size, shape, and hydrophobicity. These properties make them especially useful as fluorescent probes of heterogeneous environments, such as supramolecular host cavities, micelles, polymers and solids. This article will review the use of coumarins to probe such heterogeneous systems using fluorescence spectroscopy.

Keywords: Coumarins; Fluorescence spectroscopy; Fluorescent probes; Heterogeneous systems; Host-guest inclusion.

Introduction

Coumarins, or benzo-α-pyrones, are a very large and important family of compounds. Their defining structure consists of fused pyrone and benzene rings, with the pyrone carbonyl group at position 2 [1]; this structure is illustrated in Figure 1 for the coumarin parent molecule (IUPAC name: 2H-chromen-2-one, and also known as 1-benzopyran-2-one). Coumarins are widely occurring in nature, with coumarin itself first isolated in 1820 from a specific variety of bean, and many other coumarin derivatives found in a wide range of plants [1]. As a group, coumarins exhibit interesting fluorescence properties, which include a high degree of sensitivity to their local environment, including polarity and viscosity. This sensitivity has led to their widespread application as sensitive fluorescent probes of a wide range of systems, including homogeneous solvents and mixtures, and heterogeneous materials; the latter is the focus of this article. Specifically, the purpose of this review is two-fold: 1) to provide a detailed review of the use of coumarin fluorescence to probe the nature and properties of heterogeneous materials and systems, and 2) to provide a guide to current and future researchers studying heterogeneous and supramolecular systems to the utility of and information provided by coumarins as fluorescent probes.

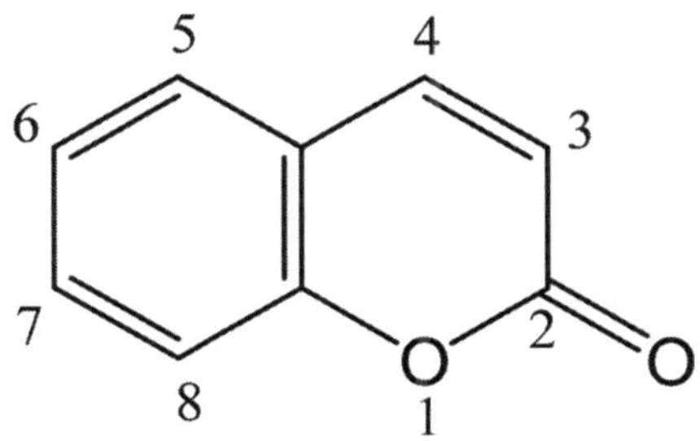

Figure 1. The chemical structure and numbering scheme of coumarin.

Numerous fluorescent coumarin derivatives have been reported, with a wide range of polarity, pH, viscosity, and other sensitivities, and varying underlying photophysical mechanisms for the observed fluorescence properties. Furthermore, thousands of papers involving some aspect of coumarin chemistry or spectroscopy have been published. Therefore, a comprehensive review of all coumarin derivatives, fluorescence properties and applications is beyond the scope of this article. Instead, the article begins with an overview of coumarin fluorescence properties using two representative groups of specific coumarin derivatives. A detailed review of the use of the fluorescence of included coumarins to study specific types of heterogeneous chemical systems and media is then presented. The use of coumarin fluorescence to probe proteins and other biochemical and biological systems will not be covered. In addition, the extensive use of coumarins as covalently attached fluorescent labels and structural components, or to generate fluorescent derivatives, will also not be reviewed, nor will the use of coumarins as fluorescent laser dyes. Thus, the scope of the review will be limited to the use of discrete coumarin fluorescent probe molecules which become included in the cavities or internal structure of hosts, discrete organized chemical structures in solution (such as micelles and polymers) or solid materials.

Coumarin Photophysics

In this section, the defining features of coumarins as fluorescent probes is discussed. In particular, the mechanism for the high degree of polarity- and viscosity-sensitivity of the fluorescence of this family of compounds is described, using specific representative coumarin derivative examples. As stated above, the purpose of this section is to provide a representative overview of coumarin photophysics, in regards to their usefulness as environmentally-sensitive fluorescent probes. In addition, specific coumarins and effects are discussed in later sections dealing with applications to specific heterogeneous systems.

In order to describe the fluorescence of coumarin probes, there are four experimental properties of interest, which can exhibit significant changes within heterogeneous systems. These are the wavelength of maximum fluorescence intensity (i.e. the wavelength of the peak of the spectrum), λF,max (alternatively the frequency of maximum intensity, νF,max), the fluorescence emission intensity at a particular wavelength, IF, the fluorescence quantum yield, NF, and the fluorescence lifetime, τF. The value of λF,max is indicative of the energy gap between the fluorescent and ground singlet states, and can undergo significant blue-shifting (to shorter wavelength/higher frequency and energy) or red-shifting (to longer wavelength/lower frequency and energy) in response to the local environment. The value of IF is indicative of the intensity of the fluorescence of the particular sample, whereas the value of NF is a measure of the efficiency of the fluorescence of the probe in this sample as a relaxation pathway, relative to all relaxation pathways, including nonradiative decay. (The value of NF is related to the integrated intensity IF taken over the entire spectrum, relative to that of a fluorescent standard). The value of τF is a measure of the lifetime of the excited state, and depends on all of the deactivation pathways available to the excited state. Both NF and τF are related to the rate constants for radiative (kr) and nonradiative (knr) decay: NF = kr /(kr + knr); τF = 1/(kr + knr). The values of all three properties IF, NF and τF can be significantly increased or decreased upon changing the local environment of the coumarin probe; the mechanisms for this will be discussed throughout this section.

A detailed spectroscopic study of coumarin itself was published in 1970 by Song and Gordon [2]. They measured fluorescence and phosphorescence spectra and lifetimes in both polar and nonpolar solvents at 77 K. They assigned the fluorescence emission to a $1(\pi,\pi^*)$ excited state, and observed a large red-shift of 30 nm in nonpolar as compared to polar solvent. Thus, the significant solvatochromism of coumarin fluorescent probes has been known for almost 40 years.

7-Aminocoumarins such as 4-methyl-7-diethylaminocoumarin (C1, shown in Figure 2a) are arguably the most important subset of coumarins, and have been the focus of intense study [3-19] and wide-spread applications as fluorescent probes. The photophysics of 7-aminocoumarins will thus be described in some detail, as a representative group. The chemical structures of some commonly-used 7-aminocoumarin fluorescent probes are shown in Figure 2. It should be noted that different authors sometimes use different coumarin numbers to describe the same coumarin derivative, thus all relevant numbers will be indicated with each coumarin structure shown. (Throughout this review, coumarin derivatives for which the chemical structure is shown in one of the figures will be indicated in bold.)

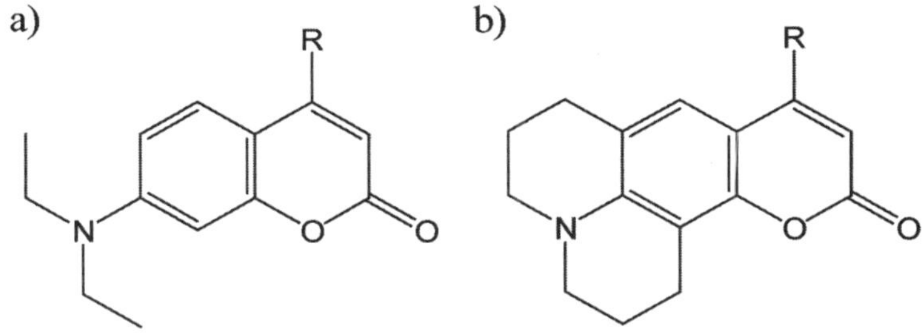

Figure 2. The chemical structure of some commonly-used 7-aminocoumarin fluorescent probes. a) R=CH3: C1, also known as C460; R=CF3: C1F, also known as C35, C152A and C481; b) R=CH3: C102, also known as C480; R=CF3: C6F, also known as C153 and C540A.

Jones et al. [4] presented an early study on the effect of solvents on three 4-trifluoromethyl substituted 7-aminocoumarins, C1F (Figure 2a), C6F (Figure 2b), and C8F. They observed a very strong red-shift in polar solvents, and were able to correlate νF,max with the solvent polarity-polarizability parameter π^*, as well as the hydrogen bonding parameter α. They also observed a significantly reduced fluorescence quantum yield NF for the non-rigid coumarin C1F with increasing polarity, which they attributed to an increased nonradiative decay rate knr, via formation of a twisted intramolecular charge-transfer (TICT) state. Jones et al. then followed up with an expanded study [5], which included 11 different 7-aminocoumarin laser dyes, and reported that hydrogen bonding is the major factor controlling TICT formation in each case, and that this effect explains the observed solvatochromism. In fact, TICT formation [20] plays the defining role in 7-aminocoumarin fluorescence properties [4-12, 14, 16, 18, 20], through increased TICT nonradiative decay in polar media. As a result of the amino group, 7-aminocoumarins also exhibit pH-dependent fluorescence. Patalakha et al. reported on the acid-base properties of a series of 7-diethylaminocoumarins with various aromatic substituents [7] or fluorine [8] at position 3, and observed very large blue shifts in the fluorescence emission of the protonated as compared to the neutral form of these probe molecules. Abdel-Mottaleb [9] studied the photophysics of both flexible and rigid 7-aminocoumarin derivatives as a function of viscosity in aqueous glycerol solutions. They measured the fluorescence depolarization rate, and correlated this with the calculated free volume fraction of the medium. They used these results to propose the use of these 7-aminocoumarins as fluorescent probes of both fluidity and polarity of the local medium. Yip et al. [10] measured the fluorescence lifetimes of coumarins C1 and C102 (Figure 2b) in a number of polar solvents, and obtained two- and three-exponential decays. They attributed these results, as well as differences in decay-associated fluorescence spectra, to an irreversible two-state solvation model.

There has been some controversy on the exact role that hydrogen bonding plays in the formation of TICT states in protic solvents. López Arbeloa et al. [11] studied the photophysics of a number of 7-aminocoumarin derivatives and invoked specific hydrogen bonding between the coumarins and solvent to explain the observed solvent dependence. Królicki et al. [13] also investigated the role of hydrogen bonding, this time in the case of the rigid 7-aminocoumarin C153 (Figure 2b), in mixed solvent systems, and observed preferential solvation in the excited state, but not the ground state. They also observed an unusual dependence of the fluorescence quantum yield on the mole fraction of methanol in methanol:toluene mixed solvent, and attributed this to specific hydrogen bonds between methanol and the coumarin probe. More recently, Moog et al. [15] extended the study of coumarin solvatochromism of coumarins C1, C120 (Figure 3a), C151, C152 and C153 by comparing three different models for treating solvent effects. They found that the multi-parameter Kamlet-Taft equation gave the best correlation for all of these coumarins, and further concluded that the effect of hydrogen bonding to solvent was a result of the increased field produced by the dipole moment of the hydrogen-bonding solvent, and not the hydrogen bonding interaction itself, in contrast to the results described above of López Arbelo et al. [11] and Królicki et al. [13]. However, Dahiya et al. [16] reported the effects of protic solvents on the coumarins C152 and C481 (Figure 2a), and concluded that the solute-solvent H-bonding interactions directly stabilize the TICT states. Most recently, Barik et al. [18] published a detailed study on the evidence for TICT-mediated nonradiative decay of coumarin C1 in high polarity protic solvents. They were able to correlate the observed Stokes shift with the solvent polarity, described using the parameter Δf. Furthermore, they saw no evidence for TICT in highly polar aprotic solvents, emphasizing the role of hydrogen bonding to the solvent. They also observed an exponential increase in the nonradiative decay rate in solvents with polarity $\Delta f > 0.28$, indicating the onset of an additional nonradiative deactivation pathway. It is clear from these studies that in the case of water, which is the predominant solvent of choice for studying heterogeneous systems in solution, hydrogen bonding plays a significant role in the formation of TICT states in 7-aminocoumarins, as will the polarity differences within the heterogeneous medium as compared to the bulk water solvent.

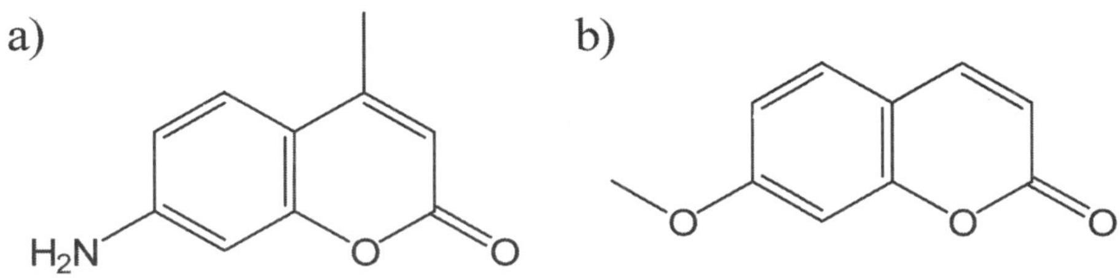

Figure 3. The chemical structures of a) coumarin derivative C120; b) 7-methoxycoumarin (7MC).
Other specific studies of 7-aminocoumarin photophysics have been reported, including the effect of substituents and concentration [12], the observation of multiple emissions from 7-aminocoumarins with heterocyclic substituents [14], and photochemical transformations upon UV irradiation [17]. In addition, Sharma et al. [19] determined the dipole moments of a number of 7-aminocoumarin dyes, both experimentally and theoretically, and found that in all cases, the dipole moment was much larger in the excited state.

It is clear from all of these studies that 7-aminocoumarins, as a group, are very sensitive fluorescent probes for the study of local environments within heterogeneous media. They have been used significantly for this purpose, as described in the following sections. The fluorescence of these probes is strongly dependent on the polarity, hydrogen bonding ability, pH and microviscosity or rotational hindrance of their local environment, and this dependence varies with the specific 7-aminocoumarin derivative used. In general, as a result of TICT state formation in polar solvents, the fluorescence emission of 7-aminocoumarins is seen to red-shift and decrease in intensity as the polarity of the medium is increased.

Another useful subset of coumarin probes is the 7-alkoxycoumarins [21, 22], such as 7¬methoxycoumarin (7MC, the structure of which is shown in Figure 3b). These coumarins exhibit a different polarity-dependent fluorescence than do 7-aminocoumarins: their fluorescence intensity increases with increasing polarity of the medium, but with negligible spectral shift [21, 22]. For example, the value of NF for 7-methoxycoumarin is 0.51 in aqueous buffer (i.e. over half of the excited molecules relax by emitting a photon), but drops to 0.033 in methanol [22]; however the wavelength of maximum emission only changes from 324.7 to 322.6 nm. This solvent dependence is a result of a completely different mechanism than the TICT formation described above for 7-aminocoumarins, and involves the changing of the energy of the $1(\pi\pi^*)$ fluorescent state relative to the closely lying first triplet $3(n\pi^*)$ state [21]. In nonpolar solvent, the $1(\pi\pi^*)$ state lies just above the $3(n\pi^*)$ state, so that the rate of nonradiative intersystem crossing (ISC) is very efficient, and NF is correspondingly low. However, in polar solvent, the $1(\pi\pi^*)$ energy is lowered below that of the triplet state, greatly decreasing the efficiency of ISC, and thus NF increases significantly. This effect of polarity on emission intensity is the exact opposite of that observed with 7-aminocoumarins.

Many other coumarin derivatives have been designed for specific fluorescence properties or sensitivities, such as cyanocoumarins [23] and 7-hydroxycoumarin-hemicyanine hybrids [24], both of which exhibit emission in the red region of the spectrum, and the pair of highly substituted coumarins shown in Figure 4a, which show fluorescence dependence solely on viscosity, rather than polarity [25]. Some other useful coumarin derivatives with specific applicability that have been prepared and studied include 3-(2'-benzimidazolyl) coumarins [26], the coumarin-based amino acid shown in Figure 4b [27], various iminocoumarins (such as those shown in Figure 4c) [28], and various biscoumarins [29, 30].

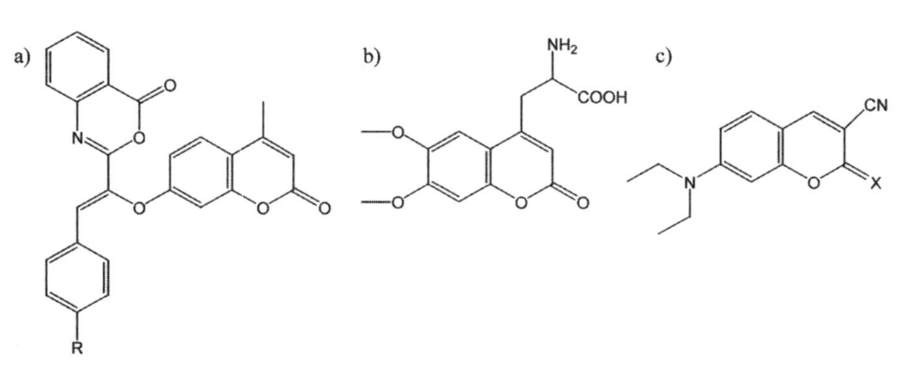

Figure 4. The chemical structure of a) a pair of specifically-designed viscosity-dependent coumarin fluorescent probes (R = OH or CO_2CH_3); b) a coumarin-based amino acid probe; c) some iminocoumarins (R = NH or $N(CO)OCH_2CH_3$).

In addition to the experimental spectroscopic studies of coumarins as described above, there have also been a number of useful theoretical studies of coumarin derivatives [9, 19, 31-33], in which the energies and electronic configurations of the excited singlet and triplet states have been calculated, in order to help to understand coumarin photophysics. The computational and theoretical approaches used include SCF-CI [9], PM3 [19], AM1 [31], PPP [32] and MM2 calculations [33].

Coumarins as Fluorescent Guests Included in Molecular Hosts

Host-guest inclusion complexes are formed when a small guest molecule becomes encapsulated within the internal cavity of a larger, cage-like host molecule. Such complexes represent one of the simplest examples of a supramolecular system, as the complex is held together only by non-covalent forces. Because of the lack of covalent binding between the host and guest, this complexation is a dynamic phenomenon, and equilibrium is established in solution between the complex and the free host and guest, as illustrated in Figure 5. The value of the binding constant, K, is the most important measurable property of the host-guest complex, and its magnitude is indicative of the total driving forces for inclusion. In order for a significant concentration of the host-guest complex to be obtained, the rate of entrance into the cavity (k_{in}) must be significantly larger that the rate of exit (k_{out}); $K = k_{in}/k_{out}$. The phenomenon of host-guest inclusion is an important aspect of the recent and growing field of supramolecular chemistry, and has found widespread and important applications, as discussed below for individual families of hosts.

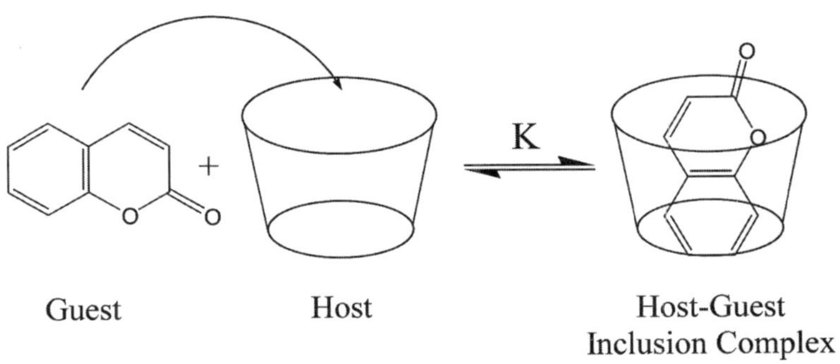

Figure 5. A representation of the inclusion of a coumarin guest inside a host cavity, forming a host-guest inclusion complex.

Host-guest inclusion complexes are usually formed in aqueous solution, as this maximizes the difference in local polarity between the relatively nonpolar internal cavity of the organic host molecule and the bulk solvent, maximizing the hydrophobic effect as a driving force for inclusion of hydrophobic guests. Thus, the guest will experience a significant lowering of the polarity of the local environment upon inclusion into the host cavity. Furthermore, the guest will be in the much more confining and restrictive cavity as opposed to being free in solution, so that guest intramolecular rotations will be expected to be significantly hindered, depending on the size and shape of the cavity. For both of these reasons, coumarins are ideal fluorescent guests to probe the nature and binding capacity of such host molecules, as both the polarity and the constriction will greatly affect the formation of TICT states, and the polarity will change the relative energy levels, all of which result in significant and easily measurable changes in the coumarin probe fluorescence.

Cyclodextrins

Cyclodextrins are cyclic oligosaccharides of glucopyranose, which through intramolecular hydrogen bonding form truncated cone-shaped structures with large internal cavities in aqueous solution [34]. The presence of the large, internal cavity makes cyclodextrins, also referred to as "molecular buckets," excellent hosts for the inclusion of a wide range of neutral and ionic guests [34], and they are by far the most widely studied and utilized molecular hosts [34, 35]. As shown in Figure 6a, there are three "native" cyclodextrins, α-, β- and γ-CD, which consist of six, seven, and eight gluocopyranose units, respectively, and hence have very different cavity sizes. The chemical structure of β-CD is shown in Figure 6b; also illustrated is the presence of the three hydroxyl groups per glucopyranose unit, which allow CDs to be readily chemically modified. The relative ease of preparing modified CDs, with specific targeted properties, has also contributed to their huge popularity as molecular hosts.

There is a relatively long and rich history of the use of the fluorescence of coumarins to investigate CDs, with the first such study reported by Takadate et al. in 1983 [36]. They studied five different 7¬substituted 4-methylcoumarins included in β-CD, and found that the fluorescence was enhanced and blue shifted for the 4-hydroxy- and 7-aminocoumarins, but that the fluorescence was quenched for 7-methoxy and 7-ethoxycoumarin. In all cases, the observed fluorescence effects were interpreted in terms of the relative polarity of the CD cavity relative to bulk water; which the authors determined to be slightly more polar than ethanol solvent, based on the measured fluorescence maxima. The authors were able to use the change in fluorescence as a function of added CD concentration to obtain the

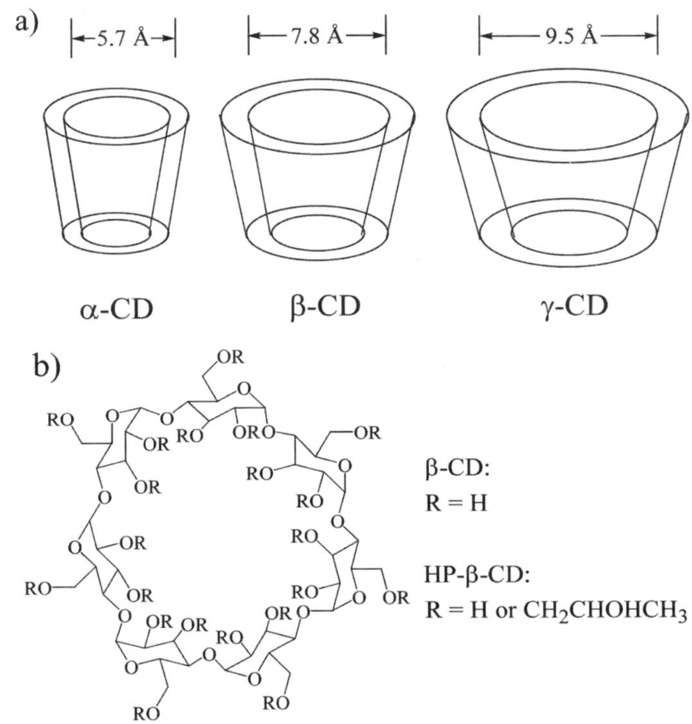

Figure 6. a) Molecular bucket depictions and cavity sizes of α-, β-, and γ-CD; b) chemical structure of native and HP-modified β-CD.

binding constant K, which ranged widely from 80 M-1 in the case of 7-methoxy-4-methylcoumarin to 893 M-1 in the case of 7-dimethylamino-4-methylcoumarin. This range in K values illustrates the impact that guest size, shape and properties (such as polarity) can have on the strength of the binding with CDs. This work was followed shortly thereafter in 1985 by a report of Scypinski and Drake [37] on the inclusion of the rigid 7-aminocoumarin derivative C540A (Figure 2b) in β- and γ-CD. Significant enhancement of C540A fluorescence was observed in both CDs (slightly larger in γ-CD), with a blue shift of 10 nm. Both observations were attributed to the formation of 1:1 CD: guest inclusion complexes, and corresponding decreased polarity within the CD cavity relative to that of water. No change in C540A fluorescence was observed in the presence of α-CD, indicating that the cavity of this CD is too small to accommodate this large coumarin guest. The values of the binding constant, K, obtained were quite small, only 54 M-1 for β-CD at 20 °C. Interestingly, two types of complexes were obtained, a "normal" and an "inverted" complex, depending on the conditions used to prepare them, with the C540A having opposite orientations within the cavity in the two types of complex. While the "normal" complex showed the fluorescence enhancement described, the "inverted" complex actually exhibited reduced fluorescence, due to enhanced quenching of the exposed guest. Hydrogen bonding between the host and guest was proposed to occur in the "normal" complex. Two other studies of the inclusion of 7-aminocoumarins in CDs have subsequently been reported. Bergmark et al. [38] showed that the co-inclusion of organic solvents with coumarins C1 and C6F in β- and γ-CD resulted in even greater enhancement of the coumarin fluorescence. They proposed that this additional enhancement occurred due to the displacement of CD cavity water by these organic solvent molecules; these water molecules directly quenched the coumarin fluorescence in solution or in CDs in the absence of organic co-solvents. However, under identical conditions, the fluorescence of coumarin C1F was found to decrease, illustrating the tremendous difference in probe fluorescence properties which can be obtained with only minor differences in the coumarin structure (in this case, replacement of a CH3 by a CF3). The presence and role of cavity water elucidated by this coumarin study is a very important property of CD cavities, and has significant effects on their host properties. Nowakowska et al. [39] showed that inclusion of 7-amino-4-methylaminocoumarin C120 into both β- and γ-CD provided significant photostabilization of this coumarin dye, illustrating a very useful application of CDs as guest stabilizers [35].

Other types of coumarins have also been used to study native CDs. Al-Kindy et al. [40] showed that both α- and β-CD significantly enhanced the fluorescence of coumarin-6-sulfonyl chloride amino acid derivatives, and that a stable 2:1 β-CD:guest inclusion complex was formed, with a very large overall 2:1 binding constant of $K = 4.7 \times 10^7$ M^{-2} for the alanine derivative. The value of the binding constant was found to depend strongly on the size of the coumarin derivative, and the possible interactions with the CD host, once again illustrating the importance of the size and fit match of guests with the CD cavity. They proposed the use of these CD-coumarin complexes as fluorescent sensors for amino acids.

Dondon and Fery-Forgues [41] studied the effect of β-CD on two 4-hydroxycoumarins substituted with heterocyclic substituents in the 3 position, HCD1 (structure shown in Figure 7) and HCD2. They found significant fluorescence enhancement upon CD inclusion, which they attributed to the constrictive effect of the CD cavity. The fluorescence spectrum of HCD1 as a function of added β–CD is also shown in Figure 7; this result is an excellent example of the fluorescence enhancement effect of host inclusion on many coumarin guests, including 7-aminocoumarins as well as the 7-hydroxycoumarin derivative shown. Significant 1:1 binding constants of 340 and 700 M-1 were obtained for the two derivatives; however, a much lower binding constant of 81 M^{-1} was obtained for the

4-hydroxycoumarin in the absence of the heterocyclic substituent, illustrating the lower affinity of CDs for phenolic guests.

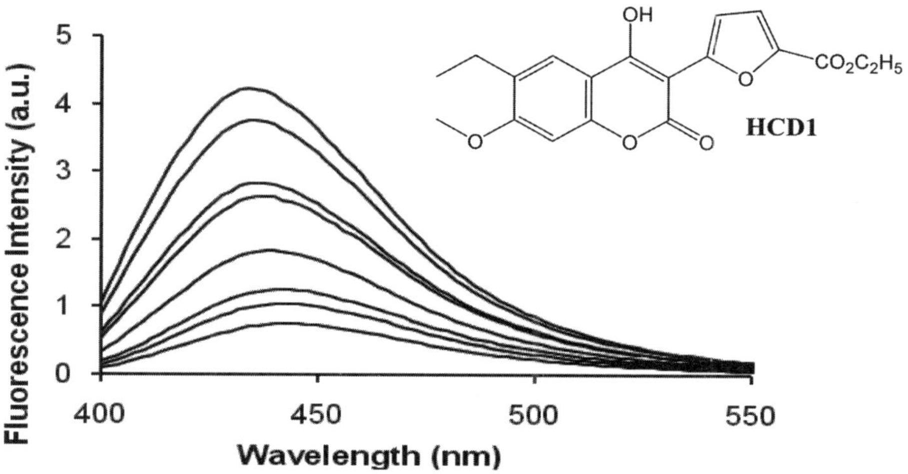

Figure 7. Structure of HCD1 and the fluorescence spectrum of its solution (1×10^{-5} M) in deionized water (1.3% ethanol) in the presence and absence of β-CD. From bottom to top: [β-CD] = 0, 1×10^{-4}, 2×10^{-4}, 1×10^{-3}, 2×10^{-3}, 4×10^{-3}, 7×10^{-3}, and 1×10^{-2} M. Spectra reproduced with permission from Figure 3 in Reference [41]. Copyright 2001 American Chemical Society.

A number of studies have been reported using coumarin guests to study dynamics within native CD cavities [42-44]. Vajda et al. [43] used femtosecond fluorescence upconversion and time-correlated single photon counting techniques to study the solvation dynamics of two coumarins, C480 (Figure 2b) and C460 (Figure 2a), in aqueous solution and included within the cavity of γ-CD, which is large enough to co-include a small number of water molecules. Solvation of C480 was found to occur on the fs timescale in pure water, but on the ps to ns timescale within γ-CD, again illustrating the highly restrictive nature of CD cavities. More recently, Bhattacharyya's group [44] studied the temperature-dependence of the anisotropy decay of C153, once again in γ-CD, the largest native CD cavity. Interestingly, they found that the C153 guests served as linkers between γ-CD hosts, generating linear "nanotube aggregates," resulting in very large steady-state and residual anisotropies. They further found a strong temperature dependence of the C153 solvation time within the CD nanotubes, which they attributed to a dynamic exchange between free and cavity-bound water molecules; this dynamic exchange of cavity water is an important feature of aqueous CD hosts.

In addition to the extensive studies of native CDs described above, a number of studies of chemically modified CDs using coumarin guest fluorescence have also been reported [45-49]. Wagner et al. [45] used the fluorescence of included 7-methoxycoumarin (7MC) to compare the host properties of native β- and γ-CDs and their hydroxypropylated (HP-β- and HP-γ-CD) derivatives. A significant reduction in 7MC fluorescence was observed in all four cases, as expected for this coumarin probe when experiencing a less polar environment. These results are illustrated in Figure 8, which shows fluorescence titration plots of the probe fluorescence intensity (F) as a function of the CD host concentration relative to that in the absence of CD (Fo), and clearly shows the decrease in 7MC fluorescence with increasing [CD].

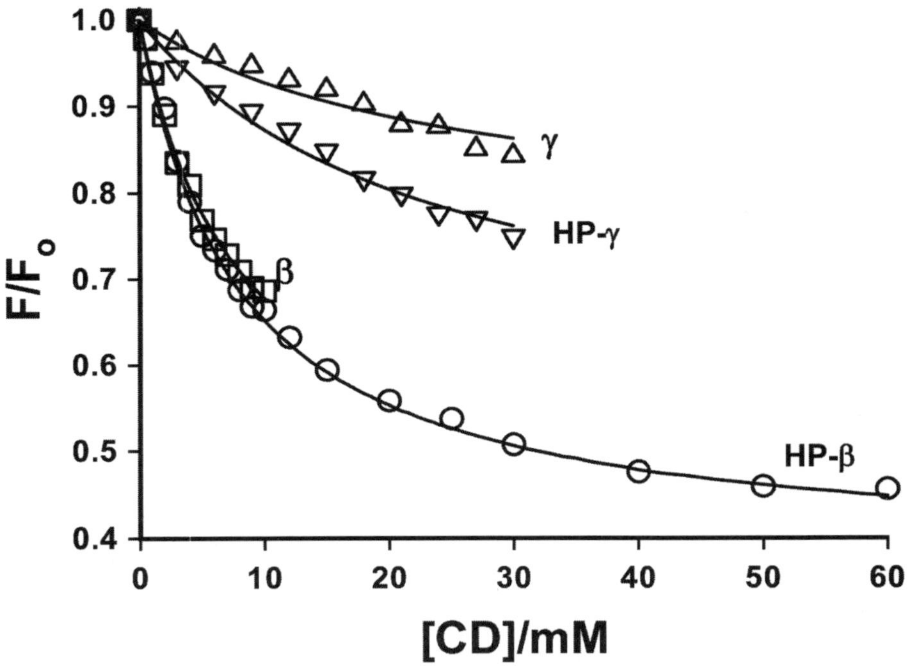

Figure 8. The effect of cyclodextrin concentration on the relative total fluorescence (F/Fo) of 7-methoxycoumarin (λex = 320 nm) for various cyclodextrins: Ö β-CD, {HP-β-CD, Δ γ-CD, ∇ HP-γ-CD; the solid lines show the fit to a 1:1 host:guest complex model. Reproduced from Reference 45 (Figure 3, © 2003 Kluwer Academic Publishers), with kind permission of Springer Science and Business Media. 1.0

Also shown in Figure 8 are the non-linear least-squares fit of the data to a 1:1 host-guest inclusion model to extract the binding constant K; relatively low values of K = 128 ± 32, 120 ± 20, 41 ± 8 and 40 ± 6 M^{-1} were obtained for β-CD, HP-β-CD, γ-CD and HP-γ-CD, respectively. Thus, the β-CD cavity provided a much better match for 7MC than did the larger γ-CD cavity, as indicated by the larger binding constants. With β-CD, there was no observed difference in either the binding constant or the degree of fluorescence suppression when compared with the modified HP-β-CD; this lack of difference can clearly be seen by the overlap of these two fluorescence titration curves in Figure 8. This lack of difference between β-CD and HP-β-CD was in contrast to the results with other fluorescent probes studied by these authors, in which a significantly less polar cavity was experienced by probes included in HP-β-CD as compared to β-CD itself, due to the extension of the cavity by the alkylhydroxy groups. This lack of effect of the HP modifying groups indicated that 7MC is well included within the β-CD cavity, and hence not affected by modifications of the CD rims. By contrast, there was a significant difference in the degree of fluorescence suppression (although not in binding constant) observed in HP-γ-CD as compared to γ-CD; in this case the much larger cavity allowed for interaction of the modifying groups with the included coumarin. Figure 8 provides a good illustration of the use of 7-alkoxy-coumarin fluorescence titrations to extract the binding constant for inclusion. (In the case of 7-amino and other coumarin probes which exhibit fluorescence enhancement upon binding into a host cavity (such as the experiment illustrated as spectra in Figure 7), the resulting fluorescence titrations would look like mirror images of those in Figure 4 about the $F/F_o = 1$ line, i.e. a curved increase (fluorescence enhancement) which plateaus at higher host concentration.)

Bhattacharyya's group extended their above-described studies on the solvation dynamics of C153 in γ-CD to methylated β-CD, and again observed interesting and significant effects of the CD cavity on the solvation dynamics, with multiple kinetic components observed [46]. Velic et al. [47, 48] used the fluorescence of coumarin probes to study the host properties of thiolated β-CD, which they subsequently attached to gold surfaces to construct fluorescent self-assembled monolayers. Most recently, Tablet and Hillebrand [49] used the inclusion of 3-carboxy-5,6-coumaric acid, a potential fluorescent marker for proteins, in native and HP-modified CDs as a model for its interactions with proteins. They observed a decrease in the coumarin fluorescence intensity, which they used to extract the binding constants for each CD. They also did molecular mechanics calculations to elucidate the structure of the CD:coumarin host:guest complex, and the contributions to the binding forces.

A few other relevant studies of coumarins included within CD cavities will also be noted here to conclude this section. Chakraborty et al. [50] used both the steady-state and time resolved fluorescence of a number of 7-aminocoumarin guests to investigate the effect of CD inclusion on photoinduced electron transfer reactions, in this case to N,N-dimethylformamide. They observed a very strong retardation of the electron transfer rate by the CD at the high free-energy region for the electron transfer, which they explained using different binding possibilities for the coumarin in the CD cavity. In addition, a number of research groups have studied solid-state CD:coumarin complexes [51-53], although the fluorescence properties were not reported.

It is clear from the wide range of studies described above that coumarin fluorescence has been successfully used to elucidate the physical and host properties of CDs, including the polarity and constriction of the internal cavity, the strength and nature of interactions between the CD and guest, and the effect of the CD host on guest reaction kinetics.

Cucurbiturils

Cucurbit[n]urils (CB[n]) are a family of macrocycles composed of n glycoluril units linked by two methylene bridges [54], as shown in Figure 9 for the parent (n=6) compound, cucurbituril. Compared to cyclodextrins, cucurbit[n]urils are extremely rigid, with well defined internal cavities, accessible through somewhat narrower carbonyl portals on both the top and bottom. The parent compound cucurbituril was first synthesized in 1904, but its structure and potential as a host compound were not elucidated until 1981 [55]. Since then, and particularly with the expansion of the family of hosts to include the n=5, 7 and 8 homologues in 2000 [56], cucurbit[n]urils have been the subject of growing interest and application.

There have been two reports on the inclusion of coumarin probes included within CB[n] hosts [57, 58]. Nau and Mohanty [57] investigated the ability of CB[7] to both stabilize and enhance the fluorescence of a number of dyes, including coumarin C102. They observed a significant effect of CB[7] inclusion on the fluorescence properties of C102, including a blue-shift from 486 to 479 nm, and an increase in both the fluorescence lifetime and relative intensity, and explained this as being a consequence of the low polarizability inside the CB[7] cavity. Barooah et al. [58] studied the binding

of a number of coumarin probes with the larger cucurbituril CB[8]. They found that most (but not all) of the coumarins formed dynamic inclusion complexes with CB[8], with varying stoichiometries, and could be made to undergo controlled photodimerizations within the CB[8] nanocontainers. They did not however report any fluorescence properties.

Compared to the extensive studies on the host properties of cyclodextrins, including the dynamics of the inclusion process itself as well as solvation within the CD cavity, the study of cucurbituril cavities using coumarin fluorescence remains vastly underexploited at this time.

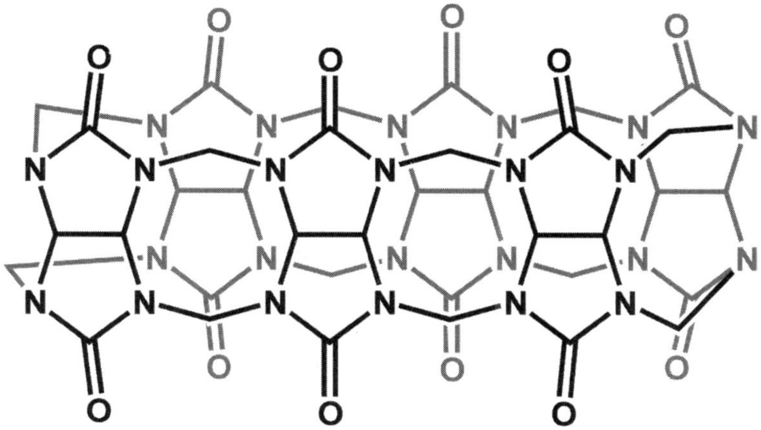

Figure 9. The cyclic structure of cucurbituril, CB[6].

Other Molecular Hosts

There have been a scattering of fluorescence studies of coumarin host:guest inclusion complexes with molecular hosts other than CDs or CB[n]s. Frauchiger et al. [59] used coumarin C102 as a dynamic probe of the local environment within amphiphilic starlike macromolecules (ASMs), which are essentially covalently-bound analogues of micelles (see next section). They have hydrophobic cores, which can encapsulate small hydrophobic molecules. The C102 fluorescence results indicated that the ASM interior is quite polar, and has a high degree of heterogeneity. Both the solvent reorientation and guest diffusion rates were found to be significantly slower than in aqueous solution, and most importantly, approximately one-tenth of the C102 guests were in microenvironments with significantly increased local friction. This latter result was used by the authors to indicate the potential use of these ASMs as drug carriers, with slow release of guests. This same research group also used coumarin fluorescence, in this case C153, to study related amphiphilic "scorpion-like" macrocycles (AScMs) as well as ASMs [60]. They used the fluorescence of encapsulated C153 to determine the core polarity and local friction, and found significant differences between ASMs and AScMs. Finally, Shirota and Segawa [61] used the time-resolved fluorescence of C153 to study the environment within crown ethers, as well as liquid oligoethylene oxides, and reported significant spectral shifting and reduced solvation times for the included C153.

Given the informative coumarin fluorescence-based studies of the cavity properties of CDs, CB[n]s and ASMs described above, there is significant (and as yet untapped) potential to use these coumarin fluorescent probes to study other discrete molecular hosts, such as calixarenes and cavitands.

Coumarins as Fluorescent Probes of Micelles

Micelles are spherical aggregates of surfactant molecules in solution. In aqueous solutions, the polar head groups form the micelle outer surface, with the organic tails oriented towards the interior, giving a relatively nonpolar core inside which hydrophobic molecules can become encapsulated. Micelles form when the surfactant concentration exceeds the critical micelle concentration (cmc). In organic solvents, the surfactants align in the opposite direction, giving reverse micelles with hydrophilic interiors. In a similar way as in the case of molecular hosts, fluorescent probes such as coumarins can be used to study the properties of micelles (such as the cmc, interior polarity and microviscosity) through measurement of micelle-induced changes to the probe fluorescent properties.

There have been a number of survey studies using coumarin fluorescence to compare micelles based on different surfactants [62-64]. In two early studies published in 1994, Marques and Marques [62] used steady state and time-resolved fluorescence of a number of different coumarin probes with a wide range of micelles, to determine which coumarin species entered the hydrocarbon core of the various micelles, while Al-Kindy et al. [63] focused on a single ionic probe, coumarin-6-sulfonyl chloride (C-6SCl), in a series of anionic and cationic micelles. Dutt published an interesting paper addressing whether the microviscosity of micelles determined using fluorescence spectroscopy is probe-dependent [64]. He used two dissimilar probes, one non-dipolar (DMDPP) and the other the dipolar coumarin C6, encapsulated within the interior of six different types of micelles; these structures are shown in Figure 10. Significantly, he found almost identical microviscosities using these two probes for each micelle type, which justifies the experimental approach which is championed in this review article!

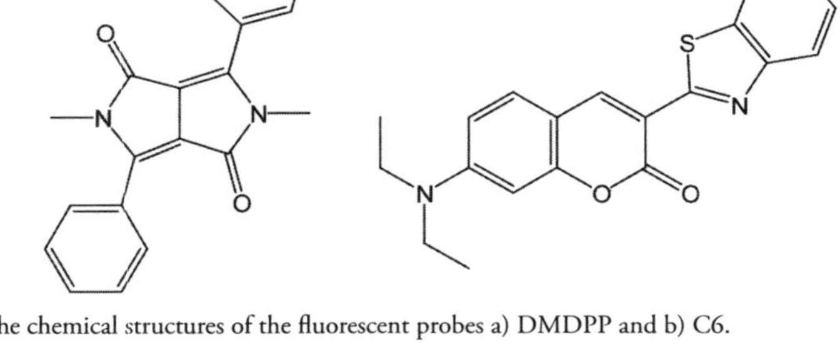

Figure 10. The chemical structures of the fluorescent probes a) DMDPP and b) C6.

In another comparative study, Hara et al. [65] looked at the pressure dependence of the solvation dynamics of coumarin C153 (one of the most widely used coumarin probes for this purpose) in neutral TX100 (see below) as compared to anionic sodium dodecyl sulphate (SDS, see below) micelles. (They had previously reported in detail the pressure dependence of the solvation dynamics of C153 in TX100 micelles [66, 67].) They observed opposite pressure-dependent spectral shifts upon increased pressure in the two micelles, namely a blue-shift in TX100 and a red-shift in SDS. Concurrently, the solvation time was found to decrease in TX100 but increase in SDS; the authors attributed these results to the different hydration structures surrounding the micelles.

In addition to these studies by Hara et al. [65-67], a number of other groups have also reported coumarin fluorescence-based studies of Triton-X [68-73], making it by far the most extensively studied type of micelle using such techniques. Carnero Ruiz et al. [68-70] reported a series of papers on Triton X-100 (TX100) micelles. They used C6 fluorescence depolarization to investigate the effect of the presence of KCl [68], ethylene glycol [69], and formamide [70] on the micelle formation, and found that electrolyte addition increased the microviscosity within the micelle whereas addition of formamide decreased it; these results were explained in terms of micellar solvation. In the case of ethylene glycol, however, solvation was not affected, but the micelle size or aggregation number decreased substantially. Kumbakhar et al. [71, 72] used dynamic Stokes' shift measurements of C153 to study solvation dynamics in both TX-100 and TX-165 micelles. They found that TX-165 micelles have a much looser Palisade layer and lower microviscosity as compared to TX-100, which they attributed to differences in micellar hydration [71]. They also found that addition of LiCl significantly slowed the hydration dynamics in these micelles, due to strong hydration of the Li+ cations. Most recently, Sarkar's group [73] studied microemulsions consisting of an ionic liquid and TX-100 micelles, using the fluorescence depolarization of C153 and C151. They found that the solvent and rotational relaxation time of C153 were not affected by addition of the ionic liquid, indicating that C153 is located at the interface of the microemulsion, whereas the relaxation times of C151 were significantly increased upon increased fraction of ionic liquid, indicating that more C151 is located in the core of the microemulsions. Thus, coumarin fluorescence depolarization studies have been successfully employed to study the nature of TX-100 micelles.

SDS micelles have also been widely studied using coumarin fluorescence [65, 74-78]. Fery-Forgues et al. [74] were able to detect a sphere-to-rod structural transition in SDS micelles, using fluorescence changes in the same 4-hydroxycoumarin derivatives they used to investigate β-CD [41], described previously. Shirota et al. [75] used picosecond fluorescence spectroscopy of C102 and C153 to study the fast solvation and orientation dynamics within SDS micelles. De Paula et al. [76] used a benzoxazolyl coumarin to measure the polarity, microviscosity, and cmc of SDS micelles. Dutt [77] also used coumarin fluorescence anisotropy measurements (in this case C6) to study the microenvironments of SDS micelles, in the presence of various organic and inorganic salts. In a vastly different SDS system, Pantano et al. used the coumarin derivative C314 to study water/air interfaces containing SDS surfactants [78].

Other types of aqueous micelles which have been investigated using coumarin fluorescence include triblock copolymer micelles, which have been extensively studied by Kumbhakar et al. [79-84], Grant et al. [85, 86] and Ghosh et al. [87] using C153 as well as other coumarins; various alkyltrimethylammonium bromide micelles [88-90]; Tween 20 [91]; and Brij-35 micelles [92].

Reverse micelles, generated in organic solvents, have also been widely investigated using coumarin fluorescence [93-108]. Levinger et al. reported on the immobilization of water at the surfactant interfaces in reverse micelles using the coumarin derivative C343 [93], and subsequently published a review of ultrafast dynamics in such systems [94]. Raju and Costa [95-96] reported a series of studies of Aerosol OT (AOT) reverse micelles, using the coumarin derivatives C35 (Figure 2b) [95], as well as C480 and a water-insoluble aminocoumarin derivative [96]. Sarkar's group [97-101] have published extensively on methanol and acetonitrile reverse micelles, using C490 [97]; C152A (Figure 2a) [98]; C153 [99]; and other coumarin dyes [100, 101]. They consistently found that the probe solvation time is strongly dependent on the ratio of polar solvent to surfactant concentration in the case of methanol, but not in the case of acetonitrile, and explained this observation in terms of the presence or absence of hydrogen bonding. A number of other papers have also been published on AOT reverse micelles studied using coumarin fluorescence properties [102-106], as well as on AOT micelle films [107, 108].

Coumarins as Fluorescent Probes of Polymer Hosts

Guest molecules can become included within the folds or pockets of polymers in solution, or within polymer thin films. Trenor et al. [109] published an excellent comprehensive review in 2004 of the use of coumarins both to study the properties of polymers using fluorescence studies, and to prepare polymers with useful optical properties, such as light harvesting. This current review will therefore only cover articles on the use of coumarins as fluorescent probes of polymer hosts which have been reported since that 2004 review.

Prabhugouda et al. [110] used the coumarin derivative C515 as an acceptor molecule to study energy transfer between dopants within polystyrene (one of the most important polymers worldwide) in aqueous solutions. Corrales et al. [111] also studied energy transfer processes within a coumarin-doped polymer, namely poly(ethylene terephthalate). In this case, the coumarin derivative C337 was used as the donor, and energy transfer to the polymer itself resulted in a strong enhancement of the polymer chemiluminescence. This emission was used to study the properties of the polymer, as it was found to be sensitive to the polymer morphology (including crystallinity) and probe mobility within the polymer.

A number of researchers have used coumarins to investigate the properties of polymer thin films [112-114]. Mason et al. [112] doped a polymer photoresist film with the pH-sensitive coumarin probe C6. They were able to measure the relative fluorescence signals from the neutral and protonated forms of C6 to determine the range of acidity and inhomogeneity within the polymer films, and to determine that proton exchange within these films happens very slowly below the glass transition temperature. Frenette et al. [113] also used C6 to study polymer resist films. They were able to determine the catalytic chain length of the prepared PMMA thin films using the coumarin fluorescence. Finally, Oh et al. [114] prepared a polymerizable coumarin derivative, and prepared fluorescent polymers with a range of practical applications in latex films.

Coumarins as Fluorescent Probes of Solid Host Materials

Two different types of solid hosts incorporating coumarin guests can be distinguished. Porous solids, such as zeolites, contain permanent cavities or channels into which guest molecules can become included. Glassy or crystalline solids, however, require that the guest be included in the solid formation process, such as crystallization, annealing or sol-gel techniques. The major difference between these two types of solid hosts involves the dynamics of the inclusion process. In the case of porous solids, a dynamic equilibrium between the included and free guest is established, much like the case of hosts in solution as described above. The binding constants are usually much higher than in solution, so that the guests are more effectively trapped, but they can still be removed. In nonporous crystalline or glassy solids, however, the guests are effectively a permanent part of the structure. This review will focus on both of these two major types of solid hosts, with sol-gel glasses being the primary type of nonporous host which has been studied using coumarin fluorescence.

Porous Solid Hosts

A number of studies have been reported using coumarin fluorescence to characterize the internal pores and channels of zeolites and related porous aluminosilicate structures. Corrent et al. [115] studied the acid-base properties of C6 (widely used as a pH-sensitive probe as described previously) within various zeolites. They were able to show that the commonly used faujasite zeolite NaY, which is generally considered to be nonacidic, in fact has acidic sites within its heterogeneous structure. In two related faujasite zeolites, HY 100 and CBV 740, the dication of C6 was observed, which was explained by the very high Bronsted and Lewis acidity, respectively, of these two zeolites. Kamijo et al. [116] used C153 fluorescence to characterize synthetic silica nanochannels prepared inside the pores of an anodic alumina membrane. They used time-resolved fluorescence to measure the solvation relaxation times of C153 co-included with various alcohols in the nanochannels; these relaxation times were found to be much longer than in bulk alcohol solvent, but that changing the bulkiness of the alcohol (e.g. decanol vs. ethanol) had no significant effect. They concluded that the alcohols are rigidly held in the silica nanochannels through an extended hydrogen bonding network.

Other porous solids have also been investigated using coumarin fluorescence, including MCM-41 and Ti-MCM-41 mesoporous molecular sieves [117, 118], which were found to have a pore size-dependent effect on the fluorescence of a number of coumarin derivatives; a pillared layer clay nanocomposite [119], which was found to enhance the fluorescence of C1; and anodic aluminum oxide films with coumarin 7 embedded in the pores [120], which exhibited an additional, long wavelength coumarin emission band.

Coumarin-Doped Glasses and Crystals

Sol-gels and related materials have been by far the most widely studied example of coumarin doped glasses. The sol-gel process is a synthetic procedure which allows for the preparation of glasses and other materials at room temperature. The preparation and properties of fluorescent probe- (including coumarin) doped sol-gel glasses [121] and nanocomposite materials [122] have been previously reviewed; therefore, only brief overview of the use of coumarin fluorescence to study the properties of such glasses will be presented here.

Takahashi et al. used coumarins C4 and C6 to study amorphous silica sol-gel glasses [123] and a sol-gel coating film [124], respectively. In both cases, significant red-shifted coumarin emission was observed, related to the probe acid-base properties. Oh et al. also used coumarin C4, in this case to probe the properties of a silica-PDMS xerogel [125]. Ferrer et al. [126] used C153 to measure the microviscosities in silica gel-glasses, while Baumann et al. [127] used this same coumarin dye to study the effects of confinement in ethanol within a sol-gel glass on the probe's rotational and solvation dynamics. Other coumarins which have been studied within sol-gel glasses include C2 [128]; C152 [129]; C307 [130, 131]; and silylated coumarin dyes [132 - 134].

Reports on the fluorescence of coumarins doped in crystalline solids have been much fewer. Ganschow et al. [135] used coumarin C40 doped in AlPO4-5 single crystal molecular sieves to investigate their properties, and found that the coumarin dye was uniformly distributed throughout the single crystals, and that these crystals had good optical properties for potential photonics applications. Similarly, Galian et al. [136] doped coumarin C6 into Photonic Crystal Fibers (PCFs) to investigate their properties and potential photonics applications.

Other Solid Host Materials

Coumarin fluorescence has also been used to study a range of other types of solid hosts, which do not fit under either of the above two categories. Aloisi et al. [137] intercalated coumarin-3-carboxylic acid with other donors and acceptors between the layers of Mg-Al hydrotalcite-like compounds to generate nanocomposite materials, and used the emission to characterize the materials themselves as well as energy transfer processes between the intercalated probes. Similarly, Fujii et al. [138] co¬intercalated coumarin probes with rhodamine 6G within the layers of a novel luminescent layered material, and used the coumarin fluorescence intensity to monitor the energy transfer to the rhodamine 6G probes. Other interesting solid materials probed using coumarin fluorescence include cellulose derivatives [139] and titania-based self-cleaning materials [140].

Other Heterogeneous Systems

In addition to the molecular, micellar, polymer and solid host systems described in detail above, there have been a few coumarin-fluorescence-based studies of other types of heterogeneous host systems. These systems include Langmuir-Blodgett films [141, 142]; self-assembled monolayers [143] and polyelectrolyte multilayer nanocontainers [144].

Summary

Coumarins as a family of compounds represent one of the most versatile and applicable family of fluorescent probes, with a wide range of sizes and hydrophobicity. Even more importantly, coumarin fluorescence shows a very broad range of responses and sensitivity to various properties of the local environment, including polarity, polarizability, microviscosity, hydrogen bonding potential and pH.

Coumarins exhibit fascinating and unique photophysical and spectroscopic properties; different derivatives can show different or even opposite behavior upon the same change in conditions. Most strikingly, 7-aminocoumarins and 7-alkoxycoumarins exhibit opposite polarity dependence: 7¬aminocoumarins show highest fluorescence intensity in nonpolar media, whereas 7-alkoxy show highest fluorescence intensity in polar media. This opposite behaviour is a result of very different underlying photophysics, and in particular different major non-radiative decay pathways for these two types of coumarins, namely TICT vs. ISC. Furthermore, this opposing polarity dependence means that a specific coumarin probe can be chosen to show a desired fluorescence change, for example either "switch on" (increased fluorescence) or "switch off" (decreased fluorescence) behavior upon inclusion in a specific cavity or region of a heterogeneous system.

A number of specific coumarin derivatives have been particularly exploited due to their unique physical and spectroscopic properties. For example, C153, a rigid 7-aminocoumarin, has been widely applied as a probe of solvation dynamics, as the lack of intramolecular rotation makes it particularly sensitive to solvent reorientation. Another example is C6, which is a pH-sensitive probe with vastly different fluorescence emission properties in its neutral and protonated form; this probe has been extensively used to investigate the acid-base properties within heterogeneous systems.

It is clear from the extensive list of studies described in this review that coumarin fluorescence can and has been used quite effectively as a probe of the physical, structural and chemical properties of a wide range of heterogeneous host systems, including molecular hosts, micelles, polymers, and porous, glassy, and crystalline solids. Researchers studying such systems should be aware of the tremendous applicability of coumarin probe fluorescence, and the extensive information which can be obtained from their use. Furthermore, as noted in various sections, coumarin fluorescence could potentially be used for other, as yet explored molecular hosts and heterogeneous systems, and should be considered as an option for all researchers preparing and investigating novel inclusion hosts and materials.

Acknowledgements

Financial support for this review was provided by the Natural Sciences and Engineering Research Council of Canada (NSERC).

References and Notes

1. Sethna, S.M.; Shah, N.M. The Chemistry of Coumarins. Chem. Rev. 1945, 36, 1–62.
2. Song, P.-S.; Gordon, W.H. III A Spectroscopic Study of the Excited States of Coumarin. J. Phys. Chem. 1970, 74, 4234–4240.
3. Woods, L.L.; Shamma, S.M. Synthesis of Substituted Coumarins with Fluorescent Properties. J. Chem. Eng. Data 1971, 16, 101–102.
4. Jones, G. II; Jackson, W.R.; Kanoktanaporn, S.; Halper, A.M. Solvent Effects on Photophysical Parameters for Coumarin Laser Dyes. Opt. Commun. 1980, 33, 315–320.
5. Jones, G. II; Jackson, W.R.; Choi, C.; Bergmark, W.R. Solvent Effects on Emission Yield and Lifetime for Coumarin Laser Dyes. Requirements for a Rotatory Decay Mechanism. J. Phys. Chem. 1985, 89, 294–300.
6. Nag, A.; Bhattacharyya, K. Role of Twisted Intramolecular Charge Transfer in the Fluorescence Sensitivity of Biological Probes: Diethylaminocoumarin Laser Dyes. Chem. Phys. Lett. 1990, 169, 12–16.
7. Patalakha, N.S.; Yufit, D.S.; Kirpichenok, M.A.; Gordeeva, N.A.; Struchkov, Yu.T.; Grandberg, I.I. Luminescence-Spectral and Acid-Base Characteristics of 3-Aryl-7-Diethylaminocoumarins. Khim. Geterot. Soed. 1991, 1, 40–46.
8. Gordeeva, N.A.; Kirpichënok, M.A.; Patalakha, N.S.; Khutorretskii, V.M.; Grandberg, I.I. Spectral, Spectral-Luminescence and Acid-Base Properties of 3-Fluoro-7–Diethylaminocoumarins. Khim. Geterot. Soed. 1991, 5, 619–624.
9. Abdel-Mottaleb, M.S.A.; Antonious, M.S.; Abo Ali, M.M.; Ismail, L.F.M.; El Sayed, B.A.; Sherief, A.M.K. Photophysics and Dynamics of Coumarin Laser Dyes and their Analytical Implications. Proc. Indian Acad. Sci. 1992, 104, 185–196.
10. Yip, R.W.; Wen, Y.-X.; Szabo, A.G. Decay Associated Fluorescence Spectra of Coumarin 1 and Coumarin 102: Evidence for a Two-State Solvation Kinetics in Organic Solvents. J. Phys. Chem. 1993, 97, 10458–10462.
11. López Arbeloa, T.; López Arbeloa, F.; Tapia, M.J.; López Arbeloa, I. Hydrogen-Bonding Effect on the Photophysical Properties of 7-Aminocoumarin Derivatives. J. Phys. Chem. 1993, 97, 4704–4707.
12. Taneja, L.; Sharma, A.K.; Singh, R.D. Study of Photophysical Properties of Coumarins: Substituent and Concentration Dependence. J. Luminesc. 1995, 63, 203–214.

13. Królicki, R.; Jarzęba, W.; Mostafavi, M.; Lampre, I. Preferential Solvation of Coumarin 153 - The Role of Hydrogen Bonding. J. Phys. Chem. A. 2002, 106, 1708–1713.

14. El-Kemary, M.; Rettig, W. Multiple Emission in Coumarins with Heterocyclic Substituents. Phys. Chem. Chem. Phys. 2003, 5, 5221–5228.

15. Moog, R.S; Kim, D.D.; Oberle, J.J.; Ostrowski, S.G. Solvent Effects on Electronic Transitions of Highly Dipolar Dyes: A Comparison of Three Approaches. J. Phys. Chem. A. 2004, 108, 9294–9301.

16. Dahiya, P.; Kumbhakar, M.; Mukherjee, T.; Pal., H. Effect of Protic Solvents on Twisted Intramolecular Charge Transfer State Formation in Coumarin-152 and Coumarin-481 Dyes. Chem. Phys. Lett. 2005, 414, 148–154.

17. Bayrakçeken, F.; Yaman, A.; Hayvali, M. Photophysical and Photochemical Study of Laser-Dye Coumarin-481 and its Photoproduct in Solution. Spectrochim. Acta A 2005, 61, 983–987.

18. Barik, A.; Kumbhakar, M.; Nath, S.; Pal, H. Evidence for the TICT Mediated Nonradiative Deexcitation Process for the Excited Coumarin-1 Dye in High Polarity Protic Solvents. Chem. Phys. 2005, 315, 277–285.

19. Sharma, V.K.; Saharo, P.D.; Sharma, N.; Rastogi, R.C.; Ghoshal, S.K.; Mohan, D. Influence of Solvent and Substituent on Excited State Characteristics of Laser Grade Coumarin Dyes. Spectrochim. Acta A 2003, 59, 1161–1170.

20. Grabowski, Z.R.; Rotkiewicz, K. Structural Changes Accompanying Intramolecular Electron Transfer: Focus on Twisted Intramolecular Charge-Transfer States and Structures. Chem. Rev. 2003, 103, 3899–4031.

21. Muthuramu, K.; Ramamurthy, V. 7-Alkoxy Coumarins as Fluorescence Probes for Microenvironments. J. Photochem. 1984, 26, 57–64.

22. Heldt, J.R.; Heldt, J.; Stoń, M.; Diehl, H.A. Photophysical Properties of 4-Alkyl- and 7–Alkoxycoumarin Derivatives. Absorption and Emission Spectra, Fluorescence Quantum Yield and Decay Time. Spectrochim. Acta A 1995, 51, 1549–1563.

23. Moeckli, P. Preparation of Some New Red Fluorescent 4-Cyanocoumarin Dyes. Dyes Pigments 1980, 1, 3–15.

24. Richard, J.-A.; Massonneau, M.; Renard, P.-Y.; Romieu, A. 7-Hydroxycoumarin-Hemicyanine Hybrids: A New Class of Far-Red Emitting Fluorogenic Dyes. Org. Lett. 2008, 10, 4175–4178.

25. Ismail, L.F.M.; Antonious, M.S.; Mohamed, H.A.; Abdel-Hay Ahmed, H. Fluorescence Properties of Some Coumarin Dyes and their Analytical Implication. Proc. Indian Acad. Sci. 1992, 104, 331–338.

26. Christie, R.M.; Liu, C.-H. Studies of Fluorescent Dyes: Part 2. An Investigation of the Synthesis and Electronic Spectral Properties of Substituted 3-(2'-Benzimidazoyl)coumarins. Dyes Pigments 2000, 47, 79–89.

27. Sui, G.; Kele, P.; Orbulescu, J.; Huo, Q.; Leblanc, R.M. Synthesis of a Coumarin Based Fluorescent Amino Acid. Lett. Pept. Sci. 2002, 8, 47–51.

28. Turki, H.; Abid, S.; El Gharbi, R.; Fery-Forgues, S. Optical Properties of New Fluorescent Iminocoumarins. Part 2. Solvatochromic Study and Comparison with the Corresponding Coumarin. C.R. Chimie 2006, 9, 1252–1259.

29. Woods, L.L.; Fooladi, M. 3,3'-Keto Biscoumarins. J. Chem. Eng. Data 1967, 12, 624–626.

30. Ammar, H.; Fery-Forgues, S.; El Gharbi, R. UV/Vis Absorption and Fluorescence Spectroscopic Study of Novel Symmetrical Biscoumarin Dyes. Dyes Pigments 2003, 57, 259–265.

31. McCarthy, P.K.; Blanchard, G.J. AM1 Study of the Electronic Structure of Coumarins. J. Phys. Chem. 1993, 97, 12205–12209.

32. Christie, R.M.; Liu, C.-H. Studies of Fluorescent Dyes: Part 1. An Investigation of the Electronic Spectral Properties of Substituted Coumarins. Dyes Pigments 1999, 42, 85–93.

33. Nemkovich, N.A.; Baumann, W.; Reis, H.; Zvinevich, Yu.V. Electro-optical and Laser Spectrofluorimetry Study of Coumarins 7 and 30: Evidence for the Existence of the Close-Lying Electronic States and Conformers. J. Photochem. Photobiol. A Chem. 1997, 109, 287–292.

34. Szejtli, J. Introduction and General Overview of Cyclodextrin Chemistry. Chem. Rev. 1998, 98, 1743–1753.

35. Hedges, A.R. Industrial Applications of Cyclodextrins. Chem. Rev. 1998, 98, 2035–2044.

36. Takadate, A.; Fujino, H.; Goya, S.; Hirayama, F.; Otagiri, M. Uekama, K.; Yamaguchi, H. Fluorescence Behavior of 7-Substituted Coumarin Derivatives by Inclusion Complexation with Beta- Cyclodextrin. Yakugaku Sasshi J. Pharm. Soc. Jap. 1983, 103, 193–197.

37. Scypinksi, S.; Drake, J.M. Photophysics of Coumarin Inclusion Complexes with Cyclodextrin. Evidence for Normal and Inverted Complex Formation. J. Phys. Chem. 1985, 89, 2432–2435.

38. Bergmark, W.R.; Davis, A.; York, C.; Macintosh, A.; Jones, G. II Dramatic Fluorescence Effects for Coumarin Laser Dyes Coincluded with Organic Solvents in Cyclodextrins. J. Phys. Chem. 1990, 94, 5020–5022.

39. Nowakowska, M.; Smoluch, M.; Sendor, D. The Effect of Cyclodextrins on the Photochemical Stability of 7-Amino-4-methylcoumarin in Aqueous Solution. J. Inclus. Phenom. Macro. Chem. 2001, 40, 213–219.
40. Al-Kindy, S.M.Z.; Suliman, F.E.O.; Al-Hamadi, A.A. Fluorescence Enhancement of Coumarin-6–sulfonyl Chloride Amino Acid Derivatives in Cyclodextrin Media. Anal. Sci. 2001, 17, 539–547.
41. Dondon, R.; Fery-Forgues, S. Inclusion Complex of Fluorescent 4-Hydroxycoumarin Derivatives with Native β-Cyclodextrin: Enhanced Stabilization Induced by the Appended Substituent. J. Phys. Chem. B 2001, 105, 10715–10722.
42. Douhal, A. Ultrafast Guest Dynamics in Cyclodextrin Nanocavities. Chem. Rev. 2004, 104, 1955–1976.
43. Vajda, S.; Jiminez, R.; Rosenthal, S.J.; Fidler, V.; Fleming, G.R.; Castner, E.W., Jr. Femtosecond to Nanosecond Solvation Dynamics in Pure Water and inside the γ-Cyclodextrin Cavity. J. Chem. Soc. Faraday Trans. 1995, 91, 867–873.
44. Roy, D.; Mondal, S.K.; Sahu, K. Ghosh, S.; Sen, P.; Bhattacharyya, K. Temperature Dependence of Anisotropy Decay and Solvation Dynamics of Coumarin 153 in γ-Cyclodextrin Aggregates. J. Phys. Chem. A 2005, 109, 7359–7364.
45. Wagner, B.D.; Fitzpatrick, S.J.; McManus, G.J. Fluorescence Suppression of 7-Methoxycoumarin upon Inclusion into Cyclodextrins. J. Inclus. Phenom. Macro. Chem. 2003, 47, 187–192.
46. Sen, P.; Roy, D.; Mondal, S.K.; Sahu, K. Ghosh, S.; Bhattacharyya, K. Fluorescence Anisotropy Decay and Solvation Dynamics in a Nanocavity: Coumarin 153 in Methyl β–Cyclodextrins. J. Phys. Chem. A 2005, 109, 9716–9722.
47. Velic, D.; Knapp, M.; Köhler, G. Supramolecular Inclusion Complexes Between a Coumarin Dye and β-Cyclodextrin, and Attachment Kinetics of a Thiolated β-Cyclodextrin to Gold Surface. J. Mol. Struct. 2001, 598, 49–56.
48. Velic, D.; Köhler, G. Supramolecular Surface Layer: Coumarin/Thiolated Cyclodextrin/Gold. Chem. Phys. Lett. 2003, 371, 483–489.
49. Tablet, C.; Hillebrand, M. Theoretical and Experimental Study of the Inclusion Complexes of the 3-Carboxy-5,6-Benzocoumarinic Acid with Cyclodextrins. Spectrochim. Acta A 2008, 70, 740–748.
50. Chakraborty, A.; Seth, D.; Chakrabarty, D.; Sarkar, N. Photoinduced Intermolecular Electron Transfer from Dimethyl Aniline to 7-Amino Coumarin Dyes in the Surface of β-Cyclodextrin. Spectrochim. Acta A 2006, 64, 801–808.

51. Moorthy, J.N.; Venkatesan, K.; Weiss, R.G. Photodimerization of Coumarins in Solid Cyclodextrin Inclusion Complexes. J. Org. Chem. 1992, 57, 3292–3297.

52. Brett, T.J.; Alexander, J.M.; Clark, J.L.; Ross, C.R. II; Harbison, G.S.; Stezowski, J.J. Chemical Insight from Crystallographic Disorder: Structural Studies of a Supramolecular β–Cyclodextrin/Coumarin Photochemical System. Chem. Commun. 1999, 1275–1276.

53. Brett, T.J.; Alexander, J.M.; Stezowski, J.J. Chemical Insight from Crystallographic Disorder-Structural Studies of Supramolecular Photochemical Systems. Part 2. The β-Cyclodextrin-4,7–Dimethylcoumarin Inclusion Complex: a New β-Cyclodextrin Dimer Packing Type, Unanticipated Photoproduct Formation, and an Examination of Guest Influence on β-CD Dimer Packing. J. Chem. Soc., Perkin Trans. 2 2000, 1095–1103.

54. Lagona, J.; Mukhopadhyay, P.; Chakrabarti, S.; Isaacs, L. The Cucurbit[n]uril Family. Angew. Chem. Int. Ed. 2005, 44, 4844–4870.

55. Freeman, W. A.; Mock, W. L.; Shih, N. Y. Cucurbituril. J. Am. Chem. Soc. 1981, 103, 7367–7368.

56. Kim, J.; Jung, I.-S.; Kim, S.-Y.; Lee, E.; Kang, J.-K.; Sakamoto, S.; Yamaguchi, K.; Kim, K. New Cucurbituril Homologues: Synthesis, Isolation, Characterization, and X-ray Crystal Structures of Cucurbit[n]uril (n=5,7, and 8). J. Am. Chem. Soc. 2000, 122, 540–541.

57. Nau, W.M.; Mohanty, J. Taming Fluorescent Dyes with Cucurbituril. Int. J. Photoenergy 2005, 7, 133–141.

58. Barooah, N.; Pemberton, B.C.; Johnson, A.C.; Sivaguru, J. Photodimerization and Complexation Dynamics of Coumarins in the Presence of Cucurbit[8]urils. Photochem. Photobiol. Sci. 2008, 7, 1473–1479.

59. Frauchiger, L.; Shirota, H.; Uhrich, K.E; Castner, E.W. Jr. Dynamic Fluorescence Probing of the Local Environments within Amphiphilic Starlike Macromolecules. J. Phys. Chem. B 2002, 106, 7463–7468.

60. Steege, K.E.; Wang, J.; Uhrich, K.E.; Castner, E.W. Jr. Local Polarity and Microviscosity in the Hydrophobic Cores of Amphiphilic Star-like and Scorpion-like Macromolecules. Macromolecules 2007, 40, 3739–3748.

61. Shirota, H.; Segawa, H. Time-Resolved Fluorescence Study on Liquid Oligo(ethylene oxide)s: Coumarin 153 in Poly(ethylene glycol)s and Crown Ethers. J. Phys. Chem. A 2003, 107, 3719–3727.

62. Marques, A.D.S.; Marques, G.S.S. Spectroscopic Studies on Coumarin in Micelles. Photochem. Photobiol. 1994, 59, 153–160.

63. Al-Kindy, S.Z.M.; El-Sherbini, S.A.; Abdel-Kader, M.H. UV-visible and Fluorescence Characteristics of the Luminescent Label Coumarin-6-sulphonyl Chloride in Homogeneous and Micellar Solutions. Anal. Chem. Acta 1994, 285, 329–333

64. Dutt, G.B. Are the Experimentally Determined Microviscosities of the Micelles Probe Dependent? J. Phys. Chem. B 2004, 108, 3651–3657.

65. Hara, K.; Baden, N.; Kajimoto, O. Pressure Effects on Water Solvation Dynamics in Micellar Media. J. Phys.: Condens. Matter 2004, 16, S1207–S1214.

66. Hara, K.; Kuwabara, H.; Kajimoto, O. Pressure Effect on Solvation Dynamics in Micellar Environment. J. Phys. Chem. A 2001, 105, 7174–7179.

67. Baden, N.; Kajimoto, O.; Hara, K. High-Pressure Studies on Aggregation Number of Surfactant Micelles Using the Fluorescence Quenching Method. J. Phys. Chem. A 2002, 106, 8621–8624.

68. Molina-Bolívar, J.A.; Aguiar, J.; Carnero Ruiz, C. Light Scattering and Fluorescence Probe Studies on Micellar Properties of Triton X-100 in KCl Solutions. Mol. Phys. 2001, 99, 1729–1741.

69. Carnero Ruiz, C.; Molina-Bolívar, J.A.; Aguiar, J.; MacIsaac, G.; Moroze, S.; Palepu, R. Thermodynamic and Structural Studies of Triton X-100 in Micelles in Ethylene Glycol-Water Mixed Solvents. Langmuir 2001, 17, 6831–6840.

70. Molina-Bolívar, J.A.; Aguiar, J.; Peula-García, J.M.; Carnero Ruiz, C. Photophysical and Light Scattering Studies on the Aggregation Behavior of Triton X-100 in Formamide-Water Mixed Solvents. Mol. Phys. 2002, 100, 3259–3269.

71. Kumbhakar, M.; Nath, S.; Mukherjee, T.; Pal, H. Solvation Dynamics in Triton X-100 and Triton X-165 Micelles: Effect of Micellar Size and Hydration. J. Chem. Phys. 2004, 121, 6026–6033.

72. Kumbhakar, M.; Goel, T.; Mukherjee, T.; Pal, H. Effect of Lithium Chloride in the Palisade Layer of the Triton X-100 Micelle: Two Sites for Lithium Ions as Revealed by Solvation and Rotational Dynamics Studies. J. Phys. Chem. A 2005, 109, 18528–18534.

73. Seth, D.; Chakroborty, A.; Setua, P.; Sarkar, N. Interaction of Ionic Liquid with Water in Ternary Microemulsions (Triton X-100/Water/1-Butyl-3-methylimidazolium Hexafluorophosphate) Probed by Solvent and Rotational Relaxation of Coumarin 153 and Coumarin 151. Langmuir 2006, 22, 7768–7775.

74. Dondon, R.; Bertorelle, F.; Fery-Forgues, S. 4-Hydroxycoumarin Derivatives in Micelles: An Approach to Detect a Structural Transition Using with Fluorescent Viscosity Probes. J. Fluoresc. 2002, 12, 163–165.

75. Shirota, H.; Tamoto, Y.; Segawa, H. Dynamic Fluorescence Probing of the Microenvironment of Sodium Dodecyl Sulfate Micelle Solutions: Surfactant Concentration Dependence and Solvent Isotope Effect. J. Phys. Chem. A 2004, 108, 3244–3252.

76. De Paula, R.; da Hora Machado, A.E.; de Miranda, J.A. 3-Benzoxazol-2-yl-7--(N,N-diethylamino)–chromen-2-one as a Fluorescence Probe for the Investigation of Micellar Microenvironments. J. Photochem. Photobiol. A Chem. 2004, 165, 109–114.

77. Dutt, G.B. Comparison of the Microenvironments of Aqueous Sodium Dodecyl Sulfate Micelles in the Presence of Inorganic and Organic Salts: A Time-Resolved Fluorescence Anisotropy Approach. Langmuir 2005, 21, 10391–10397.

78. Pantano, D.A.; Sonoda, M.T.; Skaf, M.S. Laria, D. Solvation of Coumarin 314 at Water/Air Interfaces Containing Anionic Surfactants. I. Low Coverage. J. Phys. Chem. B 2005, 109, 7365–7372.

79. Kumbhakar, M.; Goel, T.; Nath, S.; Mukherjee, T.; Pal, H. Microenvironment in the Corona Region of Triblock Copolymer Micelles: Temperature Dependent Solvation and Rotational Relaxation Dynamics of Coumarin Dyes. J. Phys. Chem. B 2006, 110, 25646–25655.

80. Kumbhakar, M.; Ganguly, R. Influence of Electrolytes on the Microenvironment of F127 Triblock Copolymer Micelles: A Solvation and Rotational Dynamics Study of Coumarin Dyes. J. Phys. Chem. B 2007, 111, 3935–3842.

81. Kumbhakar, M. Effect of Ionic Surfactants on the Hydration Behavior Triblock Copolymer Micelles: A Solvation Dynamics Study of Coumarin 153. J. Phys. Chem. B 2007, 111, 12154–12161.

82. Satpati, A.K.; Kumbhakar, M.; Nath, S.; Pal, H. Roles of Diffusion and Activation Barrier on the Appearance of Marcus Inversion Behavior: A Study of a Photoinduced Electron-Transfer Reaction in Aqueous Triblock Copolymer (P123) Micellar Solutions. J. Phys. Chem. B 2007, 111, 7550–7560.

83. Kumbhakar, M. Aggregation of Ionic Surfactants to Block Copolymer Assemblies: A Simple Fluorescence Spectral Study. J. Phys. Chem. B 2007, 111, 14250–14255.

84. Singh, P.K.; Kumbhakar, M.; Pal, H.; Nath, S.; Effect of Electrostatic Interaction on the Location of Molecular Probe in Polymer-Surfactant Supramolecular Assembly: A Solvent Relaxation Study. J. Phys. Chem. B 2008, 111, 7771–7777.

85. Grant, C.D.; DeRitter, M.R.; Steege, K.E.; Fadeeva, T.A.; Castner, E.W. Jr. Fluorescence Probing of Interior, Interfacial, and Exterior Regions in Solution

Aggregates of Poly(ethylene oxide)–Poly(propylene oxide)-Poly(ethylene oxide) Triblock Copolymers. Langmuir 2005, 21, 1745–1752.

86. Grant, C.D.; Steege, K.E.; Bunagan, M.R.; Castner, E.W. Jr. Microviscosity in Multiple Regions of Complex Aqueous Solutions of Poly(ethylene oxide)-Poly(propylene oxide)-Poly(ethylene oxide). J. Phys. Chem. B 2005, 109, 22273–22284.

87. Ghosh, S.; Dey, S.; Adhikari, A.; Mandal, U.; Bhattacharyya, K. Ultrafast Fluorescence Resonance Energy Transfer in the Micelle and the Gel Phase of a PEO-PPO-PEO Triblock Copolymer: Excitation Wavelength Dependence. J. Phys. Chem. B 2007, 111, 7085–7091.

88. Aguiar, J.; Molina-Bolívar, J.A.; Peula-García, J.M.; Carnero Ruiz, C. Thermodynamics and Micellar Properties of Tetradecyltrimethylammonium Bromide in Formamide-Water Mixtures. J. Coll. Interf. Sci. 2002, 255, 382–390.

89. Chakraborty, A.; Charkrabarty, D.; Hazra, P.; Seth, D.; Sarkar, N. Photoinduced Intermolecular Electron Transfer Between Coumarin Dyes and Electron Scavenging Solvents in Cetyltrimethylammonium Bromide (CTAB) Micelles: Evidence for Marcus Inverted Region. Chem. Phys. Lett. 2003, 382, 508–517.

90. Tamoto, Y.; Segawa, H.; Shirota, H. Solvation Dynamics in Aqueous Anionic and Cationic Micelle Solutions: Sodium Alkyl Sulfate and Alkyltrimethylammonium Bromide. Langmuir 2005, 21, 3757–4764.

91. Carnero Ruiz, C.; Molina-Bolívar, J.A.; Aguiar, J.; McIsaac, G.; Moroze, S.; Palepu, R. Effect of Ethylene Glycol on the Thermodynamic and Micellar Properties of Tween 20. Colloid Polym. Sci. 2003, 281, 531–541.

92. Chakraborty, A.; Seth, D.; Charkrabarty, D.; Setua, P.; Sarkar, N. Dynamics of Solvent and Rotational Relaxation of Coumarin 153 in Room-Temperature Ionic Liquid 1-Butyl-3–methylimidazolium Hexafluorophosphate Confined in Brij-35 Micelles: A Picosecond Time-Resolved Fluorescence Spectroscopic Study. J. Phys. Chem. A 2005, 109, 11110–11116.

93. Riter, R.E.; Willard, D.M.; Levinger, N.E. Water Immobilization at Surfactant Interfaces in Reverse Micelles. J. Phys. Chem. B 1998, 102, 2705–2714.

94. Levinger, N.E. Ultrafast Dynamics in Reverse Micelles, Microemulsions, and Vesicles. Curr. Opin. Coll. Interface Sci. 2001, 5, 118–124.

95. Raju, B.B.; Costa, S.M.B. Excited-State Behavior of 7-Dimethylaminocoumarin Dyes in AOT Reversed Micelles: Size Effects. J. Phys. Chem. B 1999, 103, 4309–4317.

96. Raju, B.B.; Costa, S.M.B. Nanosecond Time Resolved Emission Spectroscopy of Aminocoumarins in AOT Reversed Micelles. Phys. Chem. Chem. Phys. 1999, 1, 5029–5034.

97. Hazra, P.; Sarkar, N. Solvation Dynamics of Coumarin 490 in Methanol and Acetonitrile Reverse Micelles. Phys. Chem. Chem. Phys. 2002, 4, 1040–1045.

98. Hazra, P.; Charkrabarty, D.; Sarkar, N. Solvation Dynamics of Coumarin 152A in Methanol and Acetonitrile Reverse Micelles. Chem. Phys. Lett. 2002, 358, 523–530.

99. Hazra, P.; Charkrabarty, D.; Sarkar, N. Solvation Dynamics of Coumarin 153 in Aqueous and Non-Aqueous Reverse Micelles. Chem. Phys. Lett. 2003, 371, 553–562.

100. Chakraborty, A.; Seth, D.; Charkrabarty, D.; Hazra, P.; Sarkar, N. Photoinduced Electron Transfer from Dimethylaniline to Coumarin Dyes in Reverse Micelles. Chem. Phys. Lett. 2005, 405, 18–25.

101. Seth, D.; Charkrabarty, D.; Chakraborty, A.; Sarkar, N. Study of Energy Transfer from 7-Amino Coumarin Donors to Rhodamine 6G Acceptor in Non-Aqueous Reverse Micelles. Chem. Phys. Lett. 2005, 401, 546–532.

102. Yamasaki, T.; Kajimoto, O.; Hara, K. High-Pressure Studies on AOT Reverse Micellar Aggregate by Fluorescence Probe Method. J. Photochem. Photobiol. A: Chem. 2003, 156, 145–150.

103. Choudhury, S.D.; Kumbhakar, M.; Nath, S.; Sarkar, S.K.; Mukherjee, T.; Pal, H. Compartmentalization of Reactants in Different Regions of Sodium 1,4-Bis(2–ethylhexyl)sulfosuccinate/Heptane/Water Reverse Micelles and Its Influence on Bimolecular Electron-Transfer Kinetics. J. Phys. Chem. B 2007, 111, 8842–8853.

104. Mitra, R.K.; Sinha, S.S.; Pal, S.K. Temperature-Dependent Solvation Dynamics of Water in Sodium Bis(2-ethy;lhexyl)sulfosuccinate/Isooctane Reverse Micelles. Langmuir, 2008, 24, 49–56.

105. Narayanan, S.S.; Sinha, S.S.; Sarkar, R.; Pal, S.K. Picosecond to Nanosecond Reorganization of Water in AOT/Lecithin Mixed Reverse Micelles of Different Morphology. Chem. Phys. Lett. 2008, 452, 99–104.

106. Biswas, R.; Rohman, N.; Pradhan, T.; Buchner, R. Intramolecular Charge Transfer Reaction, Polarity, and Dielectric Relaxation in AOT/Water/Heptane Reverse Micelles: Pool Size Dependence. J. Phys. Chem. B 2008, 112, 9379–9388.

107. Bekiari, V.; Lianos, P. Photophysical Studies in AOT Films Deposited on Fused Silica Slides. J. Coll. Interface Sci. 1996, 183, 552–558.

108. Hof, M.; Lianos, P. Structural Studies of Thin AOT Films by Using the Polarity Fluorescent Probe Coumarin-153. Langmuir 1997, 13, 290–294.

109. Trenor, S.R.; Shultz, A.R.; Love, B.J.; Long, T.E. Coumarins in Polymers: From Light Harvesting to Photo-Cross-Linkable Tissue Scaffolds. Chem. Rev. 2004, 104, 3059–3077.

110. Prabhugouda, M.; Lagare, M.T.; Mallikarjuna, N.N.; Naidu, B.V.K.; Aminabhavi, T.M. Energy Transfer Processes Between Primary and Secondary Dopants in Polystyrene Solutions Dissolved in 1,4-Dioxane. J. Appl. Polym. Sci. 2005, 95, 336–341.

111. Corrales, T.; Abrusi, C.; Peinado, C.; Catalina, F. Fluorescent Sensor as Physical Amplifier of Chemiluminescence: Application to the Study of Poly(ethylene terephthalate). Macromolecules 2004, 37, 6596–6605.

112. Mason, M.D.; Ray, K.; Pohlers, G.; Cameron, J.F.; Grober, R.D. Probing the Local pH of Polymer Photoresist Films Using a Two-Color Single Molecule Nanoprobe J. Phys. Chem. B 2003, 107, 14219–14224.

113. Frenette, M.; Ivan, M.G.; Scaiano, J.C. Use of Fluorescent Probes to Determine Catalytic Chain Length in Chemically Amplifies Resists. Can. J. Chem. 2005, 83, 869–874.

114. Oh, J.K.; Stöeva, V.; Rademacher, J.; Farwaha, R.; Winnik, M.A. Synthesis, Characterization, and Emulsion Polymerization of Polymerizable Coumarin Derivatives. J. Polym. Sci: A: Polym. Chem. 2004, 42, 3479–3489.

115. Corrent, S.; Hahn, P.; Pohlers, G.; Connolly, T.J.; Scaiano, J.C.; Fornés, V.; Garcia, H. Intrazeolite Photochemistry. 22. Acid-Base Properties of Coumarin 6. Characterization in Solution, the Solid State, and Incorporated into Supramolecular Systems. J. Phys. Chem. B 1998, 102, 5852–5858.

116. Kamijo, T.; Yamaguchi, A.; Suzuki, S.; Teramae, N.; Itoh, T.; Ikeda, T. Solvation Dynamics of Coumarin 153 in Alcohols Confined in Silica Nanochannels. J. Phys. Chem. A 2008, 112, 11535–11542.

117. Zhao, W.; Li, D.; He, B.; Zhang, J.; Huang, J.; Zhang, L. The Photoluminescence of Coumarin Derivative Encapsulated in MCM-41 and Ti-MCM-41. Dyes Pigments 2005, 64, 265–270.

118. Li, D.; Zhao, W.; Sun, X.; Zhang, J.; Anpo, M.; Zhao, J. Photophysical Properties of Coumarin Derivative Incorporated in MCM-41. Dyes Pigments 2006, 68, 33–37.

119. Wlordarczyk, P.; Komarneni, S.; Roy, R.; White, W.B. Enhanced Fluorescence of Coumarin Laser Dye in the Restricted Geometry of a Porous Nanocomposite. J. Mater. Chem. 1996, 6, 1967–1969.

120. Gruzinskii, V.V.; Kukhto, A.V.; Mozalev, A.M.; Surganov, V.F. Luminescence Properties of Anodic Aluminum Oxide Films with Organic Luminophores Embedded into Pores. J. Appl. Spectrosc. 1997, 64, 497–502.

121. Dunn, B.; Zink, J.I. Optical Properties of Sol-Gel Glasses Doped with Organic Molecules. J. Mater. Chem. 1991, 1, 903–913.

122. Keeling-Tucker, T.; Brennan, J.D. Fluorescent Probes as Reporters on the Local Structure and Dynamics in Sol-Gel-Derived Nanocomposite Materials. Chem. Mater. 2001, 13, 3331–3350.

123. Takahashi, Y.; Shimada, R.; Maeda, A.; Kojima, K.; Uchida, K. Photophysics of Coumarin 4 Doped-Amorphous Silica Glasses Prepared by the Sol-Gel Method. J. Luminesc. 1996, 68, 187–192.

124. Takahashi, Y.; Maeda, A.; Kojima, K.; Uchida, K. Luminescence of Dyes Doped in Sol-Gel Coating Film. J. Luminesc. 2000, 87–89, 767–769.

125. Oh, E.O.; Gupta, R.K.; Whang, C.M. Effects of pH and Dye Concentration on the Optical and Structural Properties of Coumarin 4 Dye-Doped SiO_2-PDMS Xerogels. J. Sol-Gel Sci. Tech. 2003, 28, 279–288.

126. Ferrer, M.L.; del Monte, F.; Levy, D. Microviscosities of Silica Gel-Glasses and Ormosils through Fluorescence Anisotropy. J. Phys. Chem. B 2001, 103, 11076–11080.

127. Baumann, R.; Ferrante, C.; Kneuper, E.; Deeg, F.-W.; Bräuchle, C. Influence of Confinement on the Solvation and Rotational Dynamics of Coumarin 153 in Ethanol. J. Phys. Chem. A 2003, 107, 2422–2430.

128. Deshpande, A.V.; Panhalkar, R.R. Spectroscopic Properties of Coumarin 2 in HCl and HNO3 Catalyzed Sol-Gel Glasses. J. Luminesc. 2002, 96, 185–193.

129. Fukushima, M.; Yanagi, H.; Hayashu, S.; Suganuma, N.; Taniguchi, Y. Fabrication of Gold Nanoparticles and their Influence on Optical Properties of Dye-Doped Sol-Gel Films. Thin Solid Films 2003, 438-439, 39–43.

130. Unger, B.; Rurack, K.; Müller, R.; Resch-Genger, U.; Buttke, K. Effects of the Sol-Gel Processing on the Fluorescence Properties of Laser Dyes in Tetraethoxysilane Derived Matrices. J. Sol-Gel Sci. Tech. 2000, 19, 799–802.

131. Deshpande, A.V.; Kumar, U. Molecular Forms of Coumarin-307 in Sol-Gel Glasses. J. Fluoresc. 2006, 16, 679–687.

132. Suratwala, T.; Gardlund, Z.; Davidson, K.; Uhlmann, D.R. Photostability of Silylated Coumarin Dyes in Polyceram Hosts. J. Sol-Gel Sci. Tech. 1997, 8, 973–978.

133. Suratwala, T.; Gardlund, Z.; Davidson, K.; Uhlmann, D.R. Silylated Coumarin Dyes in Sol-Gel Hosts. 1. Structure and Environmental Factors of Fluorescent Properties. Chem. Mater. 1998, 10, 190–198.

134. Suratwala, T.; Gardlund, Z.; Davidson, K.; Uhlmann, D.R. Silylated Coumarin Dyes in Sol-Gel Hosts. 2. Photostability and Sol-Gel Processing. Chem. Mater. 1998, 10, 199–209.

135. Ganschow, M.; Hellriegel, C.; Kneuper, E.; Wark, M.; Thiel, C.; Schulz-Ekloff, G.; Bräuchle, C.; Wöhrle, D. Panchromatic Chromophore Mixtures in an AlPO4-5 Molecular Sieve: Spatial Separation Effects and Energy Transfer Cascades. Adv. Funct. Mater. 2004, 14, 269–276.

136. Galian, R.E.; Laferrièrre, M.; Scaiano, J.C. Doping of Photonic Crystal Fibers with Fluorescent Probes: Possible Functional Materials for Optrode Sensors. J. Mater. Chem. 2006, 16, 1697–1701.

137. Aloisi, G.G.; Costantino, U.; Elisei, F.; Latterini, L.; Natali, C.; Nocchette, M. Preparation and Photo-physical Characterization of Nanocomposites Obtained by Intercalation and Co–intercalation of Organic Chromophores into Hydrotalcite-Like Compounds. J. Mater. Chem. 2002, 12, 3316–3323.

138. Fujii, K.; Iyi, N.; Sasai, R.; Hayashi, S. Preparation of a Novel Luminous Heterogeneous System: Rhodamine/Coumarin/Phyllosilicate Hybrid with Blue Shift in Fluorescence Emission. Chem. Mater. 2008, 20, 2994–3002.

139. Fischer, K.; Prause, S.; Spange, S.; Cichos, F.; Von Borczyskowski, C. Surface Polarity of Cellulose Derivatives Observed by Coumarin 151 and Coumarin 153 As Solvatochromic and Fluorochromic Probes. J. Polymer Sci.: B: Polymer Phys. 2003, 41, 1210–1218.

140. Guan, H.; Zhu, L.; Zhou, H.; Tang, H. Rapid Probing of Photocatalytic Activity on Titania-Based Self-Cleaning Materials using 7-Hydroxycoumarin Fluorescent Probe. Anal. Chim. Acta 2008, 608, 73–78.

141. Dutta, A.K.; Ray, K.; Misra, T.N. Aggregation Induced Reabsorption of 3-(2-Benzothiazolyl)-7–Octadecyloxy Coumarin Molecules Absorbed in Langmuir-Blodgett Films: A Fluorescence Study. Solid State Commun. 1995, 94, 53–59.

142. Ray, K.; Dutta, A.K.; Misra, T.N. Spectroscopic Properties of 3-(2-Benzothiazolyl)-7–Octadecyloxy Coumarin in Langmuir-Blodgett Films. J. Luminesc. 1997, 71, 123–130.

143. Fox, M.A. Li, W.; Wooten, M.; McKerrow, A.; Whiteshell, J.K. Fluorescence Probes for Chemical Reactivity at the Interface of a Self-Assembled Monolayer. Thin Solid Films 1998, 327–329, 477–480.

144. Nicol, E.; Habib-Jiwan, J.-L.; Jonas, A.M. Polyelectrolyte Multilayers as Nanocontainers for Functional Hydrophilic Molecules. Langmuir 2003, 19, 6178–6186. Sample Availability: Not available.

Stereochemistry of 16α-Hydroxyfriedelin and 3-Oxo-16–Methylfriedel-16-Ene Established by 2D NMR Spectroscopy

Lucienir Pains Duarte, Roqueline Rodrigues Silva de Miranda, Salomão Bento Vasconcelos Rodrigues, Grácia Divina de Fátima Silva, Sidney Augusto Vieira Filho and Vagner Fernandes Knupp

ABSTRACT

Friedelin (1), 3β-friedelinol (2), 28-hydroxyfriedelin (3), 16α-hydroxyfriedelin (4), 30-hydroxyfriedelin (5) and 16α,28-dihydroxyfriedelin (6) were isolated through fractionation of the hexane extract obtained from branches of Salacia elliptica. After a week in CDCl3 solution, 16α-hydroxyfriedelin (4) reacted turning into 3-oxo-16¬methylfriedel-16-ene

(7). This is the first report of a dehydration followed by a Nametkin rearrangement of a pentacyclic triterpene in CDCl3 solution occurring in the NMR tube. These seven pentacyclic triterpenes was identified through NMR spectroscopy and the stereochemistry of compound 4 and 7 was established by 2D NMR (NOESY) spectroscopy and mass spectrometry (GC-MS). It is also the first time that all the 13C-NMR and 2D NMR spectral data are reported for compounds 4 and 7.

Keywords: Salacia elliptica; Celastraceae; 16α-Hydroxyfriedelin; 3-Oxo-16–methylfriedel-16-ene.

Introduction

The genus Salacia (Celastraceae) has a great variety of species spread throughout Brazil and other regions of South America [1]. Different bioactive compounds like salacinol [2], kotalonol [2], sesquiterpene alkaloids [3], quinonemethide triterpenes [3] and pentacyclic triterpenes (PCTT) [4] have already been isolated from Salacia sp.

From the hexane extract of Salacia elliptica branches, the following PCTT: friedelin (1), 3β–friedelinol (2), 28-hydroxyfriedelin (canophyllol, 3), 16α-hydroxyfriedelin (4), 30-hydroxyfriedelin (5) and 16α,28-dihydroxyfriedelin (celasdin-B, 6) (Figure 1) were isolated and identified by TLC comparisons with reference standards and NMR spectroscopy. Compounds 1, 2, 3, 5 and 6 have been isolated from species of the Celastraceae family [5-7]. And, this is the first report of the presence of compound 4 in Celastraceae and the isolation of compound 4, 5 and 6 from specie of the genus Salacia.

The triterpene 16α-hydroxyfriedelin (4) was previously described as constituent of Antidesma menasu Miq.ex.Tul [8]. However, to date, only 1H-NMR chemical shifts assignments of 4 have been published. The 2D NMR (HSQC, HMBC, COSY and NOESY) data are essential to elucidate the stereochemistry of PCTTs due to their complex structures [9, 10]. The analysis of 2D NMR spectral data contributed to establish the correct chemical shift assignments of all carbons and hydrogens of compound 4. After the acquisition of 1D NMR data, the CDCl3 sample solution was maintained inside the tube for a week, until the 2D NMR experiments were performed. The 2D spectral data obtained showed that compound 4 was not the same. The preliminary analysis indicated that compound 4 had been fully converted to 3-oxo-16-methylfriedel-16-ene (7). This process can be due to a dehydration accompanied by methyl migration of C-17 to C-16, which is in agreement with the Nametkin rearrangement [11, 12].

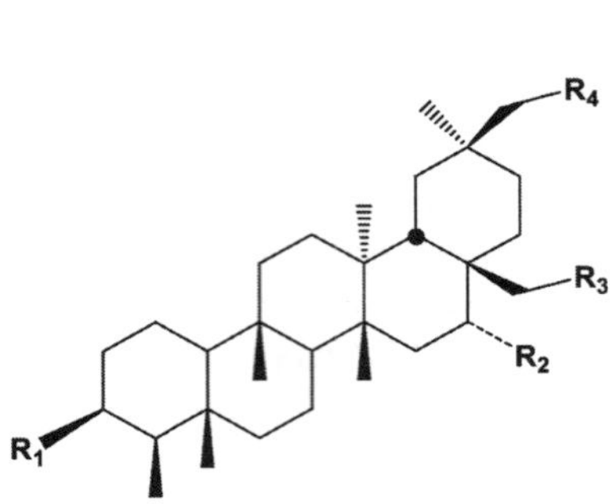

Pentacyclic triterpene	R_1	R_2	R_3	R_4
Friedelin (1)	=O	H	H	H
3β-Friedelinol (2)	OH	H	H	H
28-Hydroxyfriedelin (3)	=O	H	OH	H
16α-Hydroxyfriedelin (4)	=O	OH	H	H
30-Hydroxyfriedelin (5)	=O	H	H	OH
16α,28-Dihydroxyfriedelin (6)	=O	OH	OH	H

Figure 1. Pentacyclic triterpenes isolated from Salacia elliptica.

The triterpene 7 had already been produced by the reaction of the C-16-epimer of 4 with MsCl, but, in this case, besides compound 7, the products 3-oxo-methylfriedel-17(22)-ene and 3-oxo-16–methylfriedel-15-ene were also obtained [13].

The literature reports occurrence of olefinic and allylic hydrogen rearrangements in the presence of CDCl3, but those reactions were purposely carried out under acidic conditions to study the behavior of compounds [14-16]. In the case at hand the transformation of compound 4 into 7 is undoubtedly due to traces of DCl, which is always present in commercial CDCl3, unless the solvent is passed through basic alumina (acidity I) immediately before use.

In order to accomplish our initial aim, i.e., establish the complete chemical shifts assignments of 4, the NMR experiments were repeated, but using pyridine-D5 as solvent, and, the results showed that no rearrangement was observed.

This work describes for the first time the isolation of 16α-hydroxyfriedelin (4), 30-hydroxifriedelin (5) and 16α,28-dihydroxyfriedelin (celasdin-B, 6) from Salacia sp.; the dehydration of a PCTT (compound 4) accompanied by structural rearrangement that occurred in CDCl3, normally used as a solvent in NMR experiments, and also, the complete 2D NMR spectral data of the compound 4 and 7.

Results and Discussion

The identification of 1, 2, 3, 5 and 6 was initially developed through TLC using patterns of PCTT compounds which are commonly isolated from species of the Celastraceae family, and followed by the comparison of the NMR spectral features with literature data [5, 17]. The structural elucidation, including the stereochemistry of compounds 4 and 7, was based on the chemical shifts assignments obtained from 1D (1H, 13C and DEPT-135) and 2D (HMQC, HMBC, COSY and NOESY) NMR spectral data and mass spectrometry (GC-MS).

The 1H-NMR spectrum of 4 presented a double doublet at δH 4.25 (J=7.0 and 10.4 Hz), typical of hydrogen bounded to an oxygenated carbon, suggesting the presence of hydroxyl group in the structure. It also presented a signal at δH 2.34, characteristic of a hydrogen bonded to a carbon adjacent to a carbonyl group. The 13C-NMR of 4 presented a signal at δC 212.12, attributed to a carbonyl, confirming the presence of a ketone group, and also a signal at δC 75.70, which was attributed to a carbon bonded to a hydroxyl [17]. The 13C-NMR spectral data of 4 was compared to the data of 16β–hydroxyfriedelin [17] since, no 13C-NMR data has been reported so far for 16α-hydroxyfriedelin,. This led us to establish the structure of 4 as being 16α-hydroxyfriedelin.

Compound 4, dissolved in CDCl3, was submitted to 2D NMR experiments aiming to confirm its structure after a week inside the NMR tube. However, a quick analysis of the spectra showed that compound 4 had undergone structural modifications. The signal previously attributed to the H-C-O group hydrogen was no longer observed in the 1H-NMR spectrum. In addition, the 13C-NMR spectrum showed new signals at δC 122.57 and δC 129.50, assignable to olefinic carbons. This modified compound was numbered 7. Comparison of the spectral data of 4 and 7 suggested that 4 had undergone dehydration, probably due to the residual acidity of CDCl3, and acquired a double bound accompanied by methyl migration from C-17 to C-16. This process is the expected outcome of the so-called Nametkin rearrangement [11, 12]. To confirm the effect of the acidity of CDCl3 in this specific reaction, another 1D NMR experiment was realized with a sample of 4 dissolved in pyridine-D5. Similarly to the anterior experiment, the pyridine-D5 solution of 4 was also kept inside the NMR tube for a week and then a 2D NMR analysis was carried out. The NMR spectral data obtained using pyridine-D5 as solvent did not show any structural modifications of compound 4. The results suggest that indeed the acidity of CDCl3 was sufficient to promote the dehydration of compound 4. From the HSQC and HMBC spectra of 4 was it possible to correlate each hydrogen signal with its corresponding carbon signal. Through preliminary analysis the chemical shifts assignments of C-16 (δC 75.70, δH 4.25) and C-2 (δC 41.90, δH 2.34 and δH 2.44) were identified.

In the HMBC spectra correlations of the signal at δC 212.12 (C-3) with the signals at δH 2.34, 2.44 (H-2), 1.62, 1.86 (H-1), δH 0.95 (H-23) and at δH 2.20 (H-4) were observed. This last one presented correlations with the signals at δC 7.55 (C-23), 15.08 (C-24), 42.37 (C-5) and at δC 59.77 (C-10). This last signal correlated with δH 2.34, 2.44 (H-2), 2.20 (H-4) and δH 1.51 (H-8). The signal of C-8 (δC 50.82) presented correlation with the signals at δH 0.84 (H-25), 0.93 (H-26), 1.32 (H-11), 1.66 and δH 2.08 (H-15). The signal at δH 4.25 (H-16) correlated with the signals at δC 27.76 (C-22), 31.03 (C-28), 37.75 (C-17), 40.25 (C-14 and C-15) and δC 46.66 (C-18). This signal was correlated to the signals at δH 1.35 (H-19), 1.34 (H-27) and δH 1.37 (H-28). The signal of H-28 correlated with the δC 27.76 (C-22) and δC 37.75 (C-17). The signals at δH 1.05 and δH 1.09 presented correlations with the signal at δC 28.62 (C-20), 34.28 (C-19) and δC 34.97 (C-21). These two signals could only be attributed to H-29 and H-30. Both signals have close chemical shifts values and, for this reason, it becomes difficult to distinguish these two methyl groups through the HMBC spectrum. However, they could be distinguished and the stereochemistry of 4 established from the NOESY spectrum, since it was possible to observe NOEs between H-16 axial and H-15 equatorial, H-18, H-26 and H-28. NOEs were also observed between H-23 and H-24; and between H-24 and H-25. This last signal presented NOE with the signal of H-26 and this one correlated with H-18. It was observed correlation between H-27 and H-29 which presented correlation with H-21 equatorial. The signal of H-27 was correlated with the signal of H-8, which presented correlation with the signal of H-10. Some correlations, observed in the NOESY spectrum of compound 4, are shown in Figure 2.

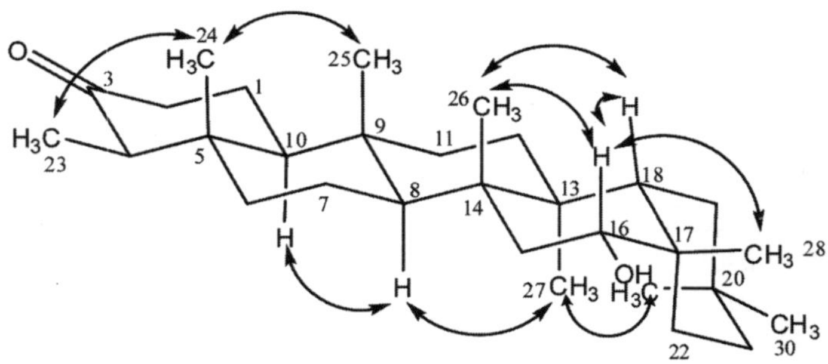

Figure 2. Some correlations observed in the NOESY spectrum of 16α-hydroxyfriedelin (4).

The 1D NMR spectral data of 16β-hydroxyfriedelin [17] are compared with those of compound 4 and the complete 1D/2D NMR data for compounds 4 and 7 are presented in Table 1.

Table 1. 1H (400 MHz) and 13C (100 MHz) NMR spectral data of 16α–hydroxyfriedelin (4) (δ values, Py-d5) and 3-oxo-16-methylfriedel-16-ene (7) (δ values, CDCl3) * (Literature data of 16α-hydroxyfriedelin) [17].

No	δ_C (lit) *	δ_C (4)	δ_H (4)	HMBC	δ_C (7)	δ_H (7)	HMBC
1	22.3	22.77	1.62 ax 1.86 eq		22.32	1.68 ax 1.98 eq	
2	41.6	41.90	2.34 eq 2.44 ax		41.55	2.39 ax 2.45 eq	
3	212.5	212.12	-	1, 2, 4, 23	213.21	-	4, 23
4	58.3	58.28	2.20, m	23, 24, 5, 10	58.25	2.27, m	
5	42.3	42.37	-		42.07	-	4, 10
6	41.4	41.54	1.54, m		41.09	1.23 eq 1.74 ax	
7	18.6	18.95	1.64, m		18.23	1.39 eq 1.48 ax	
8	53.5	50.82	1.51, m	12, 15, 25, 26	50.18	1.38, m	6, 10, 11, 25, 26
9	37.6	37.97	-		37.53	-	
10	59.7	59.77	1.56, m	2, 4, 8	59.25	1.55, m	8
11	35.8	35.64	1.32, m		35.50	1.27 ax 1.51 eq	
12	30.8	30.42	1.29 ax 1.40 eq		28.14	1.35, m	
13	39.3	39.95	-		37.29	-	8, 12, 26, 27
14	40.1	40.25	-		37.65	-	
15	44.4	40.25	1.66 ax, m 2.08 eq, m		43.07	1.52 eq 1.58 ax	
16	75.6	75.70	4.25 J = 7.0; 10.4 Hz	14, 15, 17, 18, 22, 28	122.57	-	28
17	32.1	37.75	-		129.50	-	28
18	44.8	46.66	1.63, m	19, 27, 28	40.43	1.87, m	12, 19
19	35.8	34.28	1.35, m		37.69	1.03 ax 1.29 eq	
20	28.0	28.62	-		30.00	-	
21	32.1	34.97	1.62, m		38.33	1.19 eq 1.36 ax	
22	36.0	27.76	1.97, m 2.05, m		24.60	1.91 ax 2.48 eq	

The 1H-NMR spectrum of 7 showed multiple signals in the region between δH 0.70 and δH 2.50. As mentioned, a lack of signals at the region of the H-C-O hydrogen was observed. The 13C-NMR spectrum presented a signal at δC 213.21, which was assigned to a carbonyl group, and two non-hydrogenated carbon signals at δC 122.57 and δC 129.50 that were attributed to olefinic carbons. All chemical shifts of hydrogens and carbons of compound 7 were assigned through the HMBC spectra. The signal at δC 213.21 (C-3) correlated with the signals at δH 2.27 (H-4) and δH 0.88 (H-23). This last one showed correlation with the signal at δC 42.07 (C-5), which presented correlation with δH 2.27 (H¬4) and δH 1.55 (H-10). This last one correlated with the signal at δC 50.18 (C-8). The signal of C-8 correlated with the signal at δH 1.74 (H-6), 1.55 (H-10), 1.51 and δH 1.27 (H-11), and also with the two methyl signals at δH 0.75 and 0.86, attributed to H-26 and H-25. The signal at δH 0.75 presented correlation with the signal of C-13 (δC 37.29), then this signal could only be associated to H-26, and consequently, the signal at δH 0.86 was attributed to H-25. The signal of C-13 (δC 37.29) correlated with δH 1.38 (H-8), 1.35 (H-12) and δH 0.84 (H-27) and this last one was also correlated with δC 40.43 (C-18). The signal of C-18 correlated with the signals at δH 1.35 (H-12), 1.03 and δH 1.29 (H-19). Also were observed correlations between the signals of olefinic carbon at δC 122.57 (C-16) and δC 129.50 (C-17) with the signal of methyl hydrogen at δH 1.59 attributed to H-28, because it is the only methyl group able to correlate with carbons C-16 and C-17. The signal of H-28 presented yet a correlation with the signal at δC 43.07 (C-15).

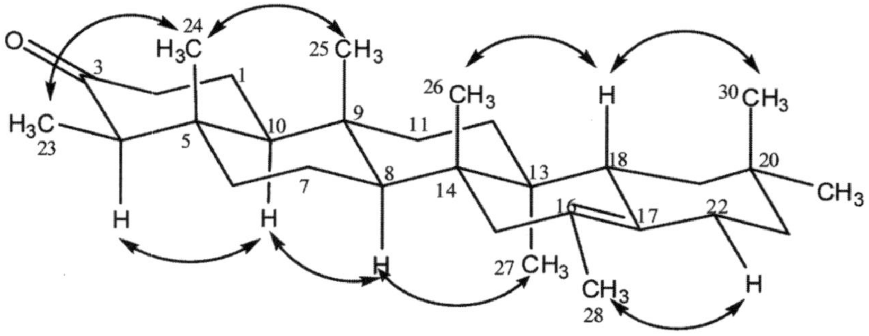

Figure 3. Some correlations observed in NOESY spectrum of 3-oxo-16-methylfriedel-16-ene (7).

The analysis of the NOESY spectrum permitted us to determine the stereochemistry of compound 7. It was possible to observe NOEs between H-30, H-18 and H-19 equatorial. These correlations indicated that the E ring has a chair conformation. NOEs were also observed between H-18 and H-26, between H-25,

H-24 and H-26, and also between H-27 and H-11 axial, H-19 axial, H-12 equatorial, H¬15 axial and H-8. It was possible to observe correlations between H-23-/H-24 and H-10/H-4, and this last one presented correlation with H-2 axial. The H-28 hydrogen presented a NOE with H-22 equatorial. Some correlations observed in the NOESY spectrum of compound 7 are shown in Figure 3. The complete 1D/2D NMR spectral data of 3-oxo-16-methylfriedel-16-ene (7) are presented in Table 1. The mass spectrum of 4 did not show the molecular ion at m/z 442, but rather showed peaks at m/z 411 (M-OH,-CH_2) and m/z 273, confirming it to be a friedelane derivative [18]. On the other hand, the mass spectrum of compound 7 showed the molecular ion at m/z 424, corroborating the molecular formula $C_{30}H_{48}O$.

By the data obtained through NMR and CG/MS was possible to confirm that the structure of PCTT 4 was modified by a process of dehydration accompanied by methyl rearrangement induced by the acidity CDCl3 that is normally used as NMR solvent, producing compound 7. To the best of our knowledge, this is the first report of the Nemetkin rearrangement of a pentacyclic triterpene dissolved in CDCl3, inside an NMR tube, and also the isolation of compounds 1 to 6 from Salacia elliptica.

Experimental

General

Melting points (uncorrected) were determined on a Mettler FP 80 HT. The IR spectra were obtained on a Perkin Elmer, Spectrum One (SN 74759) spectrophotometer. Plates of silica gel G-60 were previously activated at 100°C/30min, and developed with an acidic soln. of vanillin in perchloric acid [19] after the TLC processes. Column chromatography (CC) processes were developed using silica gel (Merck, 230-400 mesh). GC-MS analysis was carried out on a Hewlett Packard HP5890 instrument, equipped with a HP 7673 injector, HP-1 (50 m x 0.25 mm i.d. x 0.2 mm film) column and helium as mobile phase. Operating conditions: injector temperature at 300 °C, 2 μL of sample solution (10 μg/100 μL); splitless of 30s followed by split 1:40 (30 psi stream pressure). The initial oven temperature was 200 ºC/3min, followed by 10°C/min until 300ºC and with 40 min holding time. Interface: quadrupole mass spectrometer model HP 5971, electron impact ionization, 70 eV potential.

NMR Spectra

NMR spectra were recorded on a Bruker DRX400-AVANCE spectrometer operating at 400 and 100 MHz at 27 0C equipped with a direct detection 5 mm

1H/13C dual probe and a 5 mm inverse probe with z-gradient coil. The solvent was CDCl3 or pyridine-D5. Compound 4 (about 10 mg) was dissolved in 0.7 mL of CDCl3 or pyridine-D5, and transferred to a 5 mm o.d. tube. The chemical shifts are reported in ppm using TMS (0 ppm) as internal standard. One-dimensional 1H- and 13C-NMR spectra were acquired under standard conditions. Two-dimensional inverse hydrogen-detected heteronuclear shift correlation spectra were obtained by HSQC pulse sequence [1J(C, H)] and HMBC pulse sequence [nJ (C, H), n = 2 and 3], 1H homonuclear correlation spectroscopy (COSY) and homonuclear 2D¬NOESY (mixing time = 441 ms) experiment were used to confirm the assignments of all carbons and hydrogens of the compounds.

Plant Material and Compound Isolation

Leaves and branches of Salacia elliptica were collected in the Mata Samuel de Paula, Nova Lima region, Minas Gerais, Brazil, in August of 2005. They were separated, dried at room temperature (r.t.) and powdered in a mill. The branches (1158 g) were successively submitted to exhaustive extraction at r.t. with solvents of different polarity. Each solvent was removed under vacuum furnishing the hexane (6.11 g), ethyl acetate (8.16 g) and finally ethanol (144.5 g) extracts.

During the hexane removal, the formation of a white solid was observed. The solid material (0.98 g) was separated by filtration. This material was fractionated by silica gel CC eluted with dichloromethane, ethyl acetate and ethanol pure or in mixture of gradient polarity, obtaining 69 fractions. Fraction 4-5 was analyzed by TLC and GC together with PCTT standards and identified as friedelin (1). By these comparative analyses also was possible to identify fraction 6-7 as a mixture of 1 and 3β-friedelinol (2). Fraction 10-15 gave a white solid material (mp. 279.5-281.8 °C). Its 1H- and 13C-NMR data were compared with the literature data [17] and identified as 28-hydroxyfriedelin (canophyllol, 3). Fraction 18 (200 mg) was submitted to TLC, which showed the presence of only three components. It was then separated by CC using CHCl3, ethyl acetate and ethanol as eluents furnishing 16α-dihydroxyfriedelin (4, 18 mg, m.p. 238.7-244.9 oC) and compound 5 (102.5 mg, m.p. 270.6-278.7 °C), identified as 30-hydroxyfriedelin. After solvent evaporation, fraction 46-47 presented as a white solid (20.2 mg, m.p. 170-172.5 0C). By comparison of its NMR spectral data with the literature [5] this solid material was identified as 16α,28-dihydroxyfriedelin (6). For the NMR analysis, a sample of 4 (10 mg) was dissolved in CDCl3 (0.7 mL) and placed within the NMR tube. After the acquisition data of 1D NMR, the solution was kept inside the tube during a week, and then submitted to 2D NMR experiments. After CDCl3 evaporation the solid was recovered and named as 7 (melting point 240-242 oC).

Acknowledgements

The authors thank the Fundação de Amparo a Pesquisa de Minas Gerais (FAPEMIG) for financial support (Process CEX-1004/05) and the Anglogold Ashanti Ltd for the permission to collect Salacia elliptica in the Mata Samuel de Paula, Nova Lima, MG, Brazil.

References

1. Wolf, B.M.; Weisbrode, S.E. Safety evaluation of an extract from Salacia oblonga. Food Chem. Toxicol. 2003, 4, 867–870.
2. Yoshikawa, M.; Murakami, T.; Shimada, H.; Matsuda, H.; Yamahara, J.; Tanabe, G.; Muraoka, O. Salacinol, potent antidiabetic principle with unique thiosugar sulfonium sulfate structure from the ayurvedic traditional medicine Salacia reticulata in Sri Lanka and India. Tetrahedron Lett. 1997, 38, 8367–8370.
3. Jeller, A.H.; Silva, D.H.S.; Lião. L.M.; Bolzani, V.S.; Furlan; M. Antioxidant phenolic and quinonemethide triterpenes from Cheiloclinium cognatum. Phytochemistry 2004, 65, 1977–1982.
4. Deepa, M.A.; Bai, V.N.; Antibacterial activity of Salacia beddomei. Fitoterapia 2004, 75, 589–591.
5. Kuo, Y.H.; Kuo, L.M.Y. Antitumour and anti-AIDS triterpenes from Celastrus hindsii. Phytochemistry 1997, 44, 1275–1281.
6. Oliveira, M. L. G.; Duarte, L. P.; Silva, G. D. F.; Vieira Filho, S.A.; Knupp, V.F.; Alves; F.G.P. 3-Oxo-12α-hydroxyfriedelane from Maytenus gonoclada: structure elucidation by (1)H and (13)C chemical shifts assignments and 2D-NMR spectroscopy. Magn. Reson. Chem. 2007, 45, 895–898.
7. Oliveira, D.M.; Silva, G.D.F.; Duarte, L.P.; Vieira Filho, S.A. Chemical constituents isolated from roots of Maytenus acanthophylla Reissek (Celastraceae). Bioch. Syst. Ecol., 2006, 34, 661–665.
8. Rizvi, S.H.; Shoeb, A.; Kapil, R.S.; Popli, S.P. Antidesmanol - New Pentacyclic Triterpenoid From Antidesma menasu Miq Ex Tul. Experientia 1980, 36, 146–147.
9. lves, J.S.; Castro, J.C.M.; Freire, M.O.; da-Cunha; E.V.M.; Barbosa-Filho, J.M.; Silva, M.S. Complete assignment of the 1H and 13C NMR spectra of four triterpenes of the ursane, artane, lupane and friedelane groups. Magn. Reson.Chem. 2000, 38, 201–206.

10. Vieira, H.S.; Takahashi, J.A.; Gunatilaka, A.A.L.; Boaventura, M.A.D. 1H and 13C NMR signal assignments of a novel Baeyer-Villiger originated diterpene lactone. Magn. Reson. Chem. 2006, 44, 146–150.

11. Fernandez, A.H.; Cerero, S.M.; Jimenez, F.M. About the Timing of Wagner-Meerwein and Nametkin Rearrangements, 6,2-Hydride Shift, Proton Elimination and Cation Trapping in 2–Norbornyl Carbocations Tetrahedron 1998, 54, 4607–4614.

12. Starling, S.M.; Vonwiller, S.C.; Reek, J.N.H. Effect of Ortho Substituents on the Direction of 1,2–Migrations in the Rearrangement of 2-exo-Arylfenchyl Alcohols. J. Org. Chem. 1998, 63, 2262–2272.

13. Kikuchi, T.; Niwa, M.; Yokoi, T.; Kadota, S. Studies on the Neutral Constituents of Pachysandra terminalis SIEB. et ZUCC. VIII. Methyl Migration in the Dehydration Reaction of Pachysonol and Pachysandiol-B Derivatives. Chem. Pharm. Bull. 1981, 29, 1819–1826.

14. Hao; X.; Xuehui, L.; Yuxin, C.; Min, Z.; Jiaxiang, S. NMR study on the reaction of rearrangement of 17β-hydroxy-7α-methyl-19-nor-17-α-pregn-5(10)-en-20-yn-3-one. Bopuxue Zazhi 2000, 17, 17–22.

15. Lebedeva, T.L.; Vointseva, I.I.; Gilman, L.M.; Petrovskii; P.V.; Larina, T.A.; Topchiev, A.V. Solvent-induced allylic rearrangements in poly(trichlorobutadiene) chains. Russ. Chem.Bull. 1997, 46, 732–738.

16. Mitsuo, K.; Zhang, L.C.; Kabuto, C.; Sakurai, H. Synthesis and Reactions of Neutral Hypercoordinate Allylsilicon Compounds Having a Tropolonato Ligand Organometallics 1996, 15, 5335–5341.

17. Mahato, S.B.; Kundu, A.P. 13C NMR Spectra of pentacyclic triterpenoids - a compilation and some salient features Phytochemistry 1994, 37, 1517–1575.

18. Courtney, J.L.; Shannon, J.S. Studies in mass spectrometry triterpenoids: structure assignment to some friedelane derivatives. Tetrahedron Lett. 1963, 1, 13–20.

19. Matos, F.J.A. Introdução a Fitoquímica Experimental. Editora UFC: Fortaleza, Brazil, 1988; p. 126.

Copyrights

Index

C

D

E

F

N

O

P

Q

T